Practical Skills in Biology

Practical Skills in
Biology

ALLAN JONES
ROB REED
JONATHAN WEYERS

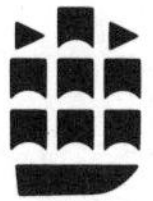

Addison Wesley Longman Limited
Edinburgh Gate, Harlow,
Essex, CM20 2JE, United Kingdom

First published 1994
Second impression 1994
Third impression 1994
Fourth impression 1995
Fifth impression 1996

British Library Cataloguing in Publication Data

A catalogue record for this book is available from the British Library

ISBN 0-582-06699-9

Library of Congress Cataloging -in-Publication Data
Jones, A. M. (Allan M.)
Practical skills in biology / Allan Jones, Robert Reed, and Jonathan Weyers.
p. cm.
Includes bibliographical references (p.) and index.
1. Biology-Methodology. I. Reed, Robert (Robert H.)
II. Weyers, Jonathan D. B. III. Title.
QH324.J64 1993
574´.078-dc20 92-36775
CIP

Produced by Longman Singapore Publishers (Pte) Ltd.
Printed in Singapore

Contents

Table of Boxes

Acknowledgements

We are grateful to the following for permission to reproduce copyright material:

Merck & Co., Inc. for Fig. 6.1 (Budavari *et al.*, 1989); Blackwell Scientific Publications for Figs 35.1 and 35.2 (Golterman *et al.*, 1978); MSE for Fig. 38.1.

Preface

All knowledge and theory in Biology has originated from practical observation and experiment. Laboratory and field work are thus important components of undergraduate training. The skills required in modern practical biology are diverse, ranging from those needed to observe, draw and record accurately to those associated with operating high-tech analytical equipment. Students are expected to design practical investigations, keep records of their work, analyse data and then present their work in written and oral forms. While some of the skills are specific, many are transferable and will be valuable in postgraduate careers. Unfortunately, practical syllabuses in biology are under increasing pressure both from within, as the knowledge base of this rapidly growing subject expands, and from without, as it competes with other subjects for limited timetable hours. Moreover, students now have increasingly diverse academic backgrounds and experiences as access to tertiary education is widened. Increasingly, they are expected to acquire complex skills at the first sitting and, as a consequence, many under-perform in practical work.

This book is designed to provide support to students (and lecturers) before, during and after practical classes. We have tried to cover most of the standard botanical, microbiological and zoological skills required in the early stages of undergraduate courses in biology and allied subjects. However, we hope the information remains useful as students progress to honours degree level and beyond. With such a wide range of skills to cover, we could not hope to provide a detailed recipe-like solution to every problem. In any case, learning general principles can give a deeper understanding of the underlying rationale of methods than learning protocols by rote. The content reflects our experiences of student needs and of the core skills which may require supplementary information. We have tried to provide the information in a concise but user-friendly manner, presenting illustrations, tips and hints, worked examples, 'how to' boxes or checklists where appropriate. Theory is given when necessary, but emphasis throughout has been placed on practical applications.

Authorship was shared equally and we have followed the convention of listing our names alphabetically. Writing this book would not have been possible without the support of our wives, Angela, Polly and Mary. We are also heavily indebted to the following colleagues and friends who read early drafts and/or presented us with much useful comment and food for thought: Gail Alexander, Steve Atkins, Janet Aucock, Olivia Bragg, Sally Brown, Cathy Caudwell, Charlie Dixon, Jackie Eager, Jennifer Gallacher, Alan Grant, Karen Gowlett-Holmes, Margaret Gruber, Bryan Harrison, David Hopkins, Steve Hubbard, Andy Johnston, Roy Oliver, Neil Paterson, John Raven, Pete Rowell, J. Andrew, C. Smith, Philip Smith, Susan Smith, Peter Sprent, Bill Tomlinson, Bob Young and Hilary-Kay Young. The following deserve to be singled out for their special assistance, often on more than one section: Richard A'Brook, Eldridge Buultjens, Hugh Ingram, Rod Herbert, Dave Holmes and Will Whitfield. Certain sections were tested out on undergraduate classes and we thank those students, too numerous to mention, who made useful comments. Despite all the above help, the responsibility for any errors is entirely ours and we would be grateful to hear of any you find.

ALLAN JONES, ROB REED AND JONATHAN WEYERS

Introduction

1 Approaching practical work

All knowledge and theory in biology has originated from practical observation and experiment: this is equally true for studies as disparate as ecology and modern genetics. Practical work forms an important part of most courses and a large percentage of your examination/assessment marks. A good practical course will give you:

Key point
Practical skills include:
- **observing**
- **measuring**
- **manipulating**
- **recording**
- **designing experiments**
- **analysing data**
- **reporting**

- experience of 'real' specimens;
- appreciation of the diversity and complexity of living things;
- understanding of experimental design;
- basic laboratory skills in observation, handling and measurement;
- technical skills needed to carry out important measurements and to use standard equipment and materials;
- transferable skills such as analysing information and preparing reports;
- better understanding of your subject.

This book aims to provide you with an easy reference source of basic practical techniques and transferable skills which will form the basis of your undergraduate practical classes. These skills will continue to be useful throughout your degree and beyond, some within science, others in any career you choose.

Being prepared

You will get most out of practicals if you prepare well. Do not go in to a practical assuming that all information and equipment will be provided. The main points to remember are:

- Read any handouts in advance: make sure you understand the purpose of the practical and the particular skills that are being taught. Does the practical relate to or expand upon a current topic in your lectures? Is there any additional preparatory reading that will help?
- Make sure you have the right equipment, and check that it is in working order.
- Know what safety hazards may be involved and any precautions you need to take.
- Take along appropriate textbooks to explain or expand points in the practical.
- Listen carefully to any preamble and note any important points; adjust your schedule/handout as necessary.
- Submit work on time if required to do so or you may lose marks.
- Catch up on any work you have missed as soon as possible and before the next practical.

2 Tools for the job

Table 2.1 Basic equipment for practical classes

Laboratory book
2H, HB and 2B pencils
Red and blue coloured pencils (for zoology)
Vinyl plastic eraser
Ruler (15 cm and 30 cm)
Pair of dividers
Draughtsman's sandpaper block
Waterproof ink, fibre-tipped pen
Safety knife/pencil sharpener
Calculator
Small set square
Flexicurve

See also pp. 108–9 for dissecting instruments.

Practical work requires a selection of good equipment (Table 2.1) and it is necessary to understand the advantages and disadvantages associated with its proper use: no student can produce work reflecting their true ability using poor tools! Clearly the range of specialized equipment is enormous but this chapter considers only items of general equipment needed to record and present your investigations.

Basic requirements

The laboratory book

A4 loose-leaf binders offer considerable flexibility since they allow insertion of laboratory handouts and a range of paper types at appropriate points in the folder. The danger of losing a page/pages from a loose-leaf system is its main drawback. Bound books should be hard-backed but those containing alternating lined/drawing or lined/graph paper tend to be wasteful.

The quality of drawing paper is important. The best paper is 80 g cartridge paper; bond (typing) paper is often used because it is cheap but the surface is very smooth, doesn't take pencil so well and is easily damaged when erasing. A variety of book and paper sizes is available — use A4 except for special purposes. A5 is too small for laboratory work,but is useful for fieldwork as it is pocketable. Use metric-scale graph paper unless there is a particular reason for choosing a different scale.

> Loose-leaf binders — to avoid losing pages use adhesive reinforcement rings.

Pencils

Good quality pencils are recommended: you will require only 2B, HB and 2H for biological work. Coloured pencils are useful for differentiating structures when interpreting complex anatomy. Mechanical ('propelling') pencils are more expensive but require no sharpening and are available in a range of hardnesses, thicknesses and colours.

Pens

Modern fibre-tipped drawing/lettering pens as made by Edding and Pentel are adaptable and reliable for general use. Roller ball and other modern pens are equally suitable for labelling and general text work. Use waterproof inks to prevent water damage.

Erasers

Vinyl plastic erasers are the most useful since they will remove most lines (except ink) from a wide variety of surfaces. A combined vinyl and ink eraser will cope with most problems but ink erasers are highly abrasive, so use them carefully and sparingly.

Associated drawing equipment

Several other items can promote better quality work.

- Templates and stencils are available in a wide variety of shapes and sizes and can be used for pencil or ink lines. There are templates for geometric shapes, flow chart symbols, chemical symbols, and specialized symbol sets for particular subjects. Those for use with drawing/lettering pens require

Sharpening pencils — frequent use of a draughtsman's sandpaper block maintains an excellent point, even on 2B pencils which blunt rather quickly.

edging strips or bevelled edges to prevent ink seepage. For drawing curves the 'Flexicurve' is generally best but for very tight curves a set of French curves may be needed.
- Rulers and set squares: the most useful rulers are made of clear plastic and should have bevelled edges if they are to be used with ink pens. Set squares are useful for hatching drawings and diagrams.
- Sharpeners: a safety knife or penknife gives best results, or you may prefer a good quality pencil sharpener.
- Dividers are valuable for taking and transferring measurements from specimens. A compass is useful to draw circular outlines.

Calculators

These range from basic machines with no pre-programmed functions and only one memory, to sophisticated programmable minicomputers with many memories. A few general tips will help when choosing your calculator:

- Power sources. Choose a battery-powered machine, preferably with a mains power option: solar types may not work properly in dim light. You will require basic functions such as logarithms, power functions, roots and basic statistical functions such as the calculation of means and standard deviations; beware, however, of the fact that some calculate population statistics, others sample statistics and some do both — check your calculator manual.
- Mode of operation. Calculators fall into two distinct groups. The older system used by e.g. Hewlett-Packard calculators is the reverse Polish notation system. To calculate the sum of two numbers, the sequence is 2 [enter] 4 + ; the answer 6 is displayed. The more expected method of calculating this equation is 2 + 4 = ; this is used by most other modern calculators. Which is better is a matter of personal preference.
- Levels of parenthesis. The number of parenthesis levels determines the complexity of the equations which your calculator can process: parentheses (brackets) are used to break up an equation into blocks which will be calculated in the correct sequence. Be aware that the position of parentheses also determines the result, e.g. $(4+3) \times 1 \neq 4 \times (3+1)$. For complex statistical calculations, six or more levels of parenthesis may be necessary so splitting the equation up on paper first can sometimes avoid difficulties.

Programmable calculators are worth considering for more advanced studies. However, it is important to note that such calculators may not be acceptable for exams. Finally, a warning: for most biologists it may be best to avoid the more complex machines full of wonderful but often unused pre-programmed functions — go for more memories and parenthesis levels or more statistical functions rather than physicochemical or mathematical constants.

Equipment for presenting more advanced practical work

Projects usually require more sophisticated presentations (see Chapters 43, 50, 53) and you may need to use a wider variety of equipment. Some of the more important types are considered briefly below.

Papers and other surfaces

Good quality paper is essential to achieve a high contrast original, often needed for photographic reproduction. The best papers, readily available from graphic arts suppliers, include Bristol Board and CS10. Tracing paper and drafting films are also suitable for good quality artwork.

Permanent markers are best for overhead transparencies — mistakes can be corrected using an ethanol swab.

Pens

Use pens with stepped nibs to prevent ink spreading beneath rulers, set squares or templates. Staedtler and Rotring are major manufacturers of such pens although disposable pens are alternatives. Special drawing equipment should be available in your department.

To prepare overhead transparencies (acetates), you need special pens, with permanent or water-soluble ink. Large areas of colour in acetates are best prepared using sheets of coloured film available for that purpose; the alternative is to use hatching techniques. Special transparencies are needed for photocopiers and laser printers.

Mounting materials

Various methods of attaching artwork to boards, etc. are available. Adhesive sprays are good for large items while cassette-based systems such as Pelikan Roll-fix® are ideal for smaller jobs. PrittStick® and rubber glues are fine for newspaper and light materials, but not for photographs and mounting board. Double-sided Sellotape® is suitable for most forms of mounting. Choice of colour for the mounting board is important and should be sympathetic with the material to be displayed.

Lettering

The most common method, and one of the best, is to use dry-transfer lettering and symbol sets such as those made by Letraset®. Images are transferred from a waxed carrier sheet to the artwork, using light pressure. They are available in sheets containing groups of letters, in numbers which reflect their frequency of usage. Dry-transfer materials are also available for covering large areas and for many specialized symbols.

Computer graphics and text

Much modern display material is now composed using various packages on personal computers. This can vary from simple text using a word processing package (Chapter 49) to complex, sophisticated graphical presentations (Chapter 50): such images are generally acceptable for display purposes or for reports. Choose simple fonts such as Helvetica for scientific reports and posters (p. 256).

Health and safety

3 The legal framework

There are several important pieces of legislation which provide the basis for safe working practice in laboratory and field.

The Health and Safety at Work, etc., Act (1974)

This requires all educational institutions to provide a working environment that is safe and without risk to health, together with appropriate information and training on safe working practices, where necessary. In turn, you must take reasonable care for the health and safety of yourself and of others, and must not misuse any safety equipment provided.

The Control of Substances Hazardous to Health (COSHH) Regulations (1988)

COSHH imposes requirements for the use of potentially harmful chemicals/microorganisms, which *must* be assessed before use to determine:

1. The intrinsic hazards, together with the maximum exposure limit (MEL), where necessary. Manufacturers' data sheets list the hazards associated with particular chemicals, e.g. toxic, corrosive, harmful and irritant substances (see Fig. 4.1). Pathogenic (disease-causing) microorganisms are categorized according to Advisory Committee on Dangerous Pathogens (ACDP) guidelines (p. 122).
2. The risks involved, taking into account the amount of substance, the manner in which it will be used and possible routes of entry into the body.
3. The persons at risk, and the ways in which they may be exposed to the substance, including the possibility of accidental exposure (e.g. spillage).
4. The measures required to prevent or control exposure. Ideally, a non-hazardous, or less hazardous alternative should be used. If this is not reasonably practicable, adequate control measures must be used, e.g. fume cupboards or other containment facilities. Personal protective equipment (e.g. lab coats, safety glasses) is necessary *in addition* to such containment measures. Safe disposal systems will be required.

Key point
It is important to distinguish between hazard and risk:
- **Hazard — the potential to cause harm.**
- **Risk — the likelihood of causing harm under a particular set of circumstances.**

i.e. risk takes into account the nature of the hazard and the possibility of exposure to it.

The assessment must be recorded, and information on the risks involved in a particular procedure must be passed on to those involved. For most biology departments COSHH risk assessments will have been carried out by a responsible person (often, the person in charge of the class). The information necessary to prevent or control exposure will be made available to all students, usually within the practical schedule for a particular exercise. Make sure you know how the system operates and that you read the appropriate information *before* you carry out any practical work with hazardous substances. You should also pay close attention to any additional safety information given by the person in charge at the beginning of the practical session, where the major risks may be emphasized. In project work, you may need to be involved in the risk assessment process, with your supervisor, since the risks associated with a particular procedure must be fully considered before practical work begins.

Most institutions will have a safety handbook — read this before you carry out practical work and abide by any instructions. This will also give names and telephone numbers of safety personnel, first aiders, hospitals, etc., and will detail the procedures for recording accidents, as required by the Recording of Injuries, Diseases and Dangerous Occurrences Regulations (1985).

Key point
The major routes of entry of harmful substances are:
- **ingestion via the mouth;**
- **inhalation into the respiratory tract;**
- **absorption/inoculation through the skin.**

4 Your responsibilities

toxic

oxidizing

corrosive

explosive

harmful, or irritant

radioactive

flammable

biological hazard (biohazard)

Fig. 4.1 Warning labels for specific chemical hazards.

Remember, it is your responsibility to understand the implications of COSHH regulations as they apply to all practical work. Follow *all* safety instructions.

Basic rules for laboratory work

- Make sure you know what to do in case of fire, including how to raise the alarm, exit routes, and where to gather on leaving the building. Remember that the most important consideration at all times is human safety: do not attempt to fight a fire unless it is safe to do so.
- Report all accidents, even those appearing insignificant. All laboratories display notices detailing the locations of first-aid kits and appropriate contacts.
- Wear protective clothing where appropriate.
- Never smoke, eat or drink in any laboratory, because of the risks of contamination by inhalation or ingestion.
- Never mouth pipette any substance. Use a pipette filler, e.g. a Propipette®, or Pi-pump®.
- Be particularly careful with glassware. Make sure you are aware of safe practices as detailed on p. 16.
- Know the international warning signs for specific chemical hazards (see Fig. 4.1).
- Use a fume cupboard for hazardous chemicals. Ensure that it is operational and open the front only as far as is necessary (many fume cupboards are marked with a maximum opening).
- Use minimum quantities of hazardous material.
- Work in a logical, tidy manner minimizing risks by thinking ahead.
- Always clear up at the end of each session. This is a very important aspect of safety, since it encourages a responsible attitude towards laboratory work.

Basic rules for fieldwork

- The objectives of the fieldwork must be understood and all potential safety hazards must be discussed and appropriate responses identified.
- Work should be carefully planned to allow for the degree of experience/training of the participants and the nature of the terrain. Do not overestimate what can be achieved!
- Physical handicaps must be notified to the course organizer so that appropriate precautions are taken.
- Carry a comprehensive first-aid kit: at least two participants should have training in first aid.
- Equipment and clothing must be suitable for all weather conditions likely to be encountered during the work.
- Never work alone without special permission from your leader.
- Be able to read a map and use a compass; your group should have both.
- Check the weather forecast before departure; be alert to changes in the weather at all times and do not hesitate to turn back if necessary.
- Leave details of intended working locations, routes and times. Never break these arrangements without informing the authorities promptly.
- Learn, and be able to use, the international distress signals (Table 4.1).

Table 4.1 International distress signals

On water
1. Use of whistles and torches
   The Morse code signal 'SOS' should be signalled as follows:
   3 SHORT blasts/flashes
   then 3 LONG blasts/flashes
   then 3 SHORT blasts/flashes
   PAUSE then repeat as often as necessary
2. Fire RED flares or ORANGE smoke
3. Raise and lower arms slowly and repeatedly
4. Wave an oar, with a clearly visible cloth tied to it, slowly from side to side

On land
1. SIX long flashes/blasts/shouts/waves in succession and repeated at about 1 min intervals

Basic laboratory techniques

5 Using glass and plastic ware

Measuring and dispensing defined volumes of solution

The equipment you should choose to measure out liquids depends on the volumes being dispensed, the accuracy required and the number of times the job must be done (Table 5.1).

Table 5.1 Criteria for choosing a method for measuring out a liquid

Method	Best volume range	Accuracy	Usefulness for repetitive measurement
Pasteur pipette	30 μl to 2 ml	Very low	Very good
Measuring cylinder	5–2 000 ml	Medium	Good
Volumetric flask	5–2 000 ml	High	Good
Burette	1–100 ml	High	Very good
Pipette/pipettor	5 μl to 25 ml	High*	Very good
Microsyringe	0.5–50 μl	High	Good
Weighing	Any (depends on accuracy of balance)	Very high	Poor

*If correctly calibrated and used properly (see p. 14).

Certain liquids may cause problems:

- High-viscosity liquids are difficult to dispense: allow time for all the liquid to transfer.
- Organic solvents may evaporate rapidly, making measurements inaccurate: work quickly; seal containers quickly.
- Solutions prone to frothing (e.g. protein and detergent solutions) are difficult to measure and dispense: try to avoid forming bubbles; do not transfer these liquids quickly.
- Suspensions (e.g. cell cultures) may sediment: thoroughly mix them before dispensing.

Pasteur pipettes

These should be used with care for hazardous solutions. Remove the tip from the solution before fully releasing pressure on the bulb: the air taken up helps prevent spillage. Squeeze gently to dispense. To avoid risk of cross-contamination, take care not to draw up solution into the bulb or to lie the pipette on its side.

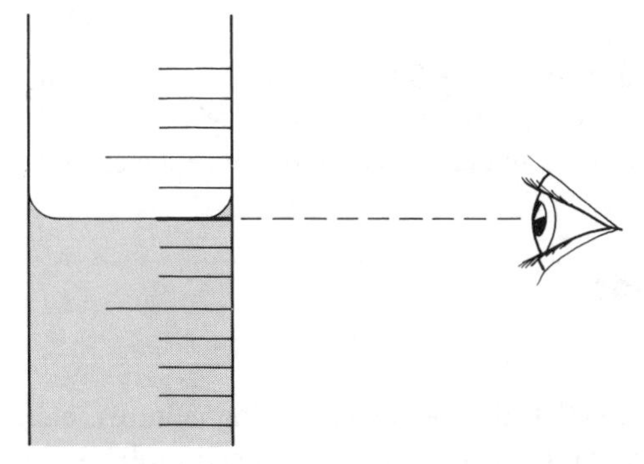

Fig. 5.1 Reading the level of the meniscus of an aqueous or organic solution.

Measuring cylinders and volumetric flasks

These must be stood on a level surface so that the scale is horizontal; you should first fill with solution from, say, a beaker until just below the desired mark; then fill slowly (e.g. using a Pasteur pipette) until the meniscus is level with the mark (Fig. 5.1). The scale measurement is taken from the bottom level of the meniscus, so make sure your eye is level with the mark. Allow time for the solution to run down the walls.

Burettes

Burettes should be mounted vertically on a clamp stand. You should first ensure the tap is closed and fill the body with solution using a funnel. Allow some liquid to fill the tubing below the tap before first use. Take a meniscus reading, noting the value. Dispense the solution via the tap and measure the new meniscus reading. The volume dispensed is the difference between the two readings.

Key point
For safety reasons, it is no longer permissible to mouth pipette — various aids are available such as the Pi-pump®.

Pipettes

These come in various designs. Take care to look at the volume scale before use: some pipettes empty from full volume to zero, others from zero to full volume; some scales refer to the shoulder of the tip, others to the tip by gravity or to the tip after blowing out.

Using a pipettor — check your technique (accuracy) by dispensing volumes of distilled water and weighing on a balance, assuming 1 mg = 1 μl = 1 mm^3. For small volumes, measure several 'squirts' together, e.g. 10 × 15 μl = 150 mg. Aim for accuracy of $\pm$1%.

Pipettors

Pipettors work by air displacement via a disposable plastic tip. If using the adjustable type, first, dial up the desired volume. Fit the tip, then press down on the trigger until a resistance (spring-loaded stop) is met. Insert the tip into the solution. Release the pressure on the trigger slowly and evenly. When the tip is full, withdraw it from the solution. To dispense, press the trigger past the first resistance until the final stop is reached. Keep the pipettor upright during use or solution from the tip may contaminate the inside.

Syringes

Syringes should be used by placing the needle in the solution and drawing the plunger up slowly to the required point on the scale. Check the barrel to make sure no air bubbles have been drawn up. Expel slowly and touch the syringe on the edge of the vessel to remove any liquid adhering to the end of the needle. Microsyringes should always be cleaned before and after use by repeatedly drawing up and expelling pure solvent. The dead space in the syringe needle can occupy up to 4% of the nominal syringe volume. A way of avoiding such problems is to fill the dead space with an inert substance (e.g. silicone oil) after sampling. An alternative is to use a syringe where the plunger occupies the needle space (available for small volumes only).

Balances

These can be used to weigh accurately how much liquid you have dispensed. Convert mass to volume using the equation:

$$\text{mass/density} = \text{volume} \qquad [5.1]$$

e.g. 0.009 kg of a liquid with a density of 1 000 kg m^{-3} = 9.0×10^{-6} m^3 (9.0 ml). Densities of common solvents can be found in Lide (1990). You will need to know the liquid's temperature, as density is temperature dependent.

Holding and storing liquids

Test tubes

These are used for colour tests, small-scale reactions, holding cultures, etc. The tube can be sterilized by heating the inside and maintained in this state with a cap or cotton wool plug.

Beakers

Beakers are used for general purposes, e.g. heating a solvent while the solute dissolves. They may have volume gradations on the side: these are *not* accurate and should only be used where approximations will suffice.

Conical (Erlenmeyer) flasks
These are used for storage of solutions; their small mouth reduces evaporation and makes them easier to seal. Volume gradations, where present, are *not* accurate.

Bottles and vials
These are used when the solution needs to be well sealed for safety, sterility or to prevent evaporation or oxidation. They usually have a screw or ground glass stopper to eliminate evaporation and contamination. Many types are available, including 'bijou', 'McCartney', 'universal', and 'Winkler'.

Storing light-sensitive chemicals — use a coloured vessel or wrap aluminium foil around a clear vessel.

You should clearly label all stored solutions (see p. 19). Seal vessels in an appropriate way, e.g. using a stopper or a sealing film such as Parafilm® or Nescofilm® to prevent evaporation. To avoid degradation store your solution in a fridge, but allow it to reach room temperature before use. Unless a solution has been sterilized or is toxic, microbes will start growing, so older solutions should be checked visually for contamination.

Creating specialized apparatus

Glassware systems incorporating ground glass connections such as Quickfit® are useful for setting up combinations of standard glass components, e.g. for chemical reactions. In project work, you may need to adapt standard forms of glassware for a special need. A glassblowing service (often available in chemistry departments) can make special items to order.

Choosing between glass and plastic

Bear in mind the following points:

Reactivity
Plastic vessels often distort at relatively low temperatures; they may be inflammable, may dissolve in certain organic solvents and may be affected by prolonged exposure to ultraviolet (UV) light. Some plasticizers have been shown to have biological activity.

Glass may adsorb ions and other molecules and then leach them into solutions, especially in alkaline conditions. Pyrex® glass is stronger than ordinary soda glass and can withstand temperatures up to 500°C.

Rigidity and resilience
Plastic vessels are not recommended where volume is critical as they may distort through time. Glass vessels are more easily broken than plastic, which is particularly important for centrifugation (see p. 192).

Opacity
Both glass and plastic absorb light in the UV range of the EMR spectrum. Quartz should be used where this is important, e.g. in cuvettes for UV spectrophotometry (see p. 196).

Disposability
Plastic items may be cheap enough to make them disposable, an advantage where there is a risk of chemical or microbial contamination.

Cleaning glass and plastic

Beware the possibility of contamination arising from prior use of chemicals or inadequate rinsing following washing. A thorough rinse with distilled or deionized water immediately before use will remove dust and other deposits. 'Strong' basic detergents (e.g. Pyroneg®) are good for solubilizing acidic deposits. If there is a risk of basic deposits remaining, use an acid wash. If there is a risk of contamination from organic deposits, a rinse with Analar® grade ethanol is recommended. Glassware can be sterilized by washing with a sodium hypochlorite bleach such as Chloros® or with sodium metabisulphite — dilute as recommended before use and rinse thoroughly with sterile water after use. Alternatively, heat glassware to at least 121 °C for 15 min in an autoclave or 160 °C for 3 h in a dry oven. Glassware that must be specially clean (e.g. for a micronutrient study) should be washed in a chromic acid bath, but this involves toxic and corrosive chemicals and should only be used under supervision.

Acid washing — use dilute acid, e.g. 100 mmol l^{-1} (100 mol m^{-3}) HCl. Rinse thoroughly at least three times with distilled or deionized water.

Safety with glass

Many minor accidents in the laboratory are due to lack of care with glassware. You should observe the following precautions.

- Wear safety glasses when there is *any* risk of glass breakage, e.g. when using low pressures or heating solutions.
- If heating glassware, use a 'soft' Bunsen flame — this avoids creating a hot spot where cracks may start. Always use tongs or special heat-resistant gloves when handling hot glassware.
- Don't use chipped or cracked glassware — it may break under very slight strains.
- Never carry large bottles by their necks — support them with a hand underneath or, better still, carry them in a basket.
- Take care when attaching tubing to glass tubes and when putting glass tubes into bungs — wear a pair of thick fabric or leather gloves if doing these tasks.
- Don't force bungs too firmly into bottles — they can be very difficult to remove. If you need a tight seal, use a screw top bottle with a rubber or plastic seal.
- Dispose of broken glass with great care — always put it into the correct bin.

6 Basic laboratory procedures

Using chemicals

Safety aspects

In practical classes, the person in charge has a responsibility to inform you of any hazards associated with the use of chemicals. In project work, your first duty when using an unfamiliar chemical is to find out about its properties, especially those relating to safety. The COSHH Regulations (1988) state that before using any chemical you must find out whether safety precautions need to be taken and complete the appropriate forms confirming that you appreciate the risks involved. Your department must provide the relevant information to allow you to do this. If your supervisor has filled out the form, read it carefully before signing.

Key safety points when handling chemicals are:

- Treat all chemicals as potentially dangerous.
- Wear a laboratory coat with buttons fastened.
- Locate relevant safety devices such as eye bath, fire extinguisher, first-aid kit, etc.
- Wear gloves and safety glasses for toxic, irritant or corrosive chemicals and carry out procedures with them in a fume cupboard.
- Use aids such as pipette fillers to minimize risk of contact.
- Extinguish all naked flames when working with flammable substances.
- Never smoke, eat or drink where chemicals are handled.
- Label solutions appropriately (see pp. 10, 19, 184).
- Report all spillages and clean them up properly.
- Dispose of chemicals in the correct manner.

The Merck Index (Budavari *et al.*, 1989) is a useful supplementary source of information on chemicals; see Fig. 6.1.

8544. Sodium Chloride. Salt; common salt. ClNa; mol wt 58.45. Cl 60.66%, Na 39.34%. NaCl. The article of commerce is also known as *table salt, rock salt* or *sea salt*. Occurs in nature as the mineral halite. Produced by mining (rock salt), by evaporation of brine from underground salt deposits and from sea water by solar evaporation: Faith, Keys & Clark's *Industrial Chemicals*. F. A. Lowenheim,M.K. Moran, Eds. (Wiley-Interscience, New York, 4th ed., 1975) pp 722-730. Comprehensive monograph: D. W. Kaufmann, *Sodium Chloride,* ACS Monograph Series no. 145 (Reinhold, New York, 1960) 743 pp.

Cubic, white crystals, granules, or powder; colorless and transparent or translucent when in large crystals. d 2.17. The salt of commerce usually contains some calcium and magnesium chlorides which absorb moisture and make it cake. mp 804° and begins to volatilize at a little above this temp. One gram dissolves in 2.8 ml water at 25°, in 2.6 ml boiling water, in 10 ml glycerol; very slightly sol in alcohol. Its soly in water is decreased by HCl and it is almost insol in concd HCl. Its aq soln is neutral. pH: 6.7-7.3. d of satd aq soln at 25° is 1.202. A 23% aq soln of sodium chloride freezes at — 20.5°C (5°F). LD_{50} orally in rats: 3.75 g/kg, Boyd. Shanas, Arch. Int. Pharmacodyn. Ther. **144**, 86 (1963).

Note: Blusalt, a brand of sodium chloride contg trace amounts of cobalt, iodine, iron, copper, manganese, zinc is used in farm animals.

Human Toxicity: Not generally considered poisonous. Accidental substitution of NaCl for lactose in baby formulas has caused fatal poisoning.

USE: Natural salt is the source of chlorine and of sodium as well as of all, or practically all, their compds, e.g., hydrochloric acid, chlorates, sodium carbonate, hydroxide, etc.; for preserving foods; manuf soap, dyes - to salt them out; in freezing mixtures; for dyeing and printing fabrics, glazing pottery, curing hides; metallurgy of tin and other metals.

THERAP CAT: Electrolyte replenisher, emetic; topical anti-inflammatory.

THERAP CAT (VET): Essential nutrient factor. May be given orally as emetic, stomachic, laxative or to stimulate thirst (prevention of calculi). Intravenously as isotonic solution to raise blood volume, to combat dehydration. Locally as wound irrigant, rectal douche.

Fig. 6.1 Example of typical *Merck Index* entry showing type of information given for each chemical.

Selection

Chemicals are supplied in various degrees of purity and this is always stated on the manufacturer's containers. Suppliers differ in the names given to the grades and there is no conformity in purity standards. Very pure chemicals cost more, sometimes a lot more, and should only be used if the situation demands. If you need to order a chemical, your department will have a defined procedure for doing this.

Using chemicals — be considerate to others: always return store room chemicals promptly to the correct place. Report when supplies are getting low to the person responsible for looking after the store.

Making up a solution of defined concentration

Box 6.1 shows the steps involved in making up a solution. The concentration you require is likely to be defined by a protocol you are following and the grade of chemical and supplier may also be specified. Success may depend on using the same source and quality, e.g. with enzyme work. To avoid waste, think carefully about the volume of solution you require, though it is always a good idea to err on the high side because you may spill some or make a mistake when dispensing it. Try to choose one of the standard volumes for measuring vessels, as this will make measuring-out easier.

If you empty an aspirator or wash bottle, fill it up from the appropriate source!

Use distilled or deionized water to make up aqueous solutions and stir to make sure all the chemical is dissolved. Magnetic stirrers are the most convenient means of doing this: carefully drop a clean magnetic stirrer bar ('flea') in the beaker, avoiding splashing; place the beaker centrally on the stirrer plate, switch on the stirrer and gradually increase the speed of stirring.

Box 6.1 How to make up an aqueous solution of known concentration

1. **Find out or decide the concentration of chemical required** and the degree of purity necessary.
2. **Decide on the volume of solution required.**
3. **Find out the molecular mass of the chemical.** This is the sum of the atomic (elemental) masses of the component element and can be found on the container. if the chemical is hydrated, i.e. has water molecules associated with it, these must be included when calculating the mass required.
4. **Work out the mass of chemical that will give the concentration desired in the volume required.** It is often easiest to think in terms of SI units.
   Suppose your protocol states that you need 100 ml of 10 mmol l^{-1} KCl.
   (a) Start by converting this to 100×10^{-6} m^3 of 10 mol m^{-3} KCl.
   (b) The required number of mol is thus $(100 \times 10^{-6}) \times (10)$.
   (c) Each mol of KCl weighs 72.56 g (the molecular mass).
   (d) Therefore you need to make up 72.56×10^{-3} g = 72.56 mg KCl to 100×10^{-6} m^3 (100 ml) with distilled water.

   See Box 7.1 for alternative methods using non-SI units.
5. **Weigh out the required mass of chemical to an appropriate accuracy.** If the mass is too small to weigh to the desired degree of accuracy, consider the following options:
   (a) Make up a greater volume of solution.
   (b) Make up a stock solution which can be diluted at a later stage.
   (c) Weigh the mass first, and calculate what volume to make the solution up to afterwards.
6. **Add the chemical to a beaker or conical flask then add a little less water than the final amount required.** If some of the chemical sticks to the paper, foil or weighing boat, use some of the water to wash it off.
7. **Stir and, if necessary, heat the solution to ensure all the chemical dissolves.** You can determine when this has happened visually by observing the disappearance of the crystals or powder.
8. **If required, check and adjust the pH of the solution when cool** (see p. 31).
9. **Make up the solution to the desired volume.** If the concentration needs to be accurate, use a volumetric flask; if a high degree of accuracy is not required, use a measuring cylinder.
   (a) Pour the solution from the beaker into the measuring vessel using a funnel to avoid spillage.
   (b) Make up the volume so that the meniscus comes up to the appropriate measurement line (p. 13). For accurate work, rinse out the original vessel and use this liquid to make up the volume.
10. **Transfer the solution to a reagent bottle or a conical flask and label the vessel clearly.**

Solubility problems — if your chemical does not dissolve after a reasonable time:
- check the limits of solubility for your compound (see *Merck Index*),
- check the pH of the solution — solubility is often low at pH extremes.

When the crystals or powder have completely dissolved, switch off and retrieve the flea with a magnet or another flea. Take care not to contaminate your solution when you do this and rinse the flea with distilled water.

'Obstinate' solutions may require heating but do this only if you know that the chemical will not be damaged at the temperature used. Use a stirrer-heater to keep the solution mixed as you heat it. Allow the solution to cool down before you measure its volume or pH as these are affected by temperature.

Stock solutions

Stock solutions are valuable when making up a range of solutions containing different concentrations of a reagent or if the solutions have some common ingredients. They also save work if the same solution is used over a prolonged period (e.g. a nutrient solution). The stock solution is more concentrated than the final requirement and is diluted as appropriate when the final solutions are made up. The principle is best illustrated with an example (Table 6.1).

Serial dilutions

These are used when a range of concentrations of a compound or microbial suspension is required in the same base medium. Two methods commonly used are doubling dilution and decimal dilution.

Table 6.1 Use of stock solutions. Suppose you need a set of solutions 10 ml in volume containing differing concentrations of KCl, with and without reagent Q. You decide to make up a stock of KCl at twice the maximum required concentration (50 mol m^{-3} = 50 mmol l^{-1}) and a stock of reagent Q at twice its required concentration. The table shows how you might use these stocks to make up the media you require. Note that the total volumes of stock you require can be calculated from the table (end column).

	Volume of stock required to make required solutions (ml)						
Stock solutions	No KCl plus Q	No KCl minus Q	15 mmol l^{-1} KCl plus Q	15 mmol l^{-1} KCl minus Q	25 mmol l^{-1} KCl plus Q	25 mmol l^{-1} KCl minus Q	Total volume of stock required (ml)
50 mmol l^{-1} KCl	0	0	3	3	5	5	16
[reagent Q] × 2	5	0	5	0	5	0	15
Water	5	10	2	7	0	5	29
Total	10	10	10	10	10	10	60

Preparing repeated serial dilutions — adding the appropriate amount of base medium to several vessels beforehand will save time when preparing a dilution series.

- Doubling dilutions: first, make up the most concentrated solution at twice the volume required. Measure out half and add a similar volume of base medium, mix thoroughly and repeat. Concentrations obtained will be 1, $\frac{1}{2}$, $\frac{1}{4}$, $\frac{1}{8}$, etc., times the original.
- Decimal dilutions: first, make up the most concentrated solution required in excess. Measure out one-tenth and add nine times as much base medium, mix thoroughly and repeat. Concentrations obtained will be 1, $\frac{1}{10}$, $\frac{1}{100}$, $\frac{1}{1000}$, etc., times the original (i.e. 10^0, 10^{-1}, 10^{-2}, 10^{-3} fold dilutions).

Solutions must be thoroughly mixed before measuring out volumes for the next dilution. Use a fresh measuring vessel for each dilution to avoid contamination, or wash your vessel thoroughly between dilutions. Clearly label the vessel containing each dilution when it is made: it is easy to get confused! When deciding on the volumes required, allow for the aliquot removed when making up the next member in the series. Remember to discard the excess from the last in the series if volumes are critical.

Storing chemicals and solutions

Labile chemicals may be stored in a fridge or freezer. Take special care when using chemicals that have been stored at low temperature: the container and its contents must be warmed up to room temperature before use, otherwise water vapour will condense on the chemical. This may render any weighing you do meaningless and it could ruin the chemical. Other chemicals may need to be kept in a desiccator, especially if they are deliquescent.

Label all stored chemicals clearly with the following information: the chemical name (if a solution, state solute(s), concentration(s) and pH if measured), the date made up, and your name.

Weighing — never weigh anything directly onto a balance's pan: you may contaminate it for other users or pick up contamination from previous users. Use a weighing boat or a slip of aluminium foil. Otherwise, choose a suitable vessel like a beaker, conical flask or aluminium tray.

Using balances

Electronic balances with digital readouts are now favoured over mechanical types: they are easy to read and their self-taring feature means the mass of the weighing boat or container can be subtracted automatically before weighing an object. The most common type offers accuracy down to 1 mg over the range 1 mg to 160 g, which is suitable for most biological applications.

To operate a standard self-taring balance:

1. Check that it is level, using the adjustable feet to centre the bubble in the spirit level (usually at the back of the machine). For accurate work, make sure a draught shield is on the balance.
2. Place an empty vessel on the balance pan and allow the reading to stabilize. *If the object is larger than the pan, take care that no part rests on the body of the balance or the draught shield as this will invalidate the reading.* Press the tare bar to bring the reading to zero.
3. Place the chemical or object carefully in the vessel (powdered chemicals should be dispensed with a suitably sized clean spatula).
4. Allow the reading to stabilize and make a note of the value.
5. If you add excess chemical, take great care when removing it. Switch off if you need to clean any deposit accidentally left on or around the balance.

Larger masses should be weighed on a top-loading balance to an appropriate degree of accuracy. Take care to note the limits for the balance: while most have devices to protect against overloading, you may damage the mechanism. In the field, spring or battery-operated balances may be preferred. Try to find a place out of the wind to use them. For extremely small masses, there are electrical balances that can weigh down to 1 μg, but these are very delicate and must be used under supervision.

Table 6.2 Suitability of devices for measuring linear dimensions

Measurement device	Suitable lengths	Degree of accuracy
Eyepiece graticule (light microscopy)	1 μm to 10 mm	0.5 μm
Vernier calipers	1–100 mm	0.1 mm
Ruler	10 mm to 1 m	1.0 mm
Tape measure	10 mm to 30 m	1.0 mm
Optical surveying devices	1 m to 100 m	0.1 m

See Box 6.2 for method of using Vernier calipers

Measuring length and area

When measuring linear dimensions, the device you need depends on the size of object you are measuring and the accuracy demanded (Table 6.2).

For many regularly shaped objects, area can be estimated from linear dimensions (see p. 225). The areas of irregular shapes can be measured with

Box 6.2 How to use Vernier calipers

(a)

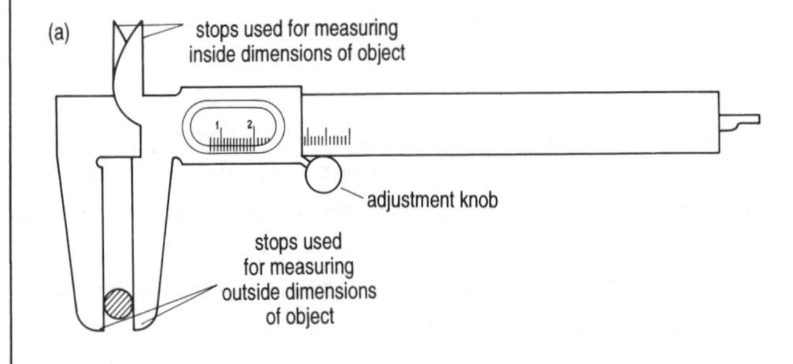

(b)

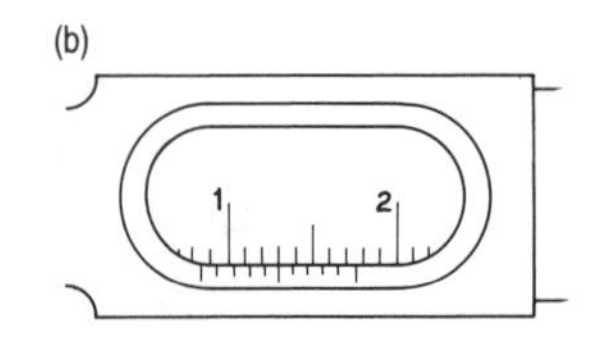

Fig. 6.2 (a) Vernier caliper. (b) Vernier measurement scales. Note that numbers on the scale refer to centimetres. Vernier scales consist of two numerical scales running side by side, the moving one being shorter with ten divisions compressed into the length for nine on the longer, static one. Use Vernier calipers to measure objects to the nearest 0.1 mm:

1. **Clamp the stops lightly over the object** as in Fig. 6.2(a), taking care not to deform it.
2. **Read off the number of whole millimetres** by taking the value on the fixed scale lying to the left of the first line on the moving (short) scale, i.e. 8 mm in Fig. 6.2(b).
3. **Read off 0.1 mm value** by finding which line in the moving scale corresponds most closely with a line on the fixed scale, i.e. 0.5 mm in Fig. 6.2(b). If the zero of the short scale corresponded to a whole number on the static scale, then record 0.0 mm as this shows fully the accuracy of the measurement.
4. **Add these numbers to give the final reading**, i.e. 8.5 mm in Fig. 6.2(b).

an optical measuring device or a planimeter. These have the benefits of speed and ease of use; instructions are machine specific. A simple 'low tech' method is to trace objects onto good quality paper or to photocopy them. If the outline is then cut round, the area can be estimated by weighing the cutout and comparing to the mass of a piece of the same paper of known area. Avoid getting moisture from the specimen onto the paper as this will affect the reading.

Measuring and controlling temperature

Heating specimens

Lighting a Bunsen — close the air hole first. Open the air hole if you need a hotter, more concentrated flame: the hottest part of the flame is just above the apex of the blue cone in its centre.

Care is required when heating specimens — there is a danger of fire whenever organic material is heated and a danger of scalding from heated liquids. Use a thermostatically controlled electric heater if possible. If using a Bunsen burner, keep the flame well away from yourself and your clothing (tie back long hair). Use a non-flammable mat beneath a Bunsen to protect the bench. Switch off when no longer required.

Ovens and drying cabinets may be used to dry specimens or glassware. They are normally thermostatically controlled. If drying organic material for dry weight measurement, do so at about 80°C to avoid caramelizing the specimen. Always state the actual temperature used as this affects results. Check that all water has been driven off by weighing until a constant mass is reached.

Cooling specimens

Heating/cooling glass vessels — take care if heating or cooling glass vessels rapidly as they may break when heat stressed. Freezing aqueous solutions in thin-walled glass vessels is risky because ice expansion may break the glass.

Fridges and freezers are used for storing stock solutions and chemicals that would either break down or become contaminated at room temperature. Normal fridge and freezer temperatures are about 4°C and −15°C respectively. Ice baths can be used when reactants must be kept at or close to 0°C. Most biology departments will have a machine which provides flaked ice for use in these baths. If common salt is mixed with ice, temperatures below 0°C can be achieved. A mixture of ethanol and solid CO_2 will provide a temperature of −72°C if required. To freeze a specimen quickly, immerse in liquid N_2 (−196°C) using tongs and wearing an apron and thick gloves, as splashes will damage your skin. Always work in a well-ventilated room.

Maintaining cultures or specimens at constant temperature

Using thermometers — some are calibrated for use in air, others require partial immersion in liquid and others total immersion — check before use. If a mercury thermometer is broken, report the spillage as mercury is a poison.

Thermostatically controlled temperature rooms and incubators can be used to maintain temperature at a desired level. Always check with a thermometer or thermograph that the thermostat is accurate enough for your study. To achieve a controlled temperature on a smaller scale, e.g. for an oxygen electrode (p. 185), use a water bath. These usually incorporate heating elements, a circulating mechanism and a thermostat. Baths for sub-ambient temperatures have a cooling element.

Controlling atmospheric conditions

Gas composition

Example
Water vapour can be removed by passing gas over dehydrated $CaCO_3$ and CO_2 may be removed by bubbling through KOH solution.

The atmosphere may be 'scrubbed' of certain gases by passing through a U-tube or Dreschel bottle containing an appropriate chemical or solution.

For accurate control of gas concentrations, use cylinders of pure gas; the contents can be mixed to give specified concentrations by controlling individual flow rates. The cylinder head regulator (Fig. 6.3) allows you to control the

Using a hydrogen cylinder — when placing a regulator on a hydrogen cylinder, note that it is tightened *anticlockwise* to avoid the chance of this potentially explosive gas being incorrectly used.

Using glassware — glass items kept at very low or high pressures should be contained within a metal cage to minimize the risk of injury.

Using a timer — always set the alarm before the critical time, so that you have adequate time to react.

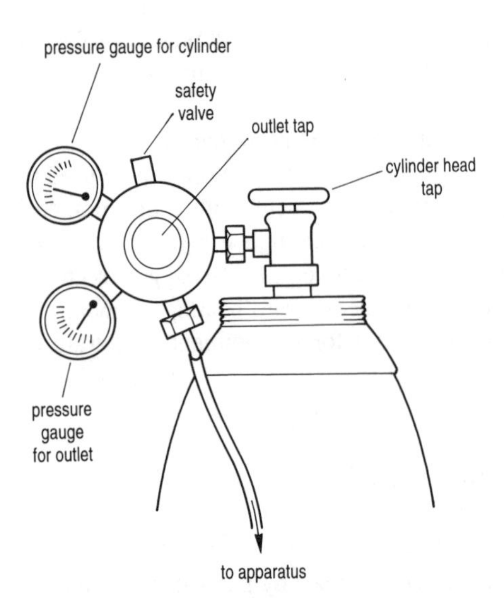

Fig. 6.3 Parts of a cylinder head regulator.

pressure (and hence flow rate) of gas; adjust using the controls on the regulator or with spanners of appropriate size. Before use, ensure the regulator outlet tap is off (turn anticlockwise), then switch on at the cylinder (turn clockwise) — the cylinder dial will give you the pressure reading for the cylinder contents. Now switch on at the regulator outlet (turn clockwise) and adjust to desired pressure/flow setting. To switch off, carry out the above directions in reverse order.

Pressure

Many forms of pumps are used to pressurize or provide partial vacuum, usually to force gas or liquid movement, each having specific instructions for use. Many laboratories are supplied with 'vacuum' (suction) and pressurized air lines that are useful for procedures such as vacuum-assisted filtration. Make sure you switch off the taps after use.

Measuring time

Many experiments and observations need to be carefully timed. Large-faced stopclocks allow you to set and follow 'experimental time' and remove the potential difficulties in calculating this from 'real time' on a watch or clock. Some timers incorporate an alarm which you can set to warn when readings or operations must be carried out; 24-h timers are available for controlling light and temperature regimes.

Miscellaneous methods for treating samples

Homogenizing

This involves breaking up and mixing specimens to give a uniform preparation. Blenders are used to homogenize animal and plant material and work best when an adequate volume of liquid is present: buffer solution may be added to specimens for this purpose. Use in short bursts to avoid overheating the motor and the sample. A pestle and mortar is used for grinding up specimens. Acid-washed sand grains can be added to help break up the tissues. For quantitative work with brittle samples, care must be taken not to lose material when the sample breaks into fragments.

Separation of components of mixtures and solutions

Particulate solids (e.g. soils) can be separated on the basis of size using sieves. These are available in stacking forms which fit on automatic shakers. Sieves with the largest pores are placed at the top and the assembly is shaken for a fixed time until the sample separates. Suspensions of solids in liquids may be separated out by centrifugation (see p. 190) or filtration. Various forms of filter paper are available having different porosities and purities. Vacuum-assisted filtration speeds up the process and is best carried out with a filter funnel attached to a filter flask. Filtration through pre-sterilized membranes with very small pores (e.g. the Millipore® type) is an excellent method of sterilizing small volumes of solution. Solvents can be removed from solutes by heating, by rotary film evaporation under low pressure and, for water, by freeze drying. The latter two are especially useful for heat-labile solutes — refer to the manufacturers' specific instructions for use.

7 Principles of solution chemistry

A solution is a homogeneous liquid, formed by the addition of solutes to a solvent (usually water in biological systems). The behaviour of solutions is determined by the type of solutes involved and by their proportions, relative to the solvent. Many laboratory exercises involve calculation of concentrations, e.g. when preparing an experimental solution at a particular concentration, or when expressing data in terms of solute concentration. Make sure that you understand the basic principles set out in this chapter before you tackle such exercises.

Solutes can affect the properties of solutions in several ways, including:

Electrolytic behaviour

This occurs where individual molecules of an electrolyte dissociate in aqueous solution to give two or more charged particles (ions). For a strong electrolyte, e.g. NaCl, dissociation is essentially complete. In contrast, a weak electrolyte, e.g. acetic acid, will be only partly dissociated, depending upon the pH and temperature of the solution (p. 32).

Osmotic effects

These are the result of solute particles lowering the effective concentration of the solvent (water). These effects are particularly relevant to biologists since membranes are far more permeable to water than to most solutes. Water moves across biological membranes from the solution with the higher effective water concentration to that with the lower effective water concentration (osmosis).

Ideal/non-ideal behaviour

This occurs because solutions of real substances do not necessarily conform to the theoretical relationships predicted for dilute solutions of so-called ideal solutes. It is often necessary to take account of the non-ideal behaviour of real solutions, especially at high solute concentrations (see Lide, 1990, for appropriate data).

Concentration

Expressing solute concentrations — you should use SI units wherever possible. However, you are likely to meet non-SI concentrations and you must be able to deal with these units too.

In SI units (p. 44), the concentration of a solute is expressed in mol m^{-3}, which is convenient for most biological purposes. The concentration of a solute is usually symbolized by square brackets, e.g. [NaCl]. Details of how to prepare a solution using SI units are given on p. 18.

A number of alternative ways of expressing the relative amounts of solute and solvent are in general use, and you may come across these terms in your practical work, or in the literature.

Molarity

Example
A 1.0 molar solution of NaCl would contain 58.44 g NaCl (the molecular mass) per litre of solution.

This is the term used to denote concentration, *C*, expressed as moles of solute per litre volume of solution (mol l^{-1} = 'molar'). This non-SI term continues to find widespread usage, in part because of the familiarity of working scientists with the term, but also because laboratory glassware is calibrated in millilitres and litres, making the preparation of molar and millimolar solutions relatively straightforward. However, the symbols in common use for molar (M) and millimolar (mM) solutions are at odds with the SI system and many people

Box 7.1 Useful procedures for calculations involving molar concentrations

1. **Preparing a solution of defined molarity.** For a solute of known molecular mass, the following relationship can be applied:

$$C = \frac{\text{mass of solute/molecular mass}}{\text{volume of solution}} \quad [7.1]$$

So, if you wanted to make up 200 ml (0.2 l) of an aqueous solution of NaCl (molecular mass 58.44 g) at a concentration of 500 mmol l^{-1} (0.5 mol l^{-1}), you could calculate the amount of NaCl required by inserting these values into eqn [7.1]:

$$0.5 = \frac{\text{mass of solute/58.44}}{0.2}$$

which can be rearranged to

$$\text{mass of solute} = 0.5 \times 0.2 \times 58.44 = 5.844 \text{ g}$$

The same relationship can be used to calculate the concentration of a solution containing a known amount of a solute, e.g. if 21.1 g of NaCl were made up to a volume of 100 ml (0.1 l), this would give

$$[\text{NaCl}] = \frac{21.1/58.44}{0.1} = 3.61 \text{ mol l}^{-1}$$

2. **Dilutions and concentrations.** The following relationship is very useful if you are diluting (or concentrating) a solution:

$$C_1 V_1 = C_2 V_2 \quad [7.2]$$

where C_1 and C_2 are the initial and final concentrations, while V_1 and V_2 are their respective volumes: each pair must be expressed in the same units. Thus, if you wanted to dilute 200 ml of 0.5 mol l^{-1} NaCl to give a final molarity of 0.1 mol l^{-1}, then, by substitution into eqn [7.2]:

$$0.5 \times 200 = 0.1 \times V_2$$

Thus V_2 = 1 000 ml (in other words, you would have to add water to 200 ml of 0.5 mol l^{-1} NaCl to give a final volume of 1 000 ml to obtain a 0.1 mol l^{-1} solution).

3. **Interconversion.** A simple way of interconverting amounts and volumes of any particular solution is to divide the amount and volume by a factor of 10^3: thus a molar solution of a substance contains 1 mol l^{-1}, which is equivalent to 1 mmol ml^{-1}, or 1 μmol μl^{-1}, or 1 nmol nl^{-1}, etc. You may find this technique useful when calculating the amount of substance present in a small volume of solution of known concentration, e.g. to calculate the amount of NaCl present in 50 μl of a solution with a concentration (molarity) of 0.5 mol l^{-1} NaCl:

   (a) this is equivalent to 0.5 μmol μl^{-1};
   (b) therefore 50 μl will contain 50 × 0.5 μmol = 25 μmol.

Alternatively, you may prefer to convert to primary SI units, for ease of calculation (see Box 6.1).

now prefer to use mol l^{-1} and mmol l^{-1} respectively, to avoid confusion. Box 7.1 gives details of some useful approaches to calculations involving molarities.

Example
A 0.5 molal solution of NaCl would contain 58.44 × 0.5 = 29.22 g NaCl per kg of water.

Molality

This is used to express the concentration of solute relative to the mass of solvent, i.e. mol kg^{-1}. Molality is a temperature-independent means of expressing solute concentration, rarely used except when the osmotic properties of a solution are of interest (p. 26).

Example
A 5% w/w sucrose solution contains 5 g sucrose and 95 g water (= 95 ml water, assuming a density of 1 g ml^{-1}).

Per cent composition (% w/w)

This is the solute mass (in g) per 100 g solution. The advantage of this expression is the ease with which a solution can be prepared, since it simply requires each component to be pre-weighed (for water, a volumetric measurement may be used, e.g. using a measuring cylinder) and then mixed together. Similar terms are parts per thousand (‰), i.e. mg g^{-1}, and parts per million (p.p.m.), i.e. μg g^{-1}.

Per cent concentration (% w/v and % v/v)

For solutes added in solid form, this is the number of grams of solute per 100 ml solution. This is more commonly used than per cent composition, since solutions can be accurately prepared by weighing out the required amount of solute and

> **Example**
> *A 5% w/v sucrose solution contains 5 g sucrose in 100 ml of solution.*
> *A 5% v/v glycerol solution would contain 5 ml glycerol and 95 ml water.*
>
> *Note that when water is the solvent this is often not specified in the expression, e.g. a 70% v/v ethanol solution contains 70% ethanol and 30% water.*

then making this up to a known volume using a volumetric flask. The equivalent expression for liquid solutes is % v/v.

The principal use of mass/mass or mass/volume terms is for solutes whose molecular mass is unknown (e.g. cellular proteins), or for mixtures of certain classes of substance (e.g. total salt in sea water). You should *never* use the per cent term without specifying how the solution was prepared, i.e. by using the qualifier w/w, w/v or v/v.

Activity (a)

This is a term used to describe the effective concentration of a solute. In dilute solutions, solutes can be considered to behave according to ideal (thermodynamic) principles, i.e. they will have an effective concentration equivalent to the actual concentration. However, in concentrated solutions ($\gtrsim 500$ mol m^{-3}), the behaviour of solutes is non-ideal, and their effective concentration (activity) will be less than the actual concentration C. The ratio between the effective concentration and the actual concentration is called the activity coefficient (γ) where

$$\gamma = \frac{a}{C} \qquad [7.3]$$

Table 7.1 Activity coefficient of NaCl solutions as a function of molality. Data from Robinson and Stokes (1970)

Molality	Activity coefficient at 25°C
0.1	0.778
0.5	0.681
1.0	0.657
2.0	0.668
4.0	0.783
6.0	0.986

Equation [7.3] can be used for SI units (mol m^{-3}), molarity (mol l^{-1}) or molality (mol kg^{-1}). In all cases, γ is a dimensionless term, since a and C are expressed in the same units. The activity coefficient of a solute is effectively unity in dilute solution, decreasing as the solute concentration increases (Table 7.1). At high concentrations of certain ionic solutes, γ may be greater than unity.

Activity is often the correct expression for theoretical relationships involving solute concentration (e.g. where a property of the solution is dependent on concentration). However, for most practical purposes, it is possible to use the *actual* concentration of a solute rather than the activity, since the difference between the two terms can be ignored for dilute solutions. The particular use of the term 'water activity' is considered below, since it is based on the mole fraction of solvent, rather than the effective concentration of solute.

> **Example**
> *A solution of NaCl with a molality of 0.5 mol kg^{-1} has an activity coefficient of 0.681 at 25°C and a molal activity of 0.5 × 0.681 = 0.340 mol kg^{-1}.*

Equivalent mass (equivalent weight)

Equivalence and normality are outdated terms, although you may come across them in older texts. They apply to certain solutes whose reactions involve the transfer of charged ions, e.g. acids and alkalis (which may be involved in H^+ or OH^- transfer), and electrolytes (which form cations and anions that may take part in further reactions). These two terms take into account the valency of the charged solutes. Thus the equivalent mass of an ion is its molecular mass divided by its valency (ignoring the sign), expressed in grams per equivalent (eq) according to the relationship:

$$\text{equivalent mass} = \frac{\text{molecular mass}}{\text{valency}} \qquad [7.4]$$

> **Examples**
> *For carbonate ions (CO_3^{2-}), with a molecular mass of 60.00 and a valency of 2, the equivalent mass is 60.00/2 = 30.00 g eq^{-1}.*
> *For sulphuric acid (H_2SO_4, molecular mass 98.08), where 2 hydrogen ions are available, the equivalent mass is 98.08/2 = 49.04 g eq^{-1}.*

For acids and alkalis, the equivalent mass is the mass of substance that will provide 1 mol of either H^+ or OH^- ions in a reaction, obtained by dividing the molecular mass by the number of available ions (n), using n as the denominator in eqn [7.4].

Normality

A 1 normal solution (1 N) is one that contains one equivalent weight of a substance per litre of solution. The general formula is:

$$\text{normality} = \frac{\text{mass of substance per litre}}{\text{equivalent weight}} \qquad [7.5]$$

> **Example**
> *A 0.5 N solution of sulphuric acid would contain 0.5 × 49.04 = 24.52 g l^{-1}.*

Osmolarity

This non-SI expression is used to describe the number of moles of osmotically active solute particles per litre of solution (osmol l^{-1}). The need for such a term arises because some molecules dissociate to give more than one osmotically active particle in aqueous solution.

> **Example**
> *Under ideal conditions, 1 mol of NaCl dissolved in water would give 1 mol of Na^+ ions and 1 mol of Cl^- ions, equivalent to a theoretical osmolarity of 2 osmol l^{-1}.*

Osmolality

This term describes the number of moles of osmotically active solute particles per unit mass of solvent (osmol kg^{-1}). For an ideal solute, the osmolality can be determined by multiplying the molality by n, the number of solute particles produced in solution (e.g. for NaCl, $n = 2$). However, for real solutes, a correction factor (the osmotic coefficient, ϕ) is used:

$$\text{osmolality} = \text{molality} \times n \times \phi \qquad [7.6]$$

> **Example**
> *A 1.0 mol kg^{-1} solution of NaCl has an osmotic coefficient of 0.936 at 25°C and an osmolality of 1.0 × 2 × 0.936 = 1.872 osmol kg^{-1}.*

If necessary, the osmotic coefficients of a particular solute can be obtained from tables (e.g. Table 7.2). Alternatively, the osmolality of a solution can be measured using an osmometer.

Table 7.2 Osmotic coefficients of NaCl solutions as a function of molality. Data from Robinson and Stokes (1970)

Molality	Osmotic coefficient at 25°C
0.1	0.932
0.5	0.921
1.0	0.936
2.0	0.983
4.0	1.116
6.0	1.271

Colligative properties and their use in osmometry

Several properties vary in direct proportion to the effective number of osmotically active solute particles per unit mass of solvent and can be used to determine the osmolality of a solution. These colligative properties include freezing point, boiling point, and vapour pressure.

An osmometer is an instrument which measures the osmolality of a solution, usually by determining the freezing point depression of the solution in relation to pure water, a technique known as cryoscopic osmometry. A small amount of sample is cooled rapidly and then brought to the freezing point (Fig. 7.1),

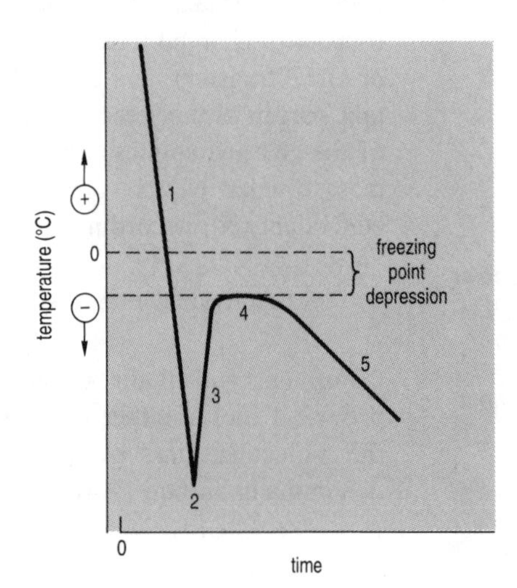

Fig. 7.1 Temperature responses of a cryoscopic osmometer. The response can be subdivided into:
1. initial supercooling
2. initiation of crystallization
3. crystallization/freezing
4. plateau, at the freezing point
5. slow temperature decrease

Using an osmometer — it is vital that the sample holder and probe are clean, otherwise small droplets of the previous sample may be carried over, leading to inaccurate measurement.

which is measured by a temperature-sensitive thermistor probe calibrated in mosmol kg^{-1}. An alternative method is used in vapour pressure osmometry, which measures the relative decrease in the vapour pressure produced in the gas phase when a small sample of the solution is equilibrated within a chamber.

Osmotic properties of solutions

Several inter-related terms can be used to describe the osmotic status of a solution. In addition to osmolality, you may come across the following.

Osmotic pressure

This is based on the concept of a membrane permeable to water, but not to solute molecules. For example, if a sucrose solution is placed on one side and pure water on the other, then a passive driving force will be created and water will diffuse across the membrane into the sucrose solution, since the effective water concentration in the sucrose solution will be lower. The tendency for water to diffuse into the sucrose solution could be counteracted by applying a hydrostatic pressure equivalent to the passive driving force. Thus, the osmotic pressure of a solution is the excess hydrostatic pressure required to prevent the net flow of water across a selectively permeable membrane from a vessel containing pure water into a vessel containing the solution. The SI unit of osmotic pressure is the pascal, Pa (= kg m^{-1} s^{-2}). Older sources may use atmospheres, or bars — conversion factors are given in Table 10.4 (p. 43). Osmotic pressure and osmolality can be interconverted using the expression 1 osmol kg^{-1} = 2.479 MPa at 25°C.

Example
A 1.0 mol kg^{-1} solution of NaCl at 25°C has an osmolality of 1.872 osmol kg^{-1} and an osmotic pressure of 1.872 × 2.479 = 4.641 MPa.

The use of osmotic pressure has been criticized as misleading, since a solution does not exhibit an 'osmotic pressure', unless it is placed on the other side of a selectively permeable membrane from pure water!

Water activity (a_w)

This is a term often used to describe the osmotic behaviour of microbial cells. It is a measure of the relative proportion of water in a solution, expressed in terms of its mole fraction, i.e. the ratio of the number of moles of water (n_w) to the total number of moles of all substances (i.e. water and solutes) in solution (n_t), taking into account the molal activity coefficient of the solvent (i.e. γ_w):

$$a_w = \gamma_w \frac{n_w}{n_t} \qquad [7.7]$$

The water activity of pure water is unity, decreasing as solutes are added. One disadvantage of a_w is the limited change which occurs in response to a change in solute concentration: a 1.0 mol kg^{-1} solution of NaCl has a water activity of 0.967.

Osmolality, osmotic pressure and water activity are measurements based solely on the osmotic properties of a solution, with no regard for any other driving forces, e.g. hydrostatic and gravitational forces. In circumstances where such other forces are important, you will need to measure a variable that takes into account these aspects of water status.

Water potential (hydraulic potential) and its applications

Water potential, Ψ_w, is the most appropriate measure of osmotic status in many areas of biology. It is a term derived from the chemical potential of water (see Nobel, 1991). It expresses the difference between the chemical potential of water in the test system and that of pure water under standard conditions and has units of pressure (i.e. Pa). It is a more appropriate term than osmotic pressure because it is based on sound theoretical principles and because it can be used to predict the direction of passive movement of water, since water will flow down a gradient of chemical potential (i.e. osmosis occurs from a solution with a higher water potential to one with a lower water potential). A solution of pure water at 20°C and at 0.1 MPa pressure (i.e. atmospheric) has a water potential of zero. The addition of solutes will lower the water potential (i.e. make it negative), while the application of pressure, e.g. from hydrostatic or gravitational forces, will raise it (i.e. make it positive).

Examples
A 1.0 mol kg^{-1} solution of NaCl has a (negative) water potential of −4.641 MPa.
A solution of pure water at 0.2 MPa pressure (about 0.1 MPa above atmospheric pressure) has a (positive) water potential of 0.1 MPa.

Often, the two principal components of water potential are referred to as the solute potential, or osmotic potential (Ψ_s, sometimes symbolized as Ψ_π or π) and the hydrostatic pressure potential (Ψ_p) respectively. For a solution at atmospheric pressure, the water potential is due solely to the presence of osmotically active solute molecules (osmotic potential) and may be calculated from the measured osmolality (osmol kg^{-1}) at 25°C, using the relationship:

$$\Psi_w \text{ (MPa)} = \Psi_s \text{ (MPa)} = -2.479 \times \text{osmolality} \qquad [7.8]$$

For aquatic microbial cells, e.g. algae, fungi and bacteria, equilibrated in their growth medium, the water potential of the external medium will be equal to the cellular water potential and the latter can be derived from the measured osmolality of the medium (eqn [7.8]). The water potential of such cells can be subdivided into two major parts, the cell osmotic potential (Ψ_s) and the cell turgor pressure (Ψ_p) as follows:

$$\Psi_w = \Psi_s + \Psi_p \qquad [7.9]$$

This equation ignores the effects of gravitational forces — for systems where gravitational effects are important, e.g. water movement in tall trees, an additional term is required (see Nobel, 1991).

To calculate the relative contribution of the osmotic and pressure terms in eqn [7.9], an estimate of the internal osmolality is required, e.g. by measuring the freezing point depression of expressed intracellular fluid. Once you have values for Ψ_w and Ψ_s, the turgor pressure can be calculated by substitution into eqn [7.9].

For terrestrial plant cells, the water potential may be determined directly using a vapour pressure osmometer, by placing a sample of the material within the osmometer chamber and allowing it to equilibrate. If Ψ_s of expressed sap is then measured, Ψ_p can be determined from eqn [7.9].

The van't Hoff relationship can be used to estimate Ψ_s, by summation of the osmotic potentials due to the major solutes, determined from their concentrations, as:

$$\Psi_s = -RTn\phi C \qquad [7.10]$$

where RT is the product of the universal gas constant and absolute temperature (2 479 J mol^{-1} at 25°C), n and ϕ are as previously defined and C is expressed in SI terms as mol m^{-3}.

8 pH and buffer solutions

Definitions
Acid — a compound that acts as a proton donor.
Base — a compound that acts as a proton acceptor.
Conjugate pair — an acid together with its corresponding base.
Alkali — a compound that liberates hydroxyl ions when it dissociates. Since hydroxyl ions are strongly basic, this will reduce the proton concentration.
Ampholyte — a compound that can act as both an acid and a base. Water is an ampholyte since it may dissociate to give a proton and a hydroxyl ion (amphoteric behaviour).

Key point
pH is strictly the negative logarithm (base 10) of H^+ activity, although in practice H^+ concentration (in kmol m^{-3}, equivalent to mol l^{-1}) is most often used. The pH scale is not strictly SI; nevertheless, it is used widely in biological science.

pH is a measure of the amount of hydrogen ions (H^+) in a solution: this affects the solubility of many substances and the activity of most biological systems, from individual molecules to whole organisms. It is usual to think of aqueous solutions as containing H^+ ions (protons), though protons actually exist in solution in their hydrated form, as hydronium ions (H_3O^+). The proton concentration $[H^+]$ is affected by several factors:

- Ionization (dissociation) of water, which liberates protons and hydroxyl ions in equal quantities, according to the reversible relationship:

$$H_2O \rightleftharpoons H^+ + OH^- \qquad [8.1]$$

- Dissociation of acids, according to the equation:

$$H-A \rightleftharpoons H^+ + A^- \qquad [8.2]$$

  where $H-A$ represents the acid and A^- is the corresponding conjugate base. The dissociation of an acid in water will increase the amount of protons, reducing the amount of hydroxyl ions as water molecules are formed (eqn [8.1]). The addition of a base (usually, as its salt) to water will decrease the amount of H^+, due to the formation of the conjugate acid (eqn [8.2]).
- Dissociation of alkalis, according to the relationship:

$$X-OH \rightleftharpoons X^+ + OH^- \qquad [8.3]$$

  where $X-OH$ represents the undissociated alkali. Since the dissociation of water is reversible (eqn [8.1]), in an aqueous solution the production of hydroxyl ions will effectively act to 'mop up' protons, lowering the proton concentration.

Many compounds act as acids, bases or alkalis: those which are almost completely ionized in solution are usually called strong acids or bases, while weak acids or bases are only slightly ionized in solution (p. 32).

In an aqueous solution, most of the water molecules are not ionized. In fact, the extent of ionization of pure water is constant at any given temperature and is usually expressed in terms of the ion product (or ionization constant) of water, K_w:

$$K_w = [H^+]\,[OH^-] \qquad [8.4]$$

where $[H^+]$ and $[OH^-]$ represent the concentration (strictly, the activity) of protons and hydroxyl ions in solution, expressed as kmol m^{-3} (or mol l^{-1}, using non-SI terminology). At 25°C, the ion product of pure water is 10^{-8} mol^2 m^{-6} (i.e. 10^{-14} mol^2 l^{-2}). This means that the concentration of protons in solution will be 10^{-4} mol m^{-3} (10^{-7} mol l^{-1}), with an equivalent concentration of hydroxyl ions (eqn [8.1]). Since these values are very low and involve negative powers of 10, it is customary to use the pH scale, where:

$$pH = -\log_{10}[H^+] \qquad [8.5]$$

where $[H^+]$ is the proton activity (see p. 25). It is satisfactory to use H^+ concentration in place of activity since they are virtually the same, given the limited dissociation of H_2O.

Table 8.1 Effects of temperature on the ionic product of water (K_w), H^+ ion concentration and pH at neutrality. Values calculated from Lide (1990)

Temp. (°C)	K_w (mol^2 m^{-6})	$[H^+]$ at neutrality (μmol m^{-3})	pH at neutrality
0	0.11×10^{-8}	33.9	7.47
4	0.17×10^{-8}	40.7	7.39
10	0.29×10^{-8}	53.7	7.27
20	0.68×10^{-8}	83.2	7.08
25	1.01×10^{-8}	100.4	7.00
30	1.47×10^{-8}	120.2	6.92
37	2.39×10^{-8}	154.9	6.81
45	4.02×10^{-8}	199.5	6.70

The value where an equal amount of H^+ and OH^- ions are present is

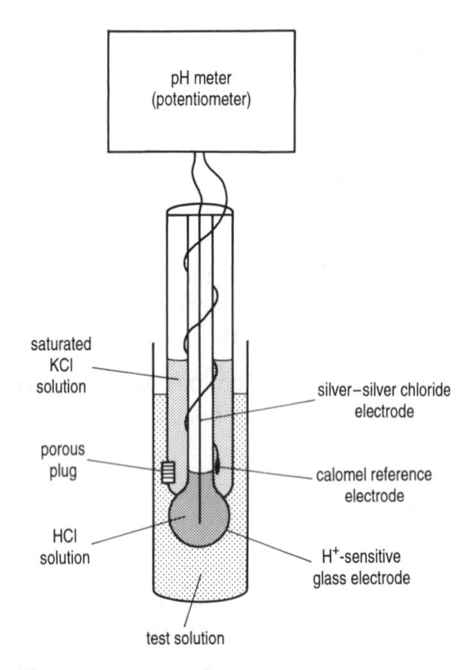

Fig. 8.1 Measurement of pH using a combination pH electrode and meter. The electrical potential difference recorded by the potentiometer is directly proportional to the pH of the test solution.

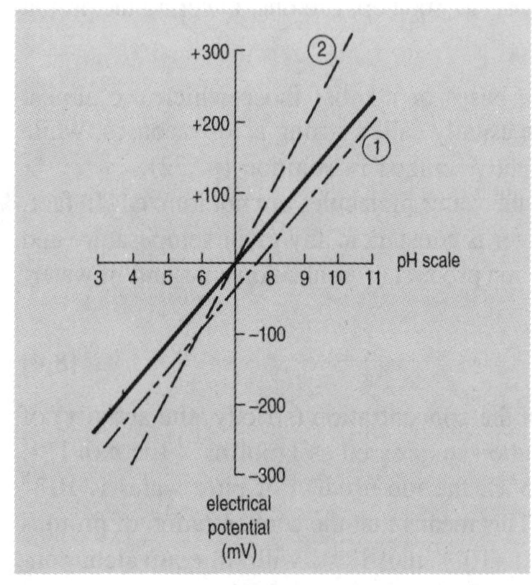

Fig. 8.2 The relationship between electrical potential and pH. The solid line shows the response of a calibrated electrode while the other plots are for instruments requiring calibration: 1 has the correct slope but incorrect isopotential point (calibration control adjustment is needed); 2 has the correct isopotential point but incorrect slope (slope control adjustment is needed).

termed neutrality: at 25°C the pH at neutrality is 7.0. At this temperature, pH values below 7.0 are acidic while values above 7.0 are alkaline. However, the pH of a neutral solution changes with temperature (Table 8.1), due to the enhanced dissociation of water with increasing temperature. This must be taken into account:

- when measuring the pH of any solution;
- when interpreting your results.

Remember that the pH scale is a logarithmic one: a solution with a pH of 3.0 is not twice as acidic as a solution of pH 6.0, but one thousand times as acidic (i.e. contains 1 000 times the amount of H^+ ions). Therefore, you may need to convert pH values into proton concentrations before you carry out mathematical manipulations (see Box 45.2). For similar reasons, it is important that pH change is expressed in terms of the original and final pH values, rather than simply quoting the difference between the values: a pH change of 0.1 has little meaning unless the initial or final pH is known.

Measuring pH

pH indicator dyes

These compounds (usually weak acids) change colour in a pH-dependent manner. They may be added in small amounts to a solution, or they can be used in paper strip form. Each indicator dye usually changes colour over a restricted pH range, typically 1–2 pH units: universal indicator dyes/papers make use of a combination of individual dyes to measure a wider pH range. Dyes are not suitable for accurate pH measurement as they are affected by other components of the solution. However, they are useful for:

- estimating the approximate pH of a solution, where the precise value is not critical;
- determining a change in pH, for example at the end-point of a titration or the production of acids during bacterial metabolism (p. 103);
- establishing the approximate pH of intracellular compartments, for example the use of neutral red as a 'vital' stain (p. 117).

pH electrodes

Accurate pH measurements can be made using a pH electrode, coupled to a pH meter. The pH electrode is usually a combination electrode, comprising two separate systems: an H^+-sensitive glass electrode and a reference electrode which is unaffected by H^+ ion concentration (Fig. 8.1). When both electrodes are immersed in a solution, a pH-dependent voltage between the two electrodes can be measured using a potentiometer (Fig. 8.2). In most cases, the pH electrode assembly (containing the glass and reference electrodes) is connected to a separate pH meter by a cable, although some hand-held instruments (pH probes) have the electrodes and meter within the same assembly, often using an H^+ sensitive field effect transistor in place of a glass electrode, to improve durability and portability.

Box 8.1 gives details of the steps involved in making a pH measurement with a glass pH electrode.

Box 8.1 Using a glass pH electrode and meter to measure the pH of a solution

The following procedure should be used whenever you make a pH measurement: consult the manufacturer's handbook for specific information, where necessary. Do not be tempted to miss out any of the steps detailed below, particularly those relating to the effects of temperature, or your measurements are likely to be inaccurate.

1. **Stir the solution thoroughly before you make any measurement**: it is often best to use a magnetic stirrer. Leave the solution for sufficient time to allow equilibration at lab temperature.
2. **Record the temperature of every solution you use**, including all calibration standards and samples, since this will affect K_w, neutrality and pH.
3. **Set the temperature compensator on the meter to the appropriate value.** This control makes an allowance for the effect of temperature on the electrical potential difference recorded by the meter: it does *not* allow for the other temperature-dependent effects mentioned elsewhere. Basic instruments have no temperature compensator, and should only be used at a specified temperature, either 20°C or 25°C, otherwise they will not give an accurate measurement.
4. **Rinse the electrode assembly with distilled water** and gently dab off the excess water onto a clean tissue: check for visible damage or contamination of the glass electrode (consult a member of staff if the glass is broken or dirty). Also check that the solution within the glass assembly is covering the metal electrode.
5. **Calibrate the instrument**: set the meter to 'pH' mode, if appropriate, and then place the electrode assembly in a standard solution of known pH, usually pH 7.00. This solution may be supplied as a liquid, or may be prepared by dissolving a measured amount of a calibration standard in water: calibration standards are often provided in tablet form, to be dissolved in water to give a particular volume of solution. Adjust the calibration control to give the correct reading. Remember that your calibration standards will only give the specified pH at a particular temperature, usually either 20°C or 25°C. If you are working at a different temperature, you must establish the actual pH of your calibration standards, either from the supplier, or from literature information.
6. **Remove the electrode assembly from the calibration solution and rinse again with distilled water**: dab off the excess water. Basic instruments have no further calibration steps (single-point calibration), while the more refined pH meters have additional calibration procedures.

   If you are using a basic instrument, you should check that your apparatus is accurate over the appropriate pH range by measuring the pH of another standard whose pH is close to that expected for the test solution. If the standard does not give the expected reading, the instrument is not functioning correctly: consult a member of staff.

   If you are using an instrument with a slope control function, this will allow you to correct for any deviation in electrical potential from that predicted by the theoretical relationship (at 25°C, a change in pH of 1.00 unit should result in a change in electrical potential of 59.16 mV) by performing a two-point calibration. Having calibrated the instrument at pH 7.00, immerse in a second standard at the same temperature as that of the first standard, usually buffered to either pH 4.00 or pH 9.00, depending upon the expected pH of your samples. Adjust the slope control until the exact value of the second standard is achieved. A pH electrode and meter calibrated using the two-point method will give accurate readings over the pH range from 3 to 11: laboratory pH electrodes are not accurate outside this range, since the theoretical relationship between electrical potential and pH is no longer valid (Fig. 8.2).
7. **Once the instrument is calibrated, measure the pH of your solution(s)**, making sure that the electrode assembly is rinsed thoroughly between measurements. You should be particularly aware of this requirement if your solutions contain organic biological material, e.g. soil, tissue fluids, protein solutions, etc., since these may adhere to the glass electrode and affect the calibration of your instrument. If your electrode becomes contaminated during use, check with a member of staff before cleaning: avoid touching the surface of the glass electrode with abrasive material. Allow sufficient time for the pH reading to stabilize in each solution before taking a measurement: for unbuffered solutions, this may take several minutes, so do not take inaccurate pH readings due to impatience!
8. **After use, the electrode assembly must not be allowed to dry out**: store in distilled water (the slight acidity of distilled water helps maintain the electrode in working condition), either by suspending the assembly in a small beaker, or by using an electrode cap, filled with distilled water.
9. **Switch the meter to zero (where appropriate), but do not turn off the power**: pH meters give more stable readings if they are left on during normal working hours.

Definition
Buffer solution — one which resists a change in H^+ concentration (pH) on addition of acid or alkali.

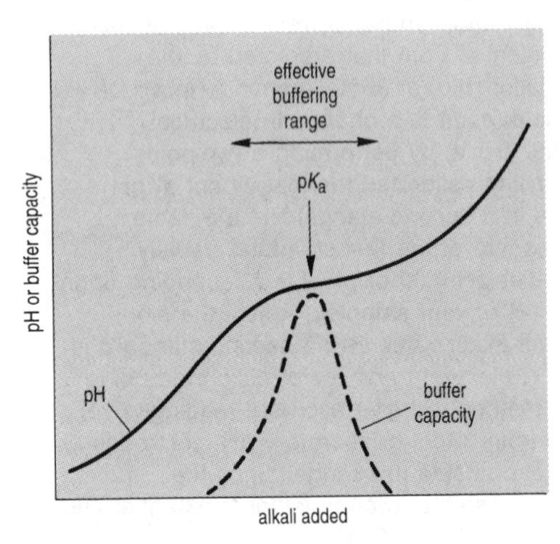

Fig. 8.3 Theoretical pH titration curve for a buffer solution. pH change is lowest and buffer capacity is greatest at the pK_a of the buffer solution.

Selecting a buffer solution — choose one whose pK_a lies within approximately one pH unit of the value you need (Tables 8.2, 8.3).

Table 8.2 pK_a values of some acids and bases used in buffer solutions. For polyprotic acids, where more than one proton may dissociate, the pK_a values are shown for each ionization step

Acid or base	pK_a value(s) at 25°C
Acetic acid	4.7
Carbonic acid	6.1, 10.2
Citric acid	3.1, 4.8, 5.4
Glycylglycine	3.1, 8.1
Phthalic acid	2.9, 5.5
Phosphoric acid	2.0, 6.7, 12.3
Succinic acid	4.2, 5.6
TRIS*	8.1

*2-amino-2-hydroxymethyl-1,3-propanediol: (tris(hydroxymethyl)aminomethane); this compound is hygroscopic and should be stored in a desiccator.

Buffers

Rather than simply measuring the pH of a solution, you may wish to *control* the pH, e.g in metabolic experiments, or in a growth medium for cell culture (p. 125). In fact, you should consider whether you need to control pH in any experiment involving a biological system, whether whole organisms, isolated cells, sub-cellular components or biomolecules. One of the most effective ways to control pH is to use a buffer solution.

A buffer solution is usually a mixture of a weak acid and its conjugate base. Added protons will be neutralized by the anionic base while a reduction in protons, e.g. due to the addition of hydroxyl ions, will be counterbalanced by dissociation of the acid (eqn [8.2]); thus the conjugate pair acts as a 'buffer' to pH change. The innate resistance of most biological fluids to pH change is due to the presence of cellular constituents that act as buffers, e.g. proteins.

Buffer capacity and the effects of pH

The extent of resistance to pH change is called the buffer capacity of a solution. The buffer capacity is measured experimentally at a particular pH by titration against a strong acid or alkali: the resultant curve will be strongly sigmoidal, with a plateau where the buffer capacity is greatest (Fig. 8.3). The mid-point of the plateau represents the pH where equal quantities of acid and conjugate base are present, and is given the symbol pK_a, which refers to the negative logarithm (to the base 10) of the acid dissociation constant, K_a, where

$$K_a = \frac{[H^+]\,[A^-]}{[HA]} \qquad [8.6]$$

By rearranging eqn [8.6] and taking negative logarithms, we obtain:

$$pH = pK_a + \log_{10}\frac{[A^-]}{[HA]} \qquad [8.7]$$

This relationship is known as the Henderson–Hasselbalch equation and it shows that the pH will be equal to the pK_a when the ratio of conjugate base to acid is unity, since the final term in eqn [8.7] will be zero. Consequently, the pK_a of a buffer solution is an important factor in determining the buffer capacity at a particular pH. In practical terms, this means that a buffer solution will work most effectively at pH values about one unit either side of the pK_a.

Biological buffers

An ideal buffer for biological purposes would possess a number of additional characteristics, including:

- impermeability to biological membranes;
- non-toxicity to metabolic and biochemical processes;
- lack of formation of insoluble complexes with cations;
- minimal effect of ionic composition or salt concentration;
- limited pH change in response to temperature.

Not all buffers meet all of the above requirements. Citric acid and phosphate buffers readily form insoluble complexes with divalent cations, while phosphate can also act as a substrate, activator or inhibitor of certain enzymes. Both of these buffers contain biologically significant quantities of cations, e.g. Na^+ or K^+. TRIS (Table 8.2) is often toxic to biological systems: due to its high

Table 8.3 pK_a values of some zwitterionic buffers at 25°C

Abbreviation	Substance	pK_a at 25°C	Useful pH range
MES	2-*N*-morpholinoethanesulphonic acid	6.1	5.5–6.7
PIPES	1,4-piperazinediethanesulphonic acid	6.8	6.1–7.5
MOPS	3-*N*-morpholinopropanesulphonic acid	7.2	6.5–7.9
HEPES	*N*-2-hydroxyethylpiperazine-*N'*-2-ethanesulphonic acid	7.5	6.8–8.2
TRICINE	*N*-2-hydroxy-1,1-bis(hydroxymethyl)ethylglycine	8.1	7.4–8.8
TAPS	2-hydroxy-1,1-bishydroxymethyl-ethylamino-1-propanesulphonic acid	8.4	7.7–9.1
CHES	2-*N*-cyclohexylaminoethanesulphonic acid	9.3	8.6–10.0
CAPS	3-cyclohexylamino-1-propanesulphonic acid	10.4	9.7–11.1

lipid solubility it can penetrate membranes, uncoupling electron transport reactions in whole cells and isolated organelles. In addition, it is markedly affected by temperature, with a tenfold increase in H^+ concentration from 4°C to 37°C. A number of zwitterionic acids have been introduced to overcome some of the disadvantages of the more traditional buffers. These newer compounds are often referred to as 'Good buffers', to acknowledge the early work of Dr N.E. Good and co-workers: HEPES (Table 8.3) is one of the most useful zwitterionic buffers, with a pK_a of 7.5 at 25°C.

These zwitterionic substances are usually added to water as the free acid: the solution must then be adjusted to the correct pH with a strong alkali, usually NaOH or KOH. Alternatively, they may be used as their sodium or potassium salts, adjusted to the correct pH with a strong acid, e.g. HCl. Consequently, you may need to consider what effects such changes in ion concentration may have in a solution where zwitterions are used as buffers. In addition, zwitterionic buffers can interfere with protein determinations (e.g. Lowry method, p. 160).

Definition
Zwitterion — a molecule possessing both positive and negative groups.

Preparing solutions of zwitterionic buffer — the acid may be relatively insoluble. Do not wait for it to dissolve fully before adding alkali to change the pH — the addition of alkali will help bring the acid into solution (but make sure it has all dissolved before the desired pH is reached).

Preparation of buffer solutions

Having selected an appropriate buffer, you will need to make up your solution to give the desired pH. You will need to consider two factors:

- The ratio of acid and conjugate base required to give the correct pH.
- The amount of buffering required; buffer capacity depends upon the absolute quantities of acid and base, as well as their relative proportions.

In most instances, buffer solutions are prepared to contain between 10 mmol l^{-1} and 200 mmol l^{-1} of the conjugate pair. While it is possible to calculate the quantities required from first principles using the Henderson–Hasselbalch equation, there are several sources which tabulate the amount of substance required to give a particular volume of solution with a specific pH value for a wide range of traditional buffers (e.g. Perrin and Dempsey, 1974). For traditional buffers, it is customary to mix stock solutions of acidic and basic components in the correct proportions to give the required pH (Table 8.4). For zwitterionic acids, the usual procedure is to add the compound to water, then bring the solution to the required pH by adding a specific amount of strong alkali or acid (obtained from tables). Alternatively, the required pH can be obtained by dropwise addition of alkali or acid, using a meter to check the pH, until the correct value is reached.

Remember that buffer solutions will only work effectively if they have sufficient buffering capacity to resist the change in pH expected during the course of the experiment. Thus a weak solution of HEPES (e.g. 10 mmol l^{-1}, adjusted to pH 7.0 with NaOH) will not be able to buffer the growth medium of a dense suspension of cells for more than a few minutes.

Finally, if you are preparing a buffer solution based on tabulated information, you should always confirm the pH with a pH meter before use.

Table 8.4 Preparation of sodium phosphate buffer solutions for use at 25°C. Prepare separate stock solutions of (a) disodium hydrogen phosphate and (b) sodium dihydrogen phosphate, both at 200 mol m^{-3}. Buffer solutions (at 100 mol m^{-3}) are then prepared at the required pH by mixing together the volume of each stock solution shown in the table, then diluting to a final volume of 100 ml using distilled or deionized water

Required pH (at 25°C)	Volume of stock (a) Na_2HPO_4 (ml)	Volume of stock (b) NaH_2PO_4 (ml)
6.0	6.2	43.8
6.2	9.3	40.7
6.4	13.3	36.7
6.6	18.8	31.2
6.8	24.5	25.5
7.0	30.5	19.5
7.2	36.0	14.0
7.4	40.5	9.5
7.6	43.5	6.5
7.8	45.8	4.2
8.0	47.4	2.6

Measures and units

9 The principles of measurement

Definition
Variable — any characteristic or property which can take one of a range of values (contrast this definition with that for a **parameter**, which is a numerical constant in any particular instance, p. 56).

The term data (singular = datum, or data value) refers to items of information, and you will use different types of data from a wide range of sources during your practical work. Consequently, it is important to appreciate the underlying features of data collection and measurement. Data gathering often involves assigning descriptive terms or numbers to a particular characteristic. A single reading, score or observation of a given variable is sometimes called a variate (i.e. a data value).

Variables

Biological variables (Fig. 9.1) can be classified as follows:

Quantitative variables

These are characteristics whose differing states can be described by means of a number. They are of two basic types:

- Continuous variables, such as length; these are usually measured against a numerical scale. Theoretically, they can take any value on the measurement scale. In practice, the number of significant figures of a measurement is directly related to the accuracy of your measuring system; for example, dimensions measured with Vernier calipers will provide readings of greater precision than a millimetre ruler (p. 20).
- Discontinuous (discrete) variables, such as the number of eggs in a nest; these are always obtained by counting and therefore the data values must be whole numbers (integers). There are no intermediate values — for example, you never find 1.25 eggs in a nest.

Measuring continuous variables — in practice, no measurement can be made with absolute accuracy, because of the limitations of the measuring device. Thus, while a given electronic timer might be able to measure to the nearest millionth of a second (μs), there remains an infinite number of intervals between each μs, unmeasurable with that instrument.

Working with discontinuous variables — note that while the original data values must be integers, derived data and statistical values do not have to be whole numbers. Thus, it is perfectly acceptable to express the *mean* number of children per family as 2.5.

Ranked variables

These provide data which can be listed in order of magnitude (i.e. ranked). A familiar example is the abundance of an organism in a sample, which is often expressed as a series of ranks, e.g. rare = 1, occasional = 2, frequent = 3, common = 4, and abundant = 5. When such data are given numerical ranks, rather than descriptive terms, they are sometimes called 'semi-quantitative data'. Note that the difference in magnitude between ranks does not need to be consistent. For example, it would not matter whether there was a one-year or a five-year gap between offspring in a family; their ranks in order of birth would be the same.

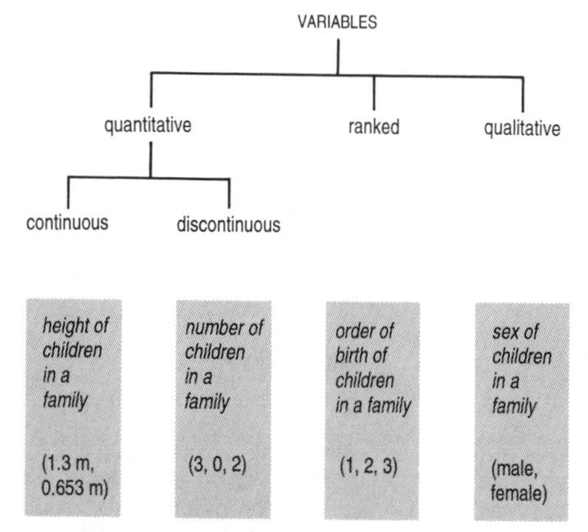

Fig. 9.1 Examples of the different types of variables as used to describe some characteristics of families.

Qualitative variables (attributes)

These are non-numerical and descriptive; they have no order of preference and therefore, are not measured on a numerical scale nor ranked in order of magnitude, but are described in terms of categories. Examples include viability (i.e. dead or alive) and shape (e.g. round, flat, elongated, etc).

Variables may be independent or dependent. Usually, the variable under the control of the experimenter (e.g. time) is the independent variable, while the variable being measured is the dependent variable (p. 50). Sometimes it is not appropriate to describe variables in this way and they are then referred to as interdependent variables (e.g. the length and breadth of an organism).

The majority of variates are recorded as direct measurements, readings or counts, but there is an important group, called derived (or computed) variates, which result from calculations based on two or more data values, e.g. ratios, percentages, indices and rates.

Measurement scales

Variables may be measured on different types of scale:

- Nominal scale: this classifies objects into categories based on a descriptive characteristic. It is the only scale suitable for qualitative data.
- Ordinal scale: this classifies by rank. There is a logical order in any number scale used.
- Interval scale: this is used for quantitative variables. Numbers on an equal unit scale are related to an arbitrary zero point.
- Ratio scale: this is similar to the interval scale, except that the zero point now represents an absence of that character (i.e. it is an absolute zero).

The measurement scale is important in determining the mathematical and statistical methods used to analyse your data. Table 9.1 presents a summary of the important properties of these scales. Note that you may be able to measure a characteristic in more than one way, or you may be able to convert data collected in one form to a different form. For instance, you might measure light in terms of the photon flux density (p. 174) between particular wavelengths of the EMR spectrum (ratio scale), or simply as 'blue' or 'red' (nominal scale); you could find out the dates of birth of individuals (interval scale) but then use this information to rank them in order of birth (ordinal scale). Where there are no other constraints, you should use a ratio scale to measure a quantitative variable, since this will allow you to use the broadest range of mathematical and statistical procedures (Table 9.1).

Examples
A nominal scale for temperature is not feasible, since the relevant descriptive terms can be ranked in order of magnitude.
An ordinal scale for temperature measurement might use descriptive terms, ranked in ascending order e.g. cold = 1, cool = 2, warm = 3, hot = 4.
The Celsius scale is an interval scale for temperature measurement, since the arbitrary zero corresponds to the freezing point of water (0°C).
The Kelvin scale is a ratio scale for temperature measurement since 0 K represents a temperature of absolute zero (for information, the freezing point of water is 273.15 K on this scale).

Table 9.1 Some important features of scales of measurement

	Measurement scale			
	Nominal	Ordinal	Interval	Ratio
Examples	Species Sexes Colour	Abundance scales Reproductive condition Optical assessment of colour development	Fahrenheit temperature scale Date (BC/AD)	Kelvin temperature scale Weight Length Response time Most physical measurements
Mathematical properties	Identity	Identity Magnitude	Identity Magnitude Equal intervals	Identity Magnitude Equal intervals True zero point
Mathematical operations	None	Rank	Rank Add Subtract	Rank Add Subtract Multiply Divide
Type of variable	Qualitative (Ranked)* (Quantitative)*	Ranked (Quantitative)*	Quantitative	Quantitative
Typical statistics used	Only those based on frequency of counts made: contingency tables, frequency distributions, etc. Chi-square test	Non-parametric methods, sign tests Mann–Whitney *U*-test	Almost all types of test, *t*-test, analysis of variance (ANOVA), etc. (check distribution before using, p. 236)	Almost all types of test, *t*-test, ANOVA, etc. (check distribution before using, p. 236)

* In some instances (see text for examples)

The features of measurement

The basic requirements for good measurements are:

- They must be related to a carefully defined problem.
- Measurements must be comparable and standardized; all measurements must be collected in a logical manner, using a carefully defined protocol.
- Replication must be adequate; a common fault is to gather an inadequate number of data values, giving an inconclusive result.
- Errors must be identified and quantified. You should consider the possibility of operator errors and other sources of error when making measurements and include descriptions of possible sources of error and estimates of their importance in any report. However, do not use 'biological variability' as an excuse for poor technique!

Examples
Sources of error include mistakes; human errors (e.g. parallax); instrumental limitations; effects of the measurement/observation; extraneous factors such as temperature change; sampling errors (e.g. unrepresentative samples); mathematical and statistical errors.

Accuracy and precision

Accuracy is the closeness of a measured or derived data value to its true value, while precision is the closeness of repeated measurements to each other (Fig. 9.2). A balance with a fault in it (i.e. a bias, see below) could give precise (i.e. very repeatable) but inaccurate (i.e. untrue) results. Unless there is bias in a measuring system, precision will lead to accuracy and it is precision that is generally the most important practical consideration, if there is no reason to suspect bias. You can investigate the precision of any measuring system by repeated measurements of individual samples.

Absolute accuracy and precision are impossible to achieve, due to both the limitations of measuring systems for continuous quantitative data and the fact that you are usually working with incomplete data sets (samples, p. 56). It is particularly important to avoid spurious accuracy in the presentation of results; include only those digits which the accuracy of the measuring system implies. This type of mistake is common when changing units (e.g. inches to metres) and in derived data, especially when calculators give results to a large number of decimal places.

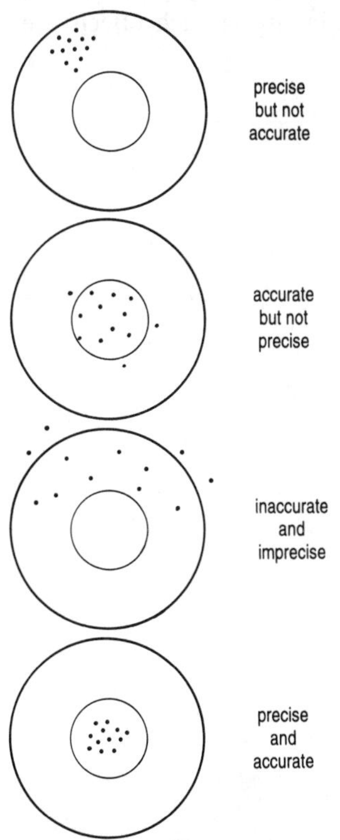

Fig. 9.2 'Target' diagrams illustrating precision and accuracy.

Bias (systematic error) and consistency

Bias is a systematic or non-random distortion and is one of the most troublesome difficulties in using numerical data. Biases may be associated with incorrectly calibrated instruments, e.g. a faulty pipettor, or with experimental manipulations, e.g. shrinkage during the preservation of a specimen. Bias in measurement can also be subjective, or personal, e.g. an experimenter's preconceived ideas about an 'expected' result.

Bias can be minimized by using a carefully standardized procedure, with fully calibrated instruments. You can investigate bias by measuring a single variable in several different ways, to see whether the same result is obtained.

Using a pilot investigation — a small-scale 'trial run' may help you to determine whether or not bias is present and to assess the errors associated with your measurements.

Measurement error

All measurements are subject to error, but the dangers of misinterpretation are reduced by recognizing and understanding the likely sources of error and by adopting appropriate protocols and calculation procedures. You must ensure that your method of obtaining data can be repeated by others.

Minimizing errors — determine early in your study what the dominant errors are likely to be and concentrate your time and effort on reducing these.

A common source of measurement error is carelessness, e.g. reading a scale in the wrong direction or parallax errors. This can be reduced greatly by careful recording and may be detected by repeating the measurement. Other errors arise from faulty or inaccurate equipment, but even a perfectly functioning machine has distinct limits to the accuracy and precision of its measurements. These limits are often quoted in manufacturer's specifications and are applicable when an instrument is new; however, you should allow for some deterioration with age. Further errors are introduced when the subject being studied is open to influences outside your control, e.g. volume may fluctuate with temperature changes or population structure may influence many variables in which you might be interested. Resolving such problems requires appropriate experimental design (Chapter 12) and sampling procedures (Chapter 13).

Working with derived data — when calculating a result based on a small difference between two nearly equal measurements, or when taking the power of a measured variable, special effort should be made to reduce measurement errors, since they will be magnified in the final value.

One major influence virtually impossible to eliminate, is the effect of the investigation itself; even putting a thermometer in a liquid may change the temperature of the liquid. The very act of measurement may give rise to a confounding variable (p. 51) — methods for minimizing such effects are given in Chapter 12.

10 SI units and their use

A single unified system of units is essential for the efficient communication of numerical data. Consistent units also simplify calculations involving several measured quantities (see pp. 184 and 222). The Système International d'Unités (SI) is an internationally ratified form of the metre–kilogram–second system of measurement and represents the accepted scientific convention for measurements of physical quantities. Although SI rules can cause difficulties because the scale of base or derived units appears inconvenient, the full benefits of using the SI depend on the strict observance of its conventions.

Most measurements consist of a number and a unit. The number expresses the ratio of the measured quantity to a fixed standard and the unit is the name for the standard measure or dimension. Certain measurements can be expressed as dimensionless ratios or logarithms (e.g. pH) and thus do not require a qualifying unit.

There are seven base units and two supplementary units in SI (Table 10.1); each of these has a specified abbreviation or symbol, both for convenience and to avoid misunderstandings. Base and supplementary units can be combined to give compound or derived units which may also be given special symbols (Table 10.2). Prefixes denote multiplication factors of 10^3 used to arrive at a simple expression for large multiples or small fractions of units (Table 10.3).

Table 10.1 The base and supplementary SI units

Measured quantity	Name of SI unit	Symbol
Base units		
Length	metre	m
Mass	kilogram	kg
Amount of substance	mole	mol
Time	second	s
Electric current	ampere	A
Temperature	kelvin	K
Luminous intensity	candela	cd
Supplementary units		
Plane angle	radian	rad
Solid angle	steradian	sr

Table 10.2 Some important derived SI units

Measured quantity	Name of unit	Symbol	Definition in base units	Alternative in derived units
Energy	joule	J	m^3 kg s^{-1}	N m
Force	newton	N	m kg s^{-2}	J m^{-1}
Pressure	pascal	Pa	kg m^{-1} s^{-2}	N m^{-2}
Power	watt	W	m^2 kg s^{-3}	J s^{-2}
Electric charge	coulomb	C	A s	J V^{-1}
Electric potential difference	volt	V	m^2 kg A^{-1} s^{-3}	J C^{-1}
Electric resistance	ohm	Ω	m^2 kg A^{-2} s^{-3}	V A^{-1}
Electric conductance	siemens	S	s^3 A^2 kg^{-1} m^{-2}	A V^{-1} or Ω^{-1}
Electric capacitance	farad	F	s^4 A^2 kg^{-1} m^{-2}	C V^{-1}
Luminous flux	lumen	lm	cd sr	
Illumination	lux	lx	cd sr m^{-2}	lm m^{-2}
Frequency	hertz	Hz	s^{-1}	
Radioactivity	becquerel	Bq	s^{-1}	
Enzyme activity	katal	kat	mol substrate s^{-1}	

Table 10.3 Prefixes used in the SI

Multiple	Prefix	Symbol	Multiple	Prefix	Symbol
10^{-3}	milli	m	10^3	kilo	k
10^{-6}	micro	μ	10^6	mega	M
10^{-9}	nano	n	10^9	giga	G
10^{-12}	pico	p	10^{12}	tera	T
10^{-15}	femto	f	10^{15}	peta	P
10^{-18}	atto	a	10^{18}	exa	E

Recommendations for describing measurements in SI units

Basic format

- Express each measurement as a number separated from its units by a space. If a prefix is required, no space is left between the prefix and the unit it refers to. Symbols for units are only written in their singular form and do not require full stops to show that they are abbreviated or that they are being multiplied together.

> ***Example***
> *10 μg is correct, while 10μg, 10 μg. and 10μ g are incorrect. 2.6 mol is right, but 2.6 mols is wrong.*

> ***Example***
> *n stands for nano and N for newtons.*

> ***Example***
> *1 982 963.192 309 kg (perhaps better expressed as 1.982 963 192 309 Gg).*

> ***Examples***
> *200 m s (metre-seconds) is completely different in meaning from 200 ms (milliseconds).*
>
> *mol m^{-3} is used rather than mol/m^3.*
>
> *Photosynthetic rate might be given in mol CO_2 (kg chlorophyll)$^{-1}$ s^{-1}.*
>
> *You might select units of Pa m^{-1} to describe a hydrostatic pressure gradient rather than kg m^{-2} s^{-1}, even though the measurements would be numerically the same and the units equivalent (see Table 10.2).*

> ***Examples***
> *10 μm is preferred to 0.000 01 m or 0.010 mm.*
>
> *1 mm^2 = 10^{-6} m^2 (not one-thousandth of a square metre).*
>
> *1 dm^3 (1 litre) is more properly expressed as 1 × 10^{-6} m^3 (see below).*
>
> *Avogadro's constant is 6.022 174 × 10^{23} mol^{-1}.*
>
> *State as MW m^{-2} rather than W mm^{-2}.*

> In this book, we use l and ml where you would normally find equipment calibrated in that way, but use SI units where this simplifies calculations.

- Give symbols and prefixes appropriate upper or lower case initial letters as this may define their meaning. Upper case symbols are named after persons but when written out in full they are not given initial capital letters.
- Show the decimal sign as a full point on the line. Some metric countries continue to use the comma for this purpose and you may come across this in overseas literature: commas should not therefore be used to separate groups of thousands. In numbers that contain many significant figures, you should separate multiples of 10^3 by spaces rather than commas.

Compound expressions for derived units

- Always separate symbols in compound expressions by a space to avoid potential confusion with prefixes.
- Express compound units using negative powers rather than a solidus (/). The solidus should be reserved for separating a descriptive label from its units (see p. 214).
- Enclose expressions being raised to a power in parentheses if this increases clarity.
- Select relevant (natural) combinations of derived and base units.

Use of prefixes

- Use prefixes to denote multiples of 10^3 (Table 10.3) so that numbers are kept between 0.1 and 1 000.
- Treat a combination of a prefix and a symbol as a single symbol. Thus, when a modified unit is raised to a power, this refers to the whole unit including the prefix.
- Avoid the prefixes deci (d) for 10^{-1} and centi (c) for 10^{-2} as they are not strictly SI.
- Express very large or small numbers as a number between 1 and 10 multiplied by a power of 10 if they are outside the range of prefixes shown in Table 10.3.
- Do not use prefixes in the middle of derived units: they should be attached only to a unit in the numerator (the exception is in the unit for mass, kg).

Conversion from other systems

For the foreseeable future, you will have to make conversions from other units to SI units, as much of the pre-SI literature quotes imperial, c.g.s. or other units. You will need to recognize these units and find the conversion factors required. Examples relevant to biology are given in Table 10.4. Table 10.5 provides values of some important physical constants in SI units.

Some implications of SI in biology

Volume

The SI unit of volume is the cubic metre, m^3, which is rather large for practical purposes. The litre (l) and the millilitre (ml) are technically obsolete, but are widely used and glassware is still calibrated using them. However, the use of these units of volume is not encouraged in formal scientific writing and constructions such as 1 × 10^{-6} m^3 (= 1 ml), or 1 mm^3 (= 1 μl) should be used.

Table 10.4 Conversion factors between some redundant units and the SI

Quantity	Old unit/symbol	SI unit/symbol	Conversion factors	
			Multiply number in SI unit by this factor for equivalent in old unit*	Multiply number in old unit by this factor for equivalent in SI unit*
Area	acre	square metre/m^2	$0.247\,105 \times 10^{-3}$	$4.046\,86 \times 10^{3}$
	hectare/ha	square metre/m^2	0.1×10^{-3}	10×10^{3}
	square foot/ft^2	square metre/m^2	10.763 9	0.092 903
	square inch/in^2	square metre/m^2	$1.550\,00 \times 10^{6}$	645.16×10^{-9}
	square yard/yd^2	square metre/m^2	1.195 99	0.836 127
Angle	degree/°	radian/rad	57.295 8	$17.453\,2 \times 10^{-3}$
Energy	erg	joule/J	10×10^{6}	0.1×10^{-6}
	kilowatt hour/kWh	joule/J	$0.277\,778 \times 10^{-6}$	3.6×10^{6}
Length	Ångstrom/Å	metre/m	10×10^{9}	0.1×10^{-9}
	foot/ft	metre/m	3.280 84	0.304 8
	inch/in	metre/m	39.370 1	25.4×10^{-3}
	mile	metre/m	$0.621\,373 \times 10^{-3}$	$1.609\,34 \times 10^{3}$
	yard/yd	metre/m	1.093 61	0.914 4
Mass	ounce/oz	kilogram/kg	$28.349\,5 \times 10^{-3}$	35.274 0
	pound/lb	kilogram/kg	2.204 62	0.453 592
	stone	kilogram/kg	0.157 473	6.350 29
	hundredweight/cwt	kilogram/kg	$19.684\,1 \times 10^{-3}$	50.802 4
	ton (UK)	kilogram/kg	$0.984\,203 \times 10^{-3}$	$1.016\,05 \times 10^{3}$
Pressure	atmosphere/atm	pascal/Pa	$9.869\,23 \times 10^{-6}$	101 325
	bar/b	pascal/Pa	10×10^{-6}	100 000
	millimetre of mercury/mmHg	pascal/Pa	$7.500\,64 \times 10^{-3}$	133.322
	torr/Torr	pascal/Pa	$7.500\,64 \times 10^{-3}$	133.322
Radioactivity	curie/Ci	becquerel/Bq	$27.027\,0 \times 10^{-12}$	37×10^{9}
Temperature	centigrade (Celsius) degree/°C	kelvin/K	°C + 273.15	K − 273.15
	Fahrenheit degree/°F	kelvin/K	(°F + 459.67) × 5/9	(K × 9/5) − 459.67
Volume	cubic foot/ft^3	cubic metre/m^3	35.314 7	0.028 316 8
	cubic inch/in^3	cubic metre/m^3	$61.023\,6 \times 10^{3}$	$16.387\,1 \times 10^{-6}$
	cubic yard/yd^3	cubic metre/m^3	1.307 95	0.764 555
	UK pint/pt	cubic metre/m^3	1 759.75	$0.568\,261 \times 10^{-3}$
	US pint/liq pt	cubic metre/m^3	2 113.38	$0.473\,176 \times 10^{-3}$
	UK gallon/gal	cubic metre/m^3	219.969	$4.546\,09 \times 10^{-3}$
	US gallon/gal	cubic metre/m^3	264.172	$3.785\,41 \times 10^{-3}$

*In the case of temperature measurements, use formulae shown.

Table 10.5 Some physical constants in SI terms

Physical constant	Symbol	Value and units
Avogadro's constant	N_A	$6.022\,174 \times 10^{23}$ mol^{-1}
Boltzmann's constant	k	1.380 626 J K^{-1}
Charge of electron	e	$1.602\,192 \times 10^{-19}$ C
Gas constant	R	8.314 43 J K^{-1} mol^{-1}
Faraday's constant	F	$9.648\,675 \times 10^{4}$ C mol^{-1}
Molar volume of ideal gas at STP	V_0	0.022 414 m^3 mol^{-1}
Speed of light *in vacuo*	c	$2.997\,924 \times 10^{8}$ m s^{-1}
Planck's constant	h	$6.626\,205 \times 10^{-34}$ J s

Mass

The SI unit for mass is the kilogram (kg) rather than the gram (g): this is unusual because the base unit has a prefix applied.

Amount of substance

You should use the mole (mol, i.e. Avogadro's constant, see Table 10.4) to express very large numbers. The mole gives the number of atoms in the atomic mass, a convenient constant. Always specify the elementary unit referred to in other situations (e.g. mol photons m^{-2} s^{-1}).

Concentration

The SI unit of concentration, mol m^{-3}, is quite convenient for biological systems. It is equivalent to the non-SI term 'millimolar' (mM) while 'molar' (M) becomes kmol m^{-3}. Note that the symbol M in the SI is reserved for mega and hence should not be used for concentrations. If the solvent is not specified, then it is assumed to be water (see Chapter 7).

Enzyme activity

The SI-derived unit is the katal (kat) which is the amount of enzyme that will transform 1 mol of substrate in 1 s (see Chapter 33).

Time

In general, use the second (s) when reporting physical quantities having a time element (e.g. give photosynthetic rates in mol CO_2 m^{-2} s^{-1}). Hours (h), days (d) and years should be used if seconds are clearly absurd (e.g., samples were taken over a 5-year period). Note, however, that you may have to convert these units to seconds when doing calculations.

Temperature

The SI unit is the kelvin, K. The degree Celsius scale has units of the same magnitude, °C, but starts at 273.15 K, the melting point of ice at STP. Temperature is similar to time in that the Celsius scale is in widespread use, but note that conversions to K may be required for calculations. Note also that you must not use the degree sign (°) with K and that this symbol must be in upper case to avoid confusion with k for kilo; however, you should retain the degree sign with °C to avoid confusion with the coulomb, C.

Definition

STP — Standard Temperature and Pressure = 293.15 K and 0.101 325 MPa.

Light

While the first six base units in Table 10.1 have standards of high precision, the SI base unit for luminous intensity, the candela (cd) and the derived units lm and lx (Table 10.2), are defined in 'human' terms. They are, in fact, based on the spectral responses of the eyes of 52 American GIs measured in 1923! Clearly, few organisms 'see' light in the same way as this sample of humans. Also, light sources differ in their spectral quality. For these reasons, it is better to use expressions based on energy or photon content (e.g. W m^{-2} or mol photons m^{-2} s^{-1}), in studies other than those on human vision. Ideally you should specify the photon wavelength spectrum involved (see Chapter 35).

Radioactivity

The SI unit is the becquerel (Bq) defined as that quantity of a radioactive substance in which one atom disintegrates per second (i.e. 1 'd.p.s.'). This replaces the curie (Ci) which equals 37 GBq (see Chapter 36).

The scientific approach

11 Making observations

Observations provide the basic information leading to the formulation of hypotheses, the first step in the scientific method (see Chapter 12). Observations are obtained either directly by our senses or indirectly through the use of instruments which extend our senses and may be either:

- Qualitative: described by words or terms rather than by numbers and including subjective descriptions in terms of variables such as colour, shape and smell; often recorded using photographs and drawings.
- Quantitative: numerical values derived from counts or measurements of a variable (see Chapter 9), frequently requiring use of some kind of instrument.

Although both types of observation are useful, try to use quantitative ones wherever appropriate.

Factors influencing the quality of observations

Perception

Observation is highly dependent upon the perception of the observer. Perception involves both visual and intuitive processes and the interpretation of what you see is very dependent upon what you already know or have seen before. Thus, two persons observing the same event may 'see' it differently, a good example of bias. This is frequently true in microscopy where experience is an important factor in interpretation.

When you start biology, your knowledge base will be limited and your experience restricted. Practical training in observation provides the opportunity to develop both aspects of your skills in a process which is effectively a positive feedback loop — the more you know/see as a result of practice, the better will your observations become.

Precision and error

Obviously very important for interpretive accuracy and with both human and non-human components. These are dealt with in Chapter 9.

Artefacts

These are artificial features introduced usually during some treatment process such as chemical fixation prior to microscopic examination. They may be included in the interpretive process if their presence is not recognized — again, prior experience and knowledge are important factors in spotting artefacts (see Chapter 29 in relation to microscopy).

Developing observational skills

Preparation — thorough theoretical groundwork before a practical class or examination is vitally important for improving the quality of your observations.

You must develop your knowledge and observational skills to benefit properly from your practical work. The only way to acquire these skills is through extensive practice.

Make sure your observations are:

- relevant, i.e. directed towards a clearly defined objective;

- accurate, i.e. related to a scale whenever possible;
- repeatable, i.e. as error free (precise) as possible.

Much biological work attempts to relate structure to function, often through careful analysis of structure at different levels. One of the best ways to develop observational skills is by making accurate drawings or diagrams, forcing you to look more carefully than is usual (see Chapter 15). Quantification of a variable helps to define observations more rigorously and is particularly important because of the inherent variation in most biological material: this is where statistics becomes a valuable tool for biologists (see Chapters 44 and 45).

Perceptual illusions — beware deceptions such as holes appearing as bumps in photographs, misinterpretations due to lack of information on scale and other simple but well-known tricks of vision.

An important observational skill to develop is the interpretation of two-dimensional images, such as sections through plant/animal material and photographs, in terms of the three-dimensional forms from which they are derived. This requires a clear understanding of the nature of the image in terms of both scale and orientation (see Chapter 29).

Counting

Counting is an observational skill that requires practice to become both accurate and efficient. It is easy to make errors or lose count when working with large numbers of objects. Use a counting aid whenever human error might be significant. There are many such aids such as tally counters, tally charts and specialized counting devices like colony counters. It is important to avoid counting items twice. For example, when counting microbial colonies on Petri plates, each colony can be marked off as counted on the base of the plate using a spirit-based marker pen.

A simple but effective aid is the use of acetate sheets as overlays for photographs, drawings and projections: by using a washable marker pen to mark each relevant object, accurate counts of cells, plants, birds, etc., are possible. Remember to mark the corners/boundaries carefully first in case the sheet slips during the operation!

Another valuable technique is to use a grid system to organize the counting procedure. This has become formalized in equipment such as the haemocytometer used for counting blood cells (see Chapter 27). Remember that you must decide on a protocol for sampling, particularly with regard to the direction of counting within the grid and for dealing with boundary overlaps to prevent double counting at the edges of the grid squares (see Chapter 13 also).

Observation and examinations

Making appropriate observations during practical examinations often causes difficulty, particularly when qualitative observations are needed, e.g. when asked to classify a specimen, giving reasons. Answering such questions clearly requires biological knowledge but also requires a strategy to provide the relevant observations. Thus for the above example, observations relevant to determining each taxonomic level should be made and recorded: set out your observations in a logical sequence so that you show the examiner how you arrived at your conclusion.

A similar approach is needed when asked to make observations related to structure or function. For example, if comments on a method of locomotion are required, do not make observations on irrelevant structures. You will obtain maximum marks only if your answers are concise and relevant (Chapter 57).

12 Scientific method and design of experiments

Definition
Hypothesis — one possible explanation for an observed event. A mechanistic hypothesis is one based on some intuition about the mechanism underlying a phenomenon.

Biology is a systematized body of knowledge derived from observation and experiment. A biologist makes observations and attempts to explain them; these tentative explanations are called hypotheses and their validity is tested by systematically forming and rejecting alternative explanations.

Some branches of biology involve observational science. Here, organisms are usually investigated in as natural a condition as possible. This is an extremely valuable form of knowledge, but it rarely explains the mechanisms of the phenomena observed. The appropriate conditions with which to test a mechanistic hypothesis may take a long time to turn up or may only occur in a location that differs in other crucial ways. In experimental science, the process of obtaining relevant conditions is speeded up and controlled by the investigator.

An experiment is a contrived situation designed to test one or more hypotheses. Any hypothesis that cannot be rejected from the results of an experiment is provisionally accepted. This 'sieve' effect leaves us with a set of current explanations for our observations. These explanations are not permanent and may be rejected on the basis of a future investigation. A hypothesis that has withstood many such tests and has been shown to allow predictions to be made is known as a theory, and a theory may generate such confidence through its predictive abilities to be known as a law (Fig. 12.1).

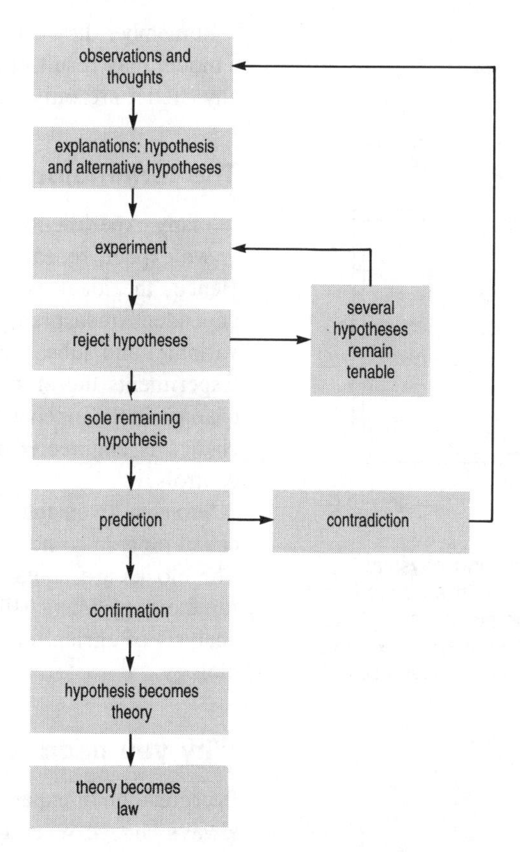

Fig. 12.1 How scientific investigations proceed.

Observations are a prelude to experimentation, but they are preconditioned by a framework of peripheral knowledge. While there is an element of luck in being at the right place and time to make important observations, as Pasteur stated, 'chance favours only the prepared mind'. A fault in scientific method is that the design of the experiment and choice of method may influence the outcome — the decisions involved may not be as objective as some scientists assume. Another flaw is that radical alternative hypotheses may be overlooked in favour of a modification to the original hypothesis, and yet just such leaps in thinking have frequently been required before great scientific advances.

Key point
The fallibility of scientific 'facts' is essential to grasp. No explanation can ever be 100% certain as it is always possible for a new alternative hypothesis to be generated. Our understanding of biology changes all the time as new observations and methods force old hypotheses to be retested.

No hypothesis can ever be rejected with certainty. Statistics allow us to quantify as vanishingly small the probability of an erroneous conclusion, but we are nevertheless left in the position of never being 100% certain that we have rejected all relevant alternative hypotheses, nor 100% certain that our decision to reject some alternative hypotheses was correct! However, despite these problems, experimental science has yielded and continues to yield many important findings.

Quantitative hypotheses, those involving a mathematical description of the system, have become very important in biology. They can be formulated concisely by mathematical models. Formulating models is often useful because it forces deeper thought about mechanisms and encourages simplification of the system. A mathematical model:

Definition
Mathematical model — an algebraic summary of the relationship between the variables in a system.

- is inherently testable through experiment;
- identifies areas where information is lacking or uncertain;
- encapsulates many observations;
- allows you to predict the behaviour of the system.

Remember, however, that assumptions and simplifications required to create a model may result in it being unrealistic. Further, the results obtained from any model are only as good as the information put into it.

The terminology of experimentation

In many experiments, the aim is to provide evidence for causality. If x causes y, we expect, repeatably, to find that a change in x results in a change in y. Hence, the ideal experiment of this kind involves measurement of y, the dependent (measured) variable, at one or more values of x, the independent variable, and subsequent demonstration of some relationship between them. Experiments therefore involve comparisons of the results of treatments — changes in the independent variable as applied to an experimental subject. The change is engineered by the experimenter under conditions that he or she controls.

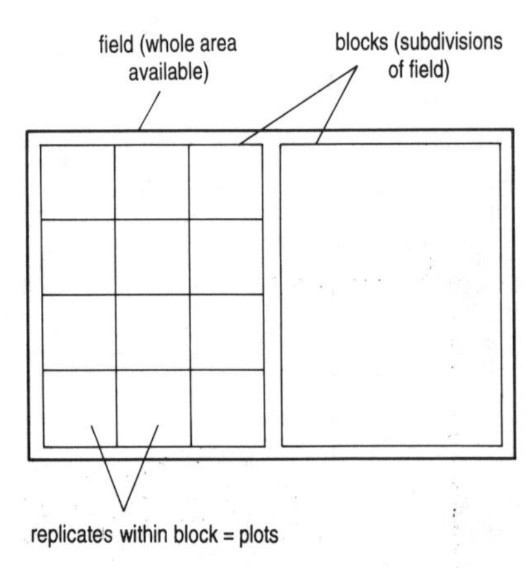

Fig. 12.2 Terminology and physical arrangement of elements in an experiment. Each block should contain the complete range of treatments (treatments may be replicated more than once in each block).

Subjects given the same treatment are known as replicates (they are also called plots in some situations). A block is a grouping of replicates or plots. The blocks are contained in a field, i.e. the whole area (or time) available for the experiment (Fig. 12.2). These terms originated from the statistical analysis of agricultural experiments, but they are now used for all areas of biology.

Why you need to control variables in experiments

Interpretation of experiments is seldom clear-cut because uncontrolled variables always change when treatments are given.

Confounding variables

These increase or decrease systematically as the independent variable increases or decreases. Their effects are known as systematic variation. This form of variation can be disentangled from that caused directly by treatments by incorporating appropriate controls in the experiment. A control is really just another treatment where a potentially confounding variable is adjusted so that its effects, if any, can be taken into account. The results from a control may therefore allow an alternative hypothesis to be rejected. There are many potential controls for any experiment.

The consequence of uncontrolled variables is that you can never be certain that the treatment, and the treatment alone, has caused an observed result. By careful design, you can, however, 'minimize the uncertainty' involved in your conclusion. Methods available include:

- Ensuring, through experimental design, that the independent variable is the only major factor that changes in any treatment.
- Incorporating appropriate controls to show that potential confounding variables have little or no effect.
- Selecting experimental subjects randomly to cancel out systematic variation arising from biased selection.
- Matching or pairing individuals among treatments so that differences in response due to their initial status are eliminated.
- Arranging subjects and treatments randomly so that responses to systematic differences in conditions do not influence the results.
- Ensuring that experimental conditions are uniform so that responses to systematic differences in conditions are minimized. When attempting this, beware 'edge effects' where subjects on the periphery of the layout receive substantially different conditions from those in the centre.

> One way you can reduce edge effects is to incorporate a 'buffer zone' of untreated subjects around the experiment proper.

Nuisance variables

These are uncontrolled variables which cause differences in the value of y independently of the value of x, resulting in random variation. Experimental biology is characterized by the high number of nuisance variables that are found and their relatively great influence on results: biological data tend to have large errors! To reduce and assess the consequences of nuisance variables, you should:

- incorporate replicates to allow random variation to be quantified;
- choose subjects that are as identical as possible;
- control random fluctuations in environmental conditions.

Constraints on experimental design

> Evaluating design constraints — a good way to do this is by processing an individual subject through the experimental procedures — a 'preliminary run' can help to identify potential difficulties.

Box 12.1 outlines the important stages in designing an experiment. At an early stage, you should find out how resources may constrain the design. For example, limits may be set by availability of subjects, cost of treatment, availability of a chemical or bench space. Logistics may be a factor (e.g. time taken to record or analyse data).

Your equipment or facilities may affect design because you cannot regulate conditions as well as you might desire. For example, you may be unable to ensure that temperature and lighting are equal over an experiment laid out in a glasshouse or you may have to accept a great deal of initial variability if your subjects are collected from the wild. This problem is especially acute for experiments carried out in the field.

Box 12.1 Checklist for designing and executing an experiment

1. **Preliminaries**
   (a) **Read background material** and decide on a precise subject area to investigate.
   (b) **Formulate a simple hypothesis to test.** It is preferable to have a clear answer to one question than to be uncertain about several questions.
   (c) **Decide which dependent variable you are going to measure and how:** is it relevant to the problem? Can you measure it accurately, precisely and without bias?
   (d) **Think about and plan the statistical analysis of your results.** Will this affect your design?
2. **Designing**
   (a) **Find out the limitations on your resources.**
   (b) **Choose treatments which alter the minimum of confounding variables.**
   (c) **Incorporate as many effective controls as possible.**
   (d) **Keep the number of replicates as high as is feasible.**
   (e) **Ensure that the same number of replicates is present in each treatment.**
   (f) **Use effective randomization and blocking arrangements.**
3. **Planning**
   (a) **List all the materials you will need.** Order any chemicals and make up solutions; grow, collect or breed the experimental subjects you require; check equipment is available.
   (b) **Organize space and/or time** in which to do the experiment.
   (c) **Account for the time taken to apply treatments and record results.** Make out a timesheet if things will be hectic.
4. **Carrying out the experiment**
   (a) **Record the results and make careful notes of everything you do** (see pages 63–66). Make additional observations to those planned if interesting things happen.
   (b) **Repeat experiment** if time and resources allow.
5. **Analysing**
   (a) **Graph data as soon as possible** (during the experiment if you can). This will allow you to visualize what has happened and make adjustments to the design (e.g. timing of measurements).
   (b) **Carry out the planned statistical analysis.**
   (c) **Jot down conclusions and new hypotheses** arising from the experiment.

Key point
When deciding the number of replicates in each treatment, try to:
- **maximize the number of replicates in each treatment;**
- **make the number of replicates even.**

Use of replicates

Replicate results show how variable the response is within treatments. They allow you to compare the differences among treatments in the context of the variability within treatments — you can do this via statistical tests such as analysis of variance (Chapter 46). Larger sample sizes tend to increase the precision of estimates of statistical parameters and increase the chances of showing a significant difference between treatments if one exists. For statistical reasons (weighting, ease of calculation, fitting data to certain tests), it is best to keep the number of replicates even. Remember that the degree of independence of replicates is important: sub-samples cannot act as replicate samples — they tell you about variability in the measurement method but not in the quantity being measured.

If the total number of replicates available for an experiment is limited by resources, you may need to compromise between the number of treatments and the number of replicates per treatment. Statistics can help here, for it is possible to work out the minimum number of replicates you would need to show a certain difference between pairs of means (say 10%) at a specified level of significance (say $P = 0.05$). For this, you need to do a preliminary investigation to estimate variability within treatments (see Sokal and Rohlf, 1981).

Box 12.2 How to use random number tables to assign subjects to positions and treatments

This is one method of many that could be used. It requires two sets of *n* random numbers — where *n* is the total number of subjects used.

1. **Number the subjects in any arbitrary order** but in such a way that you know which is which (i.e. mark or tag them).
2. **Decide how treatments will be assigned**, e.g. first five subjects selected treatment A; second five → treatment B, etc.
3. **Use the first set of random numbers in the sequence obtained to identify subjects and allocate them to treatment groups** in order of selection as decided in (2).
4. **Map the positions for subjects in the block or field. Assign numbers to these positions using the second set of random numbers**, working through the positions in some arbitrary order, e.g. top left to bottom right.
5. **Match the original numbers given to subjects with the position numbers.**

To obtain a sequence of random numbers

1. **Decide on the range of random numbers you need**, e.g. 1–20.
2. **Decide how you wish to sample the random number tables** (e.g. column by column and top to bottom) and your starting point.
3. **Moving in the selected manner, read the sequence of numbers until you come to a group that fits your needs** (e.g. in the sequence 978186, 18 represents a number between 1 and 20). Write this down and continue sampling until you get a new number. If a number is repeated, ignore it. Small numbers need to have the appropriate number of zeros preceding (e.g. 5 = 05 for a range in the tens, 21 = 021 for a range in the hundreds).
4. **When you come to the last number required, you don't need to sample any more**: simply write it down.

Example: You find the following random number sequence in a table and wish to select numbers between 1 and 10 from it.

5059146823	4862925166	**10**632**6**0**3**45
127742381**0**	**9**94**8**0**4**0**6**76	64**30**247598
8357945137	249**01**45183	59462422**08**
6588812379	2325701558	326**07**26568

Working left to right and top to bottom, the order of numbers found is 5, 10, 3, 9, 4, 6, 2, 1, 8, 7 as indicated by bold font. If the table is sampled by working right to left from bottom to top, the order is 6, 10, 7, 2, 9, 3, 4, 8, 1, 5.

Randomization of treatments

The two aspects of randomization you must consider are:

- positioning of treatments within experimental blocks;
- allocation of treatments to the experimental subjects.

For relatively simple experiments, you can adopt a completely randomized design; here, the position and treatment assigned to any subject is defined randomly. You can draw lots, use a random number generator on a calculator, or use the random number tables which can be found in most books of statistical tables (see Box 12.2).

A completely randomized layout has the advantage of simplicity but cannot show how confounding variables alter in space or time. This information can be obtained if you use a blocked design in which the degree of randomization is restricted. Here, the experimental space or time is divided into blocks, each of which accommodates the complete set of treatments (Fig. 12.2). When analysed appropriately, the results for the blocks can be compared to test for differences in the confounding variables and these effects can be separated out from the effects of the treatments. The size and shape (or timing) of the block you choose is important: besides being able to accommodate the number of replicates desired, the suspected confounding variable should be relatively uniform within the block.

> ***Example***
> *If you knew that soil type varied in a graded fashion across a field, you might arrange blocks to be long thin rectangles at right angles to the gradient to ensure conditions within the block were as even as possible.*

A Latin square is a method of placing treatments so that they appear in a balanced fashion within a square block or field. Treatments appear once in each column and row (see Fig. 12.3), so the effects of confounding variables

3×3

A	C	B
B	A	C
C	B	A

4×4

A	C	B	D
D	B	C	A
C	D	A	B
B	A	D	C

5×5

E	A	D	B	C
D	C	E	A	B
C	B	A	E	D
A	D	B	C	E
B	E	C	D	A

6×6

F	B	E	A	D	C
E	C	A	B	F	D
A	F	C	D	B	E
C	D	F	E	A	B
B	E	D	F	C	A
D	A	B	C	E	F

Fig. 12.3 Examples of Latin square arrangements for 3–6 treatments. Letters indicate treatments; the number of alternative arrangements for each size of square increases greatly as the size increases.

can be 'cancelled out' in two directions at right angles to each other. This is effective if there is a smooth gradient in some confounding variable over the field. It is less useful if the variable has a patchy distribution, where a randomized block design might be better.

Latin square designs are useful in serial experiments where different treatments are given to the same subjects in a sequence (e.g. Fig. 12.4). A disadvantage of Latin squares is the fact that the number of plots is equal to the number of replicates, so increases in the number of replicates can only be made by the use of further Latin squares. These should have different internal arrangements.

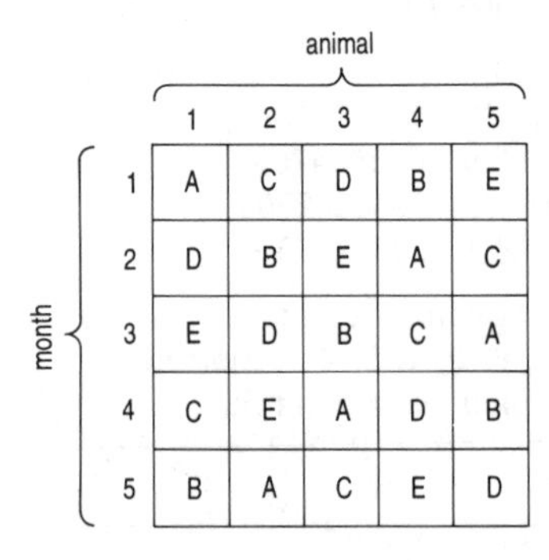

Fig. 12.4 Example of how to use a Latin square design to arrange sequential treatments. The experimenter wishes to test the effect of drugs A–E on weight gain, but only has five animals available. Each animal is fed on control diet for the first 3 weeks of each month, then on control diet plus drug for the last week. Weights are taken at start and finish of each treatment. Each animal receives all treatments.

Pairing and matching subjects

The paired comparison is a special case of blocking which you can use to remove a source of systematic variation when there are two treatments. Examples of its use are:

- 'Before and after' comparison. Here, the pairing removes variability arising from the initial state of the subjects, e.g. weight gain of mice on a diet, where the weight gain may depend on the initial weight.
- Application of a treatment and control to parts of the same subject or to closely related subjects. This allows comparison without complications arising from different origin of subjects, e.g. drug or placebo given to sibling rats, virus-containing or control solution swabbed on left or right halves of a leaf.
- Application of treatment and control under shared conditions. This allows comparison without complications arising from different environment of subjects, e.g. rats in a cage, plants in a pot.

Matched samples represent a restriction on randomization where you make a balanced selection of subjects for treatments on the basis of some attribute or attributes that may influence results, e.g. age, gender, prior history. The effect of matching should be to 'cancel out' the unwanted source(s) of variation. Disadvantages include the subjective element in choice of character(s) to be balanced, inexact matching of quantitative characteristics, the time matching takes and possible wastage of unmatched subjects.

When analysed statistically, both paired comparisons and matched samples can show up differences between treatments that might otherwise be rejected on the basis of a fully randomized design. Note that the statistical analysis of paired samples is slightly different from that of fully random or matched samples (see Sokal and Rohlf, 1981).

Definition
Interaction — where the effects of treatments given together is greater or less than the sum of their individual effects.

Multifactorial experiments

The simplest experiments are those in which one treatment (factor) is applied at a time to the subjects. This approach is likely to give clear-cut answers, but it could be criticized for lacking realism. In particular, it cannot take account of interactions among two or more conditions that are likely to occur in real life. A multifactorial experiment (Fig. 12.5) is an attempt to do this; the interactions among treatments can be analysed by specialized forms of analysis of variance.

Multifactorial experiments are economical on resources because of 'hidden replication'. This arises when two or more treatments are given to a subject because the result acts statistically as a replicate for each treatment. Choice of relevant treatments to combine is important in multifactorial experiments; for instance, an interaction may be present at certain concentrations of a chemical but not at others (perhaps because the response is saturated). It is also important that the measurement scale for the response is consistent, otherwise spurious interactions may occur. Beware when planning a multifactorial experiment that the numbers of replicates do not get out of hand: you may have to restrict the treatments to 'plus' or 'minus' the factor of interest (as in Fig. 12.5).

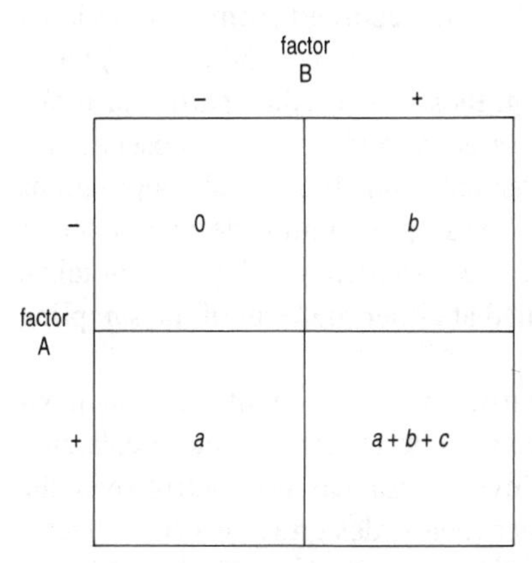

Fig. 12.5 Design of a simple multifactorial experiment. Factors A and B have effects a and b when applied alone. When both applied together, the effect is denoted by $a + b + c$. If $c = 0$, there is no interaction (e.g. $2 + 2 + c = 4$). If c is positive, there is a positive interaction (synergism) between A and B (e.g. $2 + 2 + c = 5$). If c is negative, there is a negative interaction (antagonism) between A and B (e.g. $2 + 2 + c = 3$).

Repetition of experiments

Even if you have taken great care to ensure that your experiment is well designed and statistically analysed, you are limited in the conclusions that can be made. Firstly, what you can say is valid for a particular place and time, with a particular investigator, experimental subject and method of applying treatments. Secondly, if your results were significant at the 5% level of probability, there is still an approximately one-in-twenty chance that the results did arise by chance. To guard against these possibilities, it is important that experiments are repeated. Ideally, this would be done by an independent scientist with independent materials. However, it makes sense to repeat work yourself so that you can have full confidence in your conclusions. Many scientists recommend that all experiments are carried out three times in total. This may not be possible in undergraduate practical classes or project work!

It is good practice to report how many times your experiments were repeated (in Materials and Methods); in the Results section, you need either a statement saying that the illustrated experiment is representative or one explaining the differences between results obtained.

A final piece of advice

At the start of a project, don't spend too long reading the literature and thinking about what crucial hypothesis to put to the test. Get stuck in and do an experiment. There's no substitute for 'getting your hands dirty' to stimulate new ideas.

13 Samples and sampling

Definitions
Biological population — a collection of individuals of the same species (perhaps of a given life history or sex) at a particular location at the same time.
Statistical population — the totality of individuals within a specified time or space and about which inferences are to be made, i.e. the total number of potential sampling units.
Statistical universe — the hypothetical, infinitely large, collection of individuals — applicable to experimental situations where an experiment could theoretically be repeated an infinite number of times.

It is rarely practical to make measurements using every individual in a population in laboratory or field investigations. Therefore, a sub-set (a sample) is used to estimate the values that might have been obtained had we measured every individual (the population). A sample is made up of a series of sampling units which depend on the type of variable being measured; for example, a data value for a particular variable (e.g. length) recorded from an individual sampling unit (e.g. a limpet) in a sample of n units (e.g. $n = 50$ limpets) taken from the population under investigation (e.g. those limpets on a particular rocky shore): you should be able to make an equivalent statement for any observations you make. The distinction between 'statistical' and 'biological' populations can be very important, particularly when the sampling unit is defined in terms of its area. The term 'replicate' (repeated measurement or observation taken at the same time and location) can be applied at either the level of the sampling unit or the sample.

Sampling should enable you to obtain sufficiently reliable information about the particular population in which you are interested. This chapter deals only with sampling in field work where the investigator has no control over the individuals under study. In laboratory experiments, design is the critical factor determining sampling strategy and analysis of results (see Chapter 12).

Definitions
Parameter — a numerical constant or mathematical function used to describe a particular population (e.g. the mean height of all 18-year-old females).
Statistic — an estimate of a parameter obtained from a sample (e.g. the mean height of 18-year-old females based on those in your class).

Sample data are used to make inferences about the populations from which they were drawn. Symbols are used to represent each type of sample statistic: these are given roman character symbols (e.g. $\bar{Y}$ = sample mean) while the equivalent population parameter is given a Greek symbol (e.g. μ = population or universe mean). When estimating population parameters from sample statistics, the sample size (= number of sampling units) is important: larger sample sizes usually result in greater statistical confidence. However, optimum sample size is a balance between statistical and practical considerations.

Box 13.1 Checklist for creating a sampling strategy

- Know precisely what your objectives are.
- Define the population as carefully as possible before you start.
- Carry out pilot sampling exercises before starting the main programme.
- Identify and minimize any bias in your procedures.
- Decide how you are going to analyse the data before you collect it.
- Adopt a relevant and appropriate sampling strategy.

The characteristics of a good sample

Sampling strategy (Box 13.1) has two main components:

- Selecting the sample, which involves the formulation of rules, procedures and operations (sampling protocol) by which some members of the population are included in the sample.
- Processing the data, which consists of rules and operations for calculating statistics.

Good sampling design should take into account both of these, and it should:

- relate to the objective(s) of the investigation;

- be practical and achievable;
- be cost-effective in terms of equipment and labour;
- provide estimates of population parameters that are truly representative and unbiased.

Ideally, representative samples should be:

- taken at random so that every member of the population of data has an equal chance of selection;
- large enough to give sufficient precision;
- unbiased by the sampling procedure or equipment.

These requirements may well conflict and there is rarely any unique best answer to a sampling problem.

You must provide a complete description (specification) of the population being sampled, i.e. specify the statistical population precisely. Relevant factors include location, habitat, time, age, sex, physiological condition and disease status. A poor specification may make the results difficult to interpret. When populations are to be compared, only the variable under investigation should differ between the populations if maximum significance is to be placed upon the analysis. Do not include data for which the appropriate specification is not available.

When the sample is biased, parameters cannot be realistically inferred from it and conclusions will be unreliable. To avoid bias, take great care in specification of the population and choice of sampling technique and equipment (see also Chapter 9).

Examples
Sampling bias:
If the older animals in a population are resting while the younger ones are foraging for food, a sample based on baited traps may not be representative of the entire population if the character you are measuring varies with age.

If the sex ratio in a population were known to be 2:1, a strategy which samples equal numbers of males and females might not be valid if the character of interest varies with sex.

Selection strategies for field sampling

Field sampling strategy is usually based upon a version of either point sampling, traverse (transect) sampling or quadrat sampling procedures (Fig. 13.1a–c). The distribution of sampling points or quadrats is usually based upon one of three methodological approaches:

- Simple random sampling (Fig. 13.1d);
- Systematic areal sampling (Fig. 13.1.e);
- Stratified random sampling (Fig. 13.1f and g).

The sampling protocol should be established before any investigation proceeds. Principal criteria to be determined will include:

- The dimensions and shape of the sampling unit (e.g. quadrat size);
- The number of sampling units in each sample;
- The location of sampling units within the sampling area.

One of the main factors influencing these decisions is the spatial distribution pattern of the individual objects to be sampled. Their distribution may be random (Fig. 13.2a) — surprisingly rare in nature, clumped (contagious) (Fig. 13.2b) or based upon a gradient (Fig 13.2c). The characteristics of the sampling unit must be chosen with this pattern in mind.

The size of the sampling unit

This is best considered by reference to selection of a quadrat for fieldwork sampling. Note first that the total number of available sampling units (quadrats) is the area occupied by the population being investigated (total theoretical sampling area) divided by the area of sampling units (quadrat area).

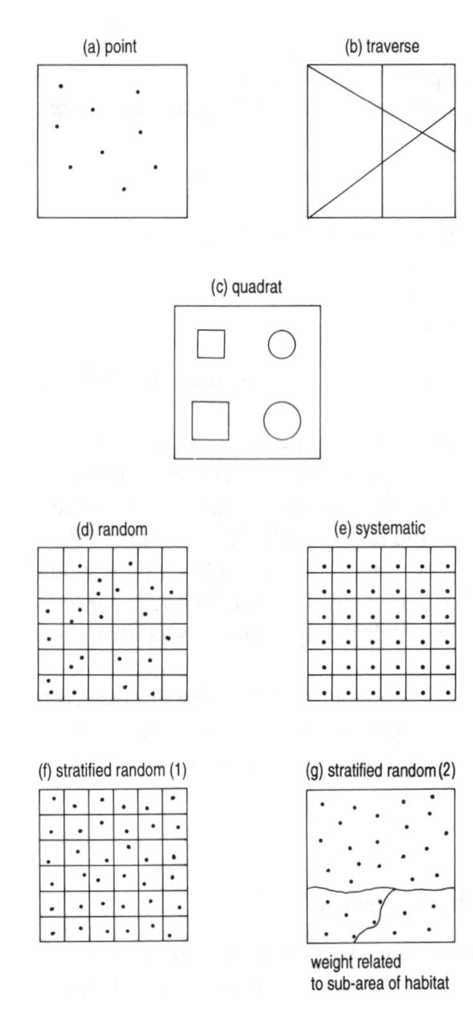

Fig. 13.1 Types of sampling strategy.

The distribution and size of the organisms to be investigated must be considered: it is obvious that you would require a different-sized sampling unit for trees in a coniferous forest than for daisies on a lawn. When the distribution is random, all quadrat sizes are equally effective for the estimation of population parameters (assuming the total number of individuals sampled is equal). If the distribution is clumped, a smaller unit size is usually more effective than a larger one. Too large a sampling unit may obscure the true nature of a clumped distribution. Alternatively, you may wish to exclude a clumped distribution from your investigation and thus will choose a relatively large sampling unit. If the distribution is graded, then sample size is less important and non-random methods such as systematic sampling or the use of transects/traverses may be appropriate.

Small sample units are often advantageous because more small units can usually be taken for the same amount of labour, thereby increasing precision. Many small units cover a wider range of the habitat than a few large units, so the catch can be more representative. However, sampling error at the edge of the quadrat is proportionally greater as sample size diminishes: establish a protocol for dealing with items which overlap the edge of the sampling unit, e.g. include those on the top and right hand edges and exclude bottom and left hand edges.

The final choice of quadrat size is always a compromise between statistical and practical requirements.

The number of sampling units in each sample

When small numbers of sampling units are used, this can give rise to imprecise estimates of population parameters. This is especially true if the spatial distribution of individuals is clumped (contagious), as is often the case. However, taking very large numbers of replicate units, e.g. > fifty sampling units per sample, usually presents an impractical workload. A general method for determining a suitable number of sampling units, usually performed as a pilot study, is as follows:

1. Take five sampling units at random: calculate the arithmetic mean of the measured variable, e.g. abundance or dimensions of a particular organism.
2. Take five more units at random: calculate the mean for the ten units collected.
3. Continue sampling in five-unit steps: plot the cumulative mean against sample number; when the mean fluctuates within acceptable limits, e.g. ± 5%, a suitable number of sampling units has been reached (Fig. 13.3).

It may be impractical for you to calculate cumulative means during sampling and, therefore, this procedure is limited in application. Methods exist for the calculation of the number of sampling units required for specific levels of precision (Elliott, 1971).

The position of the sampling units

In simple random sampling the total area to be investigated is divided into sampling units of an appropriate size, based upon a grid. Grid references are obtained using random numbers (p. 53). This procedure is repeated for each sampling unit until the required sample size has been obtained. Every potential sampling unit in the population thus has an equal chance of selection.

Stratified random sampling is often preferable if randomly selected units could miss out a distinct area and is particularly useful in habitats with clearly defined micro-habitats, e.g. a tidal seashore. The population is divided into

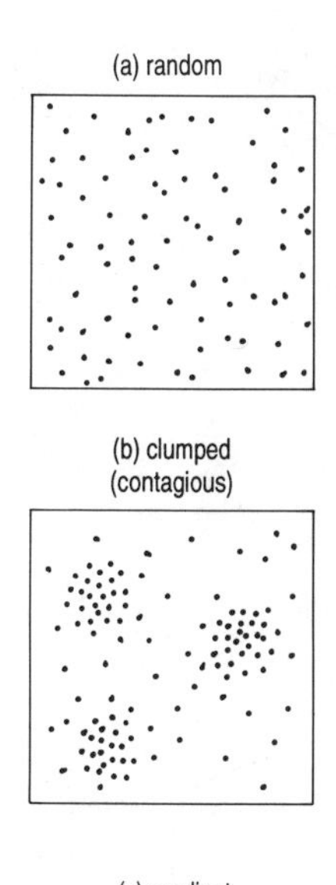

(c) gradient

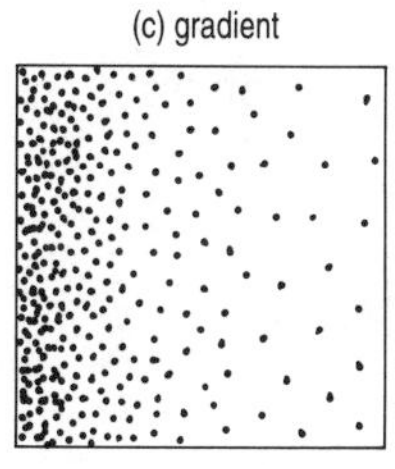

Fig. 13.2 Types of distribution: (a) random; (b) clumped (contagious); (c) gradient.

several distinct sub-populations or strata e.g. organisms within well-defined areas of known size. Sampling units are then allocated randomly within each stratum. You can analyse data from different strata by a one-way analysis of variance (ANOVA, see Chapter 46) with sources of variation between and within strata. Strata are usually sampled in proportion to their area (Fig. 13.1g).

Systematic sampling may be required for determining the position of organisms within the habitat (mapping). The location of the first sampling unit is chosen at random and then subsequent units are selected at fixed distances from it: the advantages are simplicity and regular distribution of the sampling units. The disadvantages are firstly, that the sample can be very biased if the interval between units coincides with a periodic spatial distribution of the population and secondly, that there is no reliable method for estimating the standard error of the sample mean.

The process of determining a strategy for sampling is outlined in Fig. 13.4.

Sampling in time

Sampling in both laboratory and field over a period of time has its own problems. If you are examining a process that has a regularly fluctuating nature (e.g. with a periodicity governed by day and night, or high and low tide), then the frequency of sampling has to be determined with that periodicity in mind depending upon your objectives. A pilot study is usually appropriate for such investigations. The existence of non-obvious periodic phenomena must always be considered. In certain experiments, where the effect of time is best regarded as a logarithmic phenomenon, e.g. a toxicology study, sampling intervals should be spaced according to a logarithmic series.

Sub-sampling

Sometimes sampling units may be too large to allow complete analysis of the sample but reducing the size of the sampling unit may not be possible after the collection has been made. An alternative solution is to use sub-sampling (2-stage sampling): the original sampling unit is then termed the primary unit and the second-stage units referred to as sub-units. If organisms are randomly dispersed within the primary unit before the sub-units are taken and only a small proportion of the total catch is removed, the counts for sub-units should be distributed according to a Poisson series (see Chapter 46): this is testable

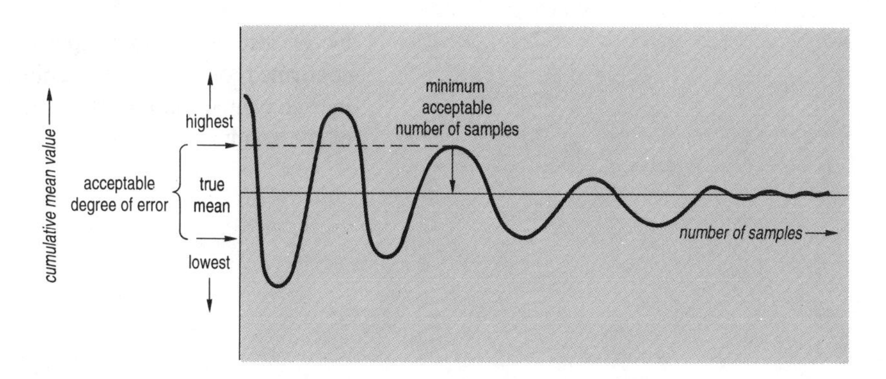

Fig. 13.3 Determination of the number of sampling units by pilot investigation.

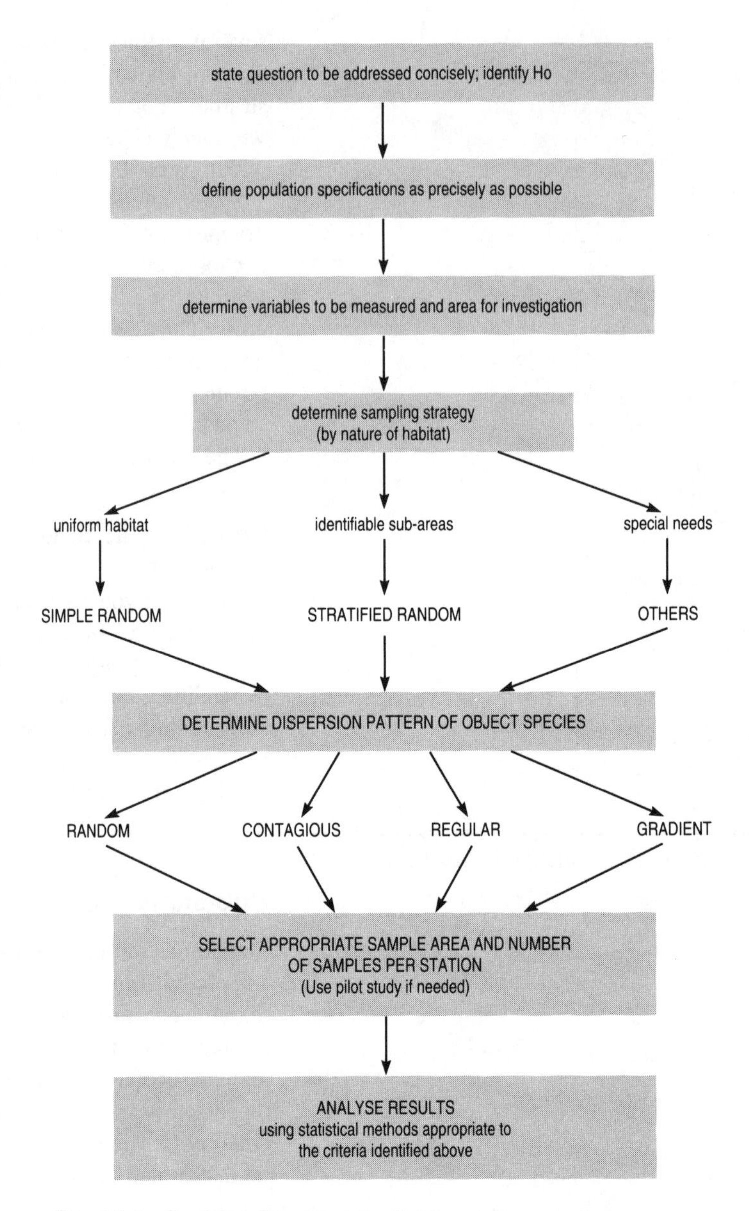

Fig. 13.4 Decision flowchart for field sampling studies.

by χ^2 test on a sub-sample of at least 5 sub-units. If this distribution is confirmed, only single sub-units need to be taken to estimate the total numbers in each primary unit — the precision of this estimate then depends on the size of the count.

Keeping a record

14 Making notes of practical work

When carrying out advanced lab work or research projects, you will need to master the important skill of managing data and observations and learn how to keep a record of your studies in a lab book. This is important for the following reasons:

- An accurate and neat record helps when using information later, perhaps for exam purposes or when writing a report.
- It allows you to practise important skills such as scientific writing, drawing diagrams, preparing graphs and tables and interpreting results.
- Analysing and writing-up your data as you go along prevents a backlog at the end of your study time.
- You can show your work to a future employer to prove you have developed the skills necessary for writing-up properly; in industry, this is vital so that others in your team can interpret and develop your work.

A good set of lab notes should:

- outline the purpose of your experiment or observation;
- set down all the information required to describe your materials and methods;
- record all relevant information about your results or observations and provide a visual representation of the data;
- note your immediate conclusions and suggestions for further experiments.

Collecting and recording primary data

Recording primary data — never be tempted to jot down data on scraps of paper: you may lose them, or forget what individual values mean.

Individual observations (e.g. laboratory temperature) can be recorded in the text of your notes, but tables are the most convenient way to collect large amounts of information. When preparing a table, you should:

1. Give each table a concise title. A numbered code can be useful for cross-referencing at a later stage.
2. Decide on the number of variables to be measured and their relationship with each other and lay out the table appropriately:
   (a) The first column of your table should show values of the independent (controlled) variable, with subsequent columns for the individual (measured) values for each replicate or sampling unit.
   (b) If several variables are measured for the same organism or sampling unit, each should be given a row.
   (c) In time-course studies, put the replicates as columns grouped according to treatment, with the rows relating to different times of measurement.
3. Make sure the arrangement reflects the order in which the values will be collected. Your table should be designed to make the recording process as straightforward as possible, to minimize the possibility of mistakes. For final presentation, a different arrangement may be best (Chapter 43).
4. Make sure there is sufficient space in each column for the values — if in doubt, err on the generous side.
5. Consider whether additional columns are required for subsequent calculations. Create a separate column for each mathematical manipulation, so the step-by-step calculations are clearly visible. Use a

computer spreadsheet (p. 250) if you are manipulating lots of data.

6. Use a pencil to record data so that mistakes can be easily corrected.
7. Take sufficient time to record quantitative data unambiguously — use large clear numbers, making sure that individual numerals cannot be confused.
8. Record numerical data to an appropriate number of significant figures, reflecting the accuracy and precision of your measurement (p. 39). Do not round off data values, as this might affect the subsequent analysis of your data.
9. Record discrete or grouped data as a tally chart (see p. 211), each row showing the possible values or classes of the variable. Providing tally marks are of consistent size and spacing, this method has the advantage of providing an 'instant' frequency distribution chart.
10. Prepare duplicated recording tables if your experiments or observations will be repeated.
11. Explain any unusual data values or observations in a footnote. Don't rely on your memory.

Suggested method for recording details of project work

The recommended system is one where you make a dual record.

Primary record

Choosing a notebook — a spiral-bound notebook is good for making a primary record — it lies conveniently open on the bench and provides a simple method of dealing with major mistakes!

The primary record is made at the bench or in the field. In this, you must concentrate on the detail of materials, methods and results. Include information that would not be used elsewhere, but which might prove useful in error tracing: for example, if you make a note of exactly how a solution was made up (precise volumes and weights used rather than concentration alone), this could reveal whether a miscalculation had been the cause of a rogue result. Note the origin, type and state of the chemicals and organism(s) used. Make rough diagrams to show the exact arrangement of replicates, the position of equipment, etc. If you are forced to use loose paper to record data, make sure each sheet is dated and taped to your lab book or that they are collected in a ring binder or attached together with a treasury tag. The same applies to traces, printouts and graphs.

The basic order of the primary record should mirror that of a research report (see p. 271), including: the title and the date of the experiment, a brief introduction, comprehensive materials and methods, the data collected and short conclusions.

Secondary record

You should make a secondary record concurrently or later at a desk in a bound book and it ought to be neater, in both organization and presentation. This book will be used when discussing results with your supervisor, and when writing up a report or thesis, and may be used as part of your course assessment. While these notes should retain the essential features of the primary record, they should be more concise and the emphasis should move towards analysis of the experiment. Outline the aims more carefully at the start and link the experiment to others in a series (e.g. 'Following the results of Expt. D24, I decided to test whether . . .'). You should present data in an easily digested form, e.g. as tables of means or as summary graphs. Use appropriate statistical tests (p. 234) to support your analysis of the results. The choice of a bound book ensures that data are not easily lost.

Points to note

The dual method of recording deals with the inevitable untidiness of notes taken at the bench or in the field; these often have to be made rapidly, in awkward positions and in a generally complex environment. Writing a second, neater version forces you to consider again details that might have been overlooked in the primary record. Having two versions of your work means vital information is automatically duplicated in case of loss or damage.

The diary aspect of the record can be used to establish precedence (e.g. for patentable research where it can be important to 'minute' where and when an idea arose and whose it was); for error tracing (e.g. you might be able to find patterns in the work affecting the results); or even for justifying your activities to a supervisor.

If you find it difficult to decide on the amount of detail required in Materials and Methods, the basic ground rule is to record enough information to allow a reasonably competent scientist to repeat your work exactly. You must tread a line between the extremes of pedantic, irrelevant detail and the omission of information essential for a proper interpretation of the data — better perhaps to err on the side of extra detail to begin with. An experienced worker can tell you which subtle shifts in technique are important (e.g. batch numbers for an important chemical, or when a new stock solution is made up and used). Many important scientific advances have been made because of careful observation and record taking and because coincident data were recorded that did not seem of immediate value.

When creating a primary record, take care not to lose any of the information content of the data: for instance, if you only write down means and not individual values, this may affect your ability to carry out subsequent statistical analyses.

There are numerous ways to reduce the labour of keeping a record. Don't repeat Materials and Methods for a series of similar experiments; use devices such as 'method as for Expt. B4'. A photocopy might suffice if a method is being copied from a text or article (check with supervisor). To save time, make up and copy a checklist in which details such as chemical batch numbers can be entered.

Always analyse and think about data immediately after collecting them as this may influence your subsequent activities. Particularly valuable is a graphical indication of what has happened. Carry out statistical analyses before moving on to the next experiment because apparent differences among treatments may not turn out to be statistically significant when tested. Write down any conclusions you make while analysing your data: sometimes those which seem obvious at the time of doing the work are forgotten when the time comes to write up a report or thesis. Likewise, ideas for further studies may occur to you and a note of these may prove valuable later. Even if your experiment appears to be a failure, suggestions as to the likely causes might prove useful.

Special requirements for fieldwork

The main problems you will encounter in the field are the effects of the weather while taking a primary record and the distance you might be from a suitable place to make a neat secondary record. Wind, rain and cold temperatures are not conducive to neat note-taking and you should be prepared for the worst possible conditions at all times. Make sure your clothing allows you to feel comfortable while recording data.

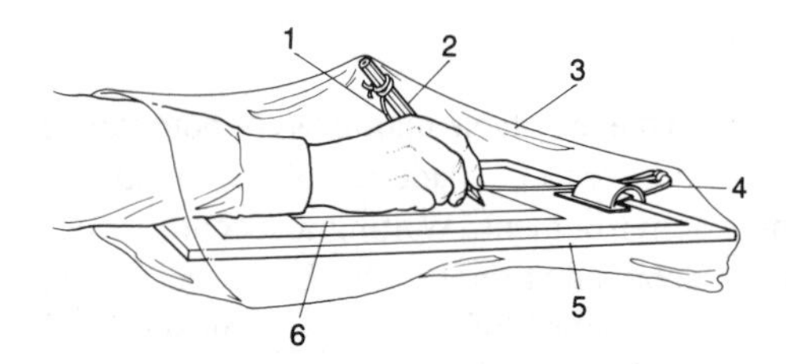

Fig. 14.1 Simple but effective method for keeping notes dry in the field.
1. string attaching pencil to clip
2. pencil (not pen) writes on damp paper
3. transparent plastic bag
4. bulldog clip
5. hardboard (at least 31 × 22 cm for A4 paper)
6. record sheet.

Waterproof notebooks — notebooks with waterproof paper are available (for example the 'Aquascribe' brand supplied by Hawkins and Mainwaring, Ltd, Newark, Notts). These are relatively expensive, but you can re-use them by rubbing out.

The simplest method of protecting a field notebook is to enclose it in a clear polythene bag large enough for you to take notes inside (Fig. 14.1). Alternatively, you could use a clipboard with a waterproof cover to shield your notes or a special notebook with a waterproof cover. When selecting a field notebook, choose a small size — the dimensions of outside pockets may dictate the upper size limit.

If recording results and observations outdoors:

- Use a pencil as ink pens such as ball-points smudge in wet conditions, are temperamental in the cold and may not work at awkward angles. Don't forget to take a sharpener.
- Prepare well to enhance the speed and quality of your field note-taking — the date and site details can be written down before setting out and tables can be made out ready for data entry.
- Transcribe field notes to a duplicate primary record at your base each time you return there. There is a very real risk of your losing or damaging a field notebook. Also, poor weather may prevent full note-taking and the necessary extra details should be written up while fresh in your memory.
- Consider using a tape recorder rather than a notebook, in which case transcription into written form should also take place while your memory is fresh in case the sound quality is poor.
- Use photographs to set data in context, when appropriate. Develop photographs as soon as possible to check their suitability. Consider using a Polaroid® camera when the suitability of a record must be guaranteed.

Field data may be logged automatically, stored temporarily in the instrument's electronic memory, and downloaded to a portable computer ('data logger') when convenient. The information is then transferred to a data bank back at base. If you are using this system, back up each period's data as soon as possible — if the recording instrument's memory is cleared or overwritten after reading there may be no recourse if the logging machine fails.

Using communal records

If working with a research team, you may need to use their communal databases. These avoid duplication of effort and ensure uniformity in techniques. They may also form part of the legal safety requirements for lab work. You will be expected to use the databases carefully and contribute to them properly. They might include:

- a shared notebook of common techniques (like how to make up media or buffer solutions);
- a set of step-by-step instructions for use of equipment. Manuals are often complex and poorly written and it is useful to simplify them, incorporating any differences in procedure adopted by the group;
- an alphabetical list of suppliers of equipment and consumables (perhaps held on a card index system);
- a list of chemicals required by the group and where they are stored;
- the COSHH risk assessment sheets for dangerous procedures (p. 9);
- the record book detailing the use of radioisotopes and their disposal.

15 Drawing and diagrams

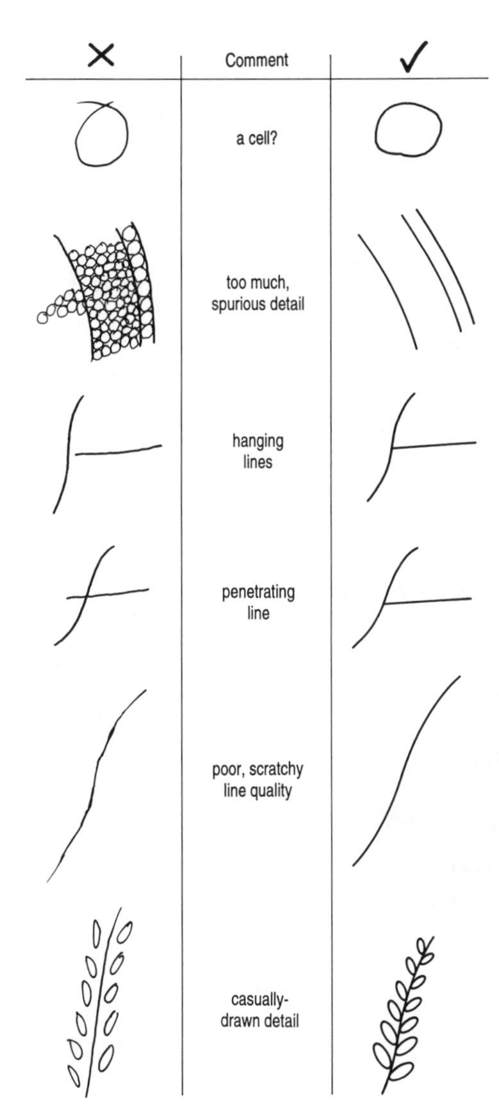

Fig. 15.1 Examples of common errors in biological drawings.

Drawing apparatus — it is often best to use a longitudinal section rather than a perspective drawing (Fig. 15.2).

The study of biology requires considerable time spent observing structures and discovering their interrelationships and functions. The accurate drawing of biological objects is a skill which has always been an important component of biological communication. Although drawing has largely been superseded in research studies by techniques such as photography, it occupies an important place in biological teaching because of its role in developing observation skills. The basic skills of drawing can be learned and are associated with the more important skill of observation; poor drawing (Fig. 15.1) is usually associated with poor observation.

The illustration of a specimen, idea or concept requires both thought and practice. To produce good illustrations (figures), both planning and careful execution are needed. The quality and type of drawing/diagram required will vary with the stage of your course: during the early years, most of your practical requirements will be line drawings and diagrams from observations made as directed, while in later years, during the course of projects, etc., you may be required to produce high quality drawings and illustrations using more sophisticated drawing aids (see Briscoe, 1990).

Basic types of figures

- A drawing is strictly a detailed and accurate representation of a specimen requiring no previous biological knowledge and could best be done by an artist. This is never a requirement for normal practical work.
- A diagram, however, needs to be accurate in its general proportions but is very stylized and only shows the important features: therefore, it requires biological knowledge in order to select items for inclusion and to decide what detail to ignore. You should be able to prepare a variety of diagrams (which may be referred to as drawings since the distinction is frequently ignored) during your course.

Figures are particularly appropriate for:

- Equipment, techniques or procedures — often shown in line drawings or flow charts e.g. Fig. 15.2.
- Pathways and hypotheses e.g. Fig. 15.3, which may be expressed in flow charts and schematic diagrams.
- Anatomical and histological diagrams and drawings, e.g. Figures 15.4 to 15.7.
- Tracings and experimental records which provide verification of data.
- Data, often best represented by graphs, histograms, etc. (see Chapter 42).

Line drawings/diagrams

These are usually done in pencil in practical notes. Use pen and ink to create line drawings for posters, project reports, etc.; such diagrams should be in black and white only and are necessarily stylized. Tones of grey can be achieved by stippling, fine contour lines, hatching or cross-hatching, but never by shading. A tissue diagram or map (e.g. Figures 15.4 and 15.5) is a special form where only the boundaries between tissues are marked and *no* cellular details are included.

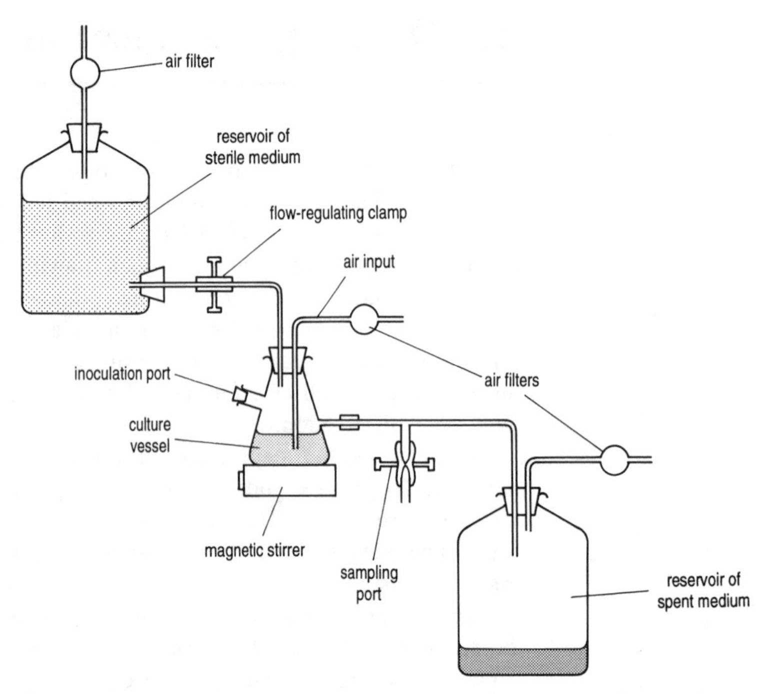

Fig. 15.2 Example of a two-dimensional lab equipment diagram of components of a chemostat.

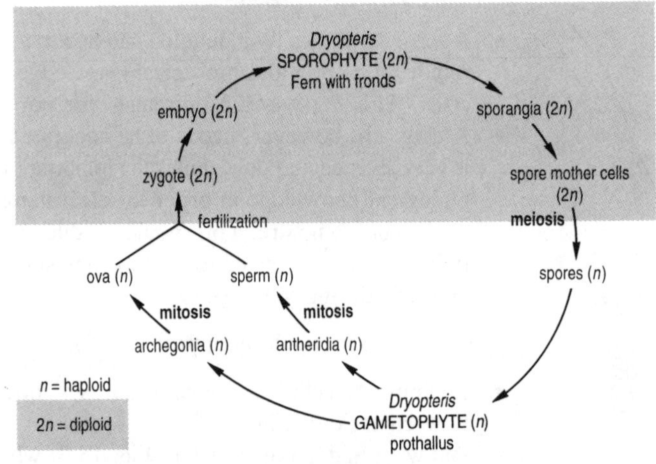

Fig. 15.3 Diagrammatic life cycle of *Dryopteris filix-mas*.

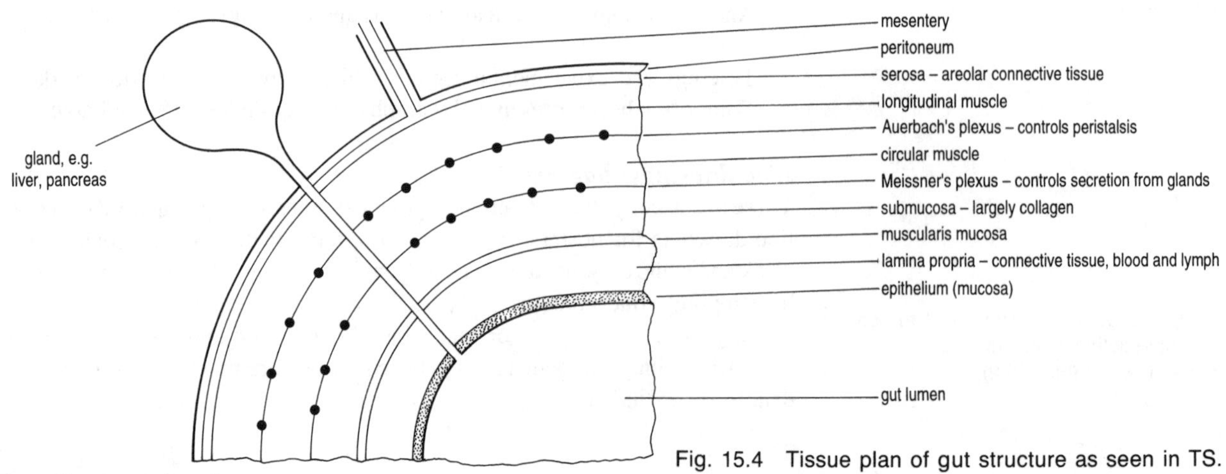

Fig. 15.4 Tissue plan of gut structure as seen in TS.

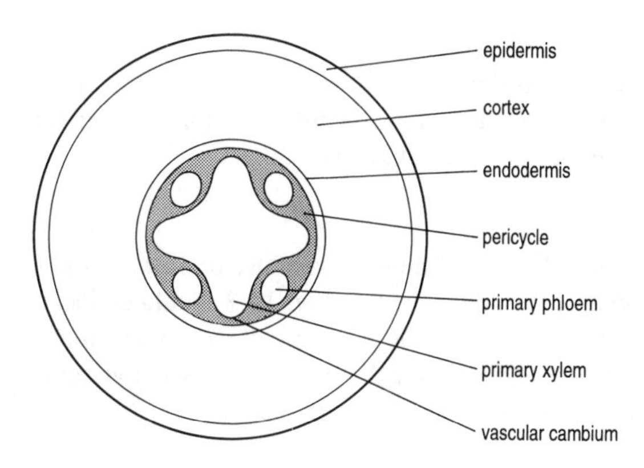

Fig. 15.5 Tissue plan diagram of a transverse section (TS) through a dicotyledon.

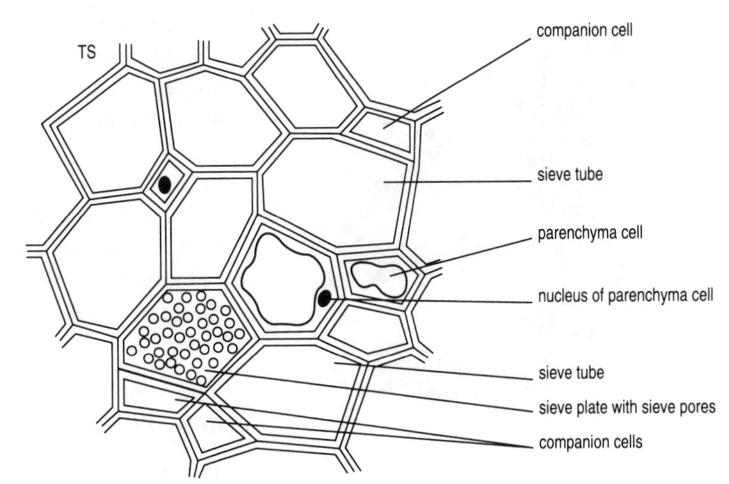

Fig. 15.6 Line diagram/drawing of TS of phloem of *Helianthus* stem.

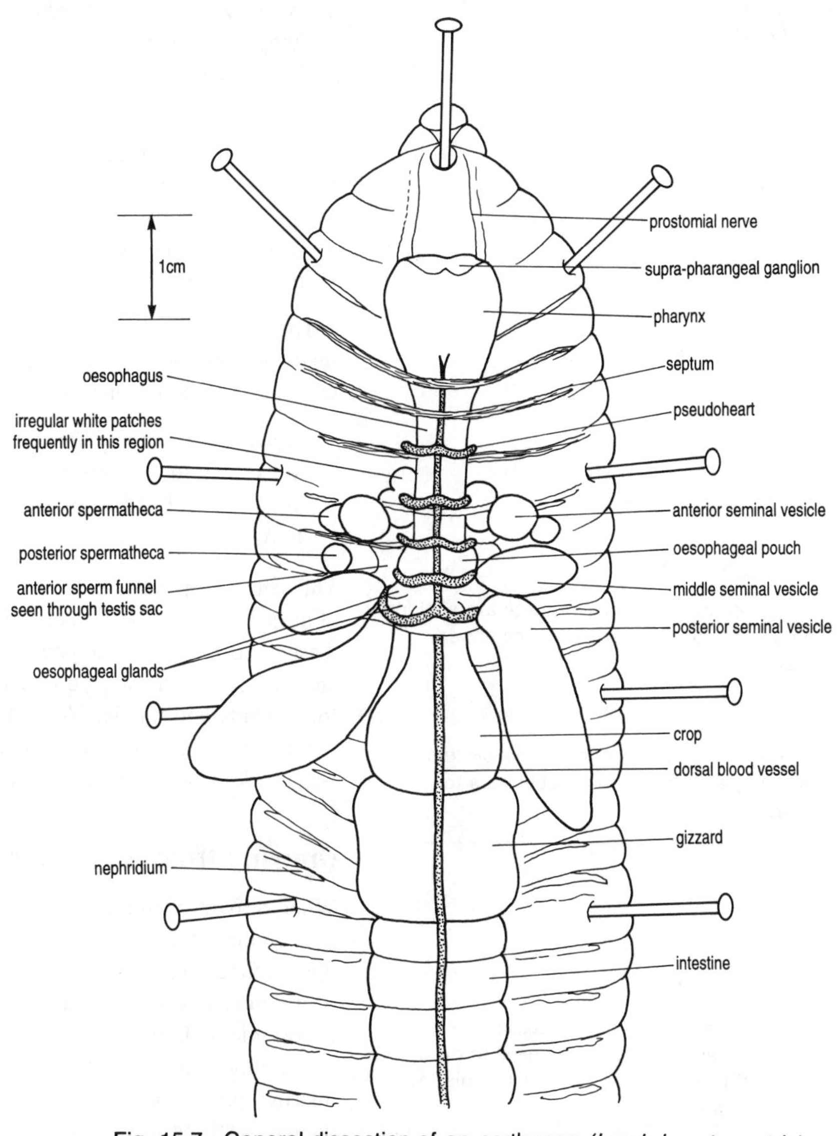

Fig. 15.7 General dissection of an earthworm (*Lumbricus terrestris*), dorsal view.

Fig. 15.8 Leafy shoot and fruits of *Ilex aquifolium* (holly).

Continuous tone drawings

These are usually done in pencil and fine detail and realism are possible using this technique (e.g. Fig. 15.8). These drawings are only used for presentation work and are not normally expected from students.

Charts

These may contain pictures, words, and/or numbers and use outlines or maps to show sequences and pathways. A good chart e.g. Fig. 15.3 should delineate and organise information: to be effective it must be clearly and simply organized. Make use of boxes or circles, simple drawings, arrows and varieties of size and style of labels to make it interesting and clear.

Improving your drawing and diagrams

The development of this skill, like any other, requires much practice and concentration: how many people do you know who have become good musicians without much practice? Remember that the requirements of biological drawing are not those of artistic representation and, therefore, the level of skill required is easier to achieve. You are required only to produce accurate and clear line drawings and diagrams. The main problems are:

- getting the correct size/scale of drawing;
- placing it correctly within the space available;
- getting the proportions of the drawing correct;
- achieving accuracy and consistency in the lines drawn;
- making labelling neat and accurate.

The first stage of any drawing is to decide exactly what to draw — this may seem obvious but until you have focused your thoughts, only a vague idea will be present in your mind. You must answer the following questions:

- Why you are doing the drawing?
- What should go into it?
- What magnification or reduction is required?
- How should it be laid out in the space available?

This often requires the use of biological knowledge such as the recognition of dorsal and ventral or anterior and posterior orientation (see Fig. 29.1). Once these decisions are made you are then in a position to determine the placement and size of your drawing on the paper; remember to include the space required for legends and labels. Your drawing should be as large as practicable.

Box 15.1 summarizes the steps to be followed once these decisions have been made, while Box 15.2 presents a general strategy for approaching drawing.

Using construction lines — these are vital for producing consistently well-proportioned drawings.

Labelling diagrams — do not mix systems of labelling and do not allow label lines to cross.

Drawing from a microscope

When drawing from a microscope, position the paper on the same side of the microscope as your drawing hand and use the 'opposite' eye for examining the specimen; thus for a right-handed person, the paper is placed on the right of the microscope and you use your left eye. If you keep both eyes open, it is possible with practice to learn to draw and see the page with one eye while continuing to observe the specimen with the other. When using a binocular microscope, use only one of the eyepieces. For more advanced work, projection devices such as the *camera lucida* may be available.

Drawing sections — if your section is symmetrical, you might only need to draw one half providing that you explain this in your legend.

(a)

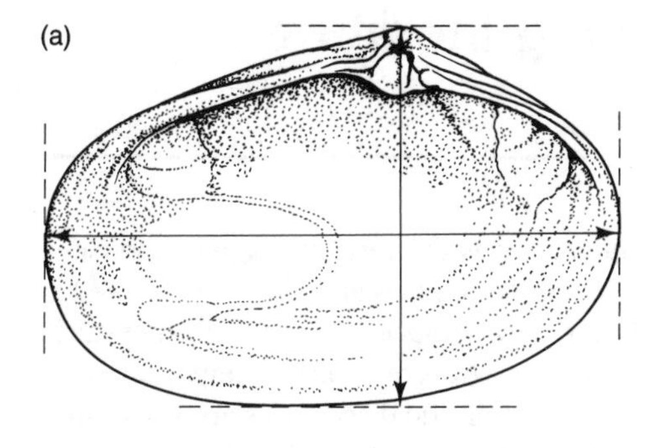

(b)

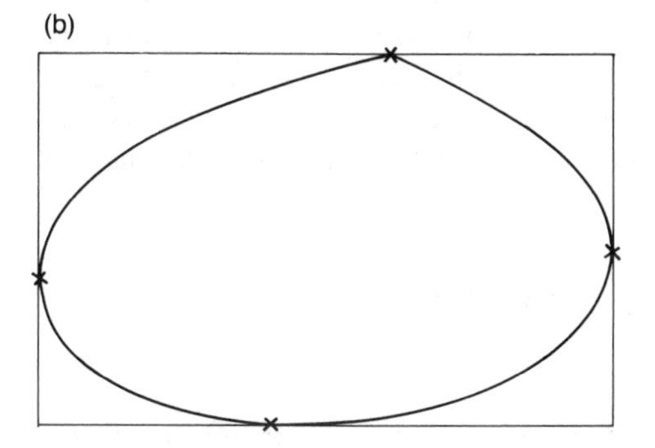

Fig. 15.9 Drawing an object. (a) determine linear dimensions, (b) construct a frame and outline.

Box 15.1 The steps required to produce a good biological drawing/diagram

1. **Draw a faint rectangle in pencil** to mark the boundaries.
2. **Draw very faint 'construction lines'** using a 2H pencil with a sharp point to allow you to get the basic proportions and outlines correct before progressing: they can be erased once the basic drawing is complete. To lay in construction lines, use a ruler or pair of dividers to determine the actual proportions of the object to be drawn (Fig. 15.9a) and then, using these dimensions, construct a scaled frame to allow further important reference points to be located (Fig. 15.9b).
3. **Draw the main outlines faintly** with your 2H pencil. When satisfied, go over the lines with an HB or 2B pencil with a sharp point. Draw firm, continuous lines, not hesitant, scratchy ones and make sure that junctions between lines are properly drawn. Figure 15.1 shows some of the more common errors found in student drawings. If you need to distinguish between different regions within your drawing, use hatching or stippling techniques (but not shading) and avoid trying to draw spurious detail such as large numbers of cells (see Fig. 15.1).
4. **Complete your drawing by adding labels**. This requires you to interpret your observations and helps you to remember what structures look like. Careful and accurate labelling is as important as the drawing itself and should be done clearly and neatly using radiating or horizontal lines ending in arrowheads or large dots to indicate precise label references. Labels should be written clearly and so as they can be read without needing to turn the paper. Annotations (short explanatory notes in brackets below the labels) are strongly recommended.
5. **Add a scale or magnification factor (see Chapter 29), a title and a legend.** The legend should include all relevant information including:
   - **(a)** the approximate magnification,
   - **(b)** the species,
   - **(c)** a full taxonomy if appropriate,
   - **(d)** details of the preparation of the subject, e.g. TS, whole mount, ventral dissection etc., and any stains used.

Figures 15.2 to 15.7 present examples of the way drawings should be presented.

Box 15.2 A checklist for making a good drawing

1. **Decide exactly what you are going to draw and why.**
2. **Decide how large the drawing should be.**
3. **Decide where you are going to place the drawing on the page.**
4. **Make sure that your equipment (pencils, ruler, dividers, etc.) is prepared and to hand.**
5. **Start drawing:**
   - **(a)** Draw what you see, *not* what you expect to see.
   - **(b)** Use carefully measured construction lines to provide the correct proportions.
   - **(c)** Avoid shading.
   - **(d)** Avoid excess detail, especially in tissue diagrams.
   - **(e)** Use conventions where appropriate, e.g. cut edges represented by a double line.
6. **Label the drawing/diagram carefully and comprehensively.**
7. **Give the drawing a full title, scale and legend:** include organism, classification, part drawn, orientation, stains, magnification etc.

16 Finding and using literature references

The basic unit of scientific communication is the research paper, published in a scientific journal (see p. 272). These periodicals are generally devoted to a narrow subject area and collectively they comprise the primary literature. Certain prestigious general journals (e.g. *Nature*) contain important new advances from a wide subject area. New scientific discoveries are sometimes reported in specialized books (known as monographs if they cover a single topic), but the research may not be subject to refereeing as with the primary literature. Reviews encapsulate research work in a defined field and are a good way of getting orientated with the literature in an unfamiliar field. Papers read at scientific meetings may be published as abstracts. These can lack detail but may later appear in the literature as full papers.

How to find relevant literature and information

For essays and revision

You are unlikely to delve into the primary literature for these purposes — books and reviews are likely to be much more readable! If a lecturer or tutor specifies a particular book, then it should not be difficult to find out where it is shelved in your library, as most now have a computerized index system and their staff will be happy to assist with any queries. If you want to find out which books your library holds on a specified topic, use the system's subject index. You will also be able to search by author or by key words. Browsing the shelves may turn up extremely interesting material but it will only let you see those books not currently out on loan, often the older or less useful ones.

There are two main systems used by libraries to classify books: the Dewey Decimal system and the Library of Congress system. Libraries differ in the precise way in which they use these systems, so enquire at your library for a full explanation. To show how these systems work, the book *The Selfish Gene* by Richard Dawkins (1976; Oxford University Press) is likely to be classified in the following ways:

Dewey decimal system: 591.51

where	591	refers to zoology
	591.5	refers to ecology of animals
	591.51	refers to habits and behaviour patterns

Library of Congress system: QL751

where	Q	refers to science
	QL	refers to zoology
	QL75	refers to animal behaviour
	QL751	refers to general works and treatises

Depending on the library, these marks may be followed by further letters and/or numbers designating shelving position and edition number.

For literature surveys and project work

You will probably need to consult the primary literature. If you are starting a new research project or writing a report from scratch, you can build up a core of relevant papers by using the following methods:

Especially useful to biologists is the Biosis Previews Database based on *Biological Abstracts* accessed via Dialog Information Services, Inc. (Palo Alto, California, USA). Accessing this database is a skilled job involving the use of a specific database language, so this is likely to be handled by a trained librarian.

Several databases are now produced on CD-ROM for student use (e.g. Medline, Applied Science and Technology Index, etc.).

- Asking around: supervisors or their postgraduate students will almost certainly be able to supply you with a reference or two that will start you off.
- Searching a computer database: a very good way to start a reference collection, although a charge is often made for access and sending out a listing of the papers selected (your library may or may not pass this on to you). Computer databases cover a wide area and can be particularly useful for finding articles published in foreign languages or in obscure journals.
- Consulting the bibliography of other papers in your collection — an important way of finding the key papers in your field. In effect, you are taking advantage of the fact that another researcher has already done all the hard work!
- Referring to 'current awareness' journals: these are useful for keeping you up to date with current research; they usually provide a listing of article details (title, authors, source, author address) arranged by subject and cross-referenced by subject and author. Current awareness journals save time searching through contents pages and cover a wider range of primary journals than could ever be available in any one library. They are usually updated on a monthly or bimonthly basis. Examples include:
  (a) The *Current Contents* series published by the Institute of Scientific Information, Philadelphia, USA, which reproduces the contents pages of journals of a particular subject area and presents an analysis by author and subject.
  (b) The *Current Advances* series published by Pergamon Press, Oxford, UK, which subdivides papers by subject within research areas and cross-references by subject and author.
  (c) Specialist abstracting journals, e.g. *Biological Abstracts*, in which each paper's abstract is also reproduced. Papers may be cross-referenced according to various taxa, which is useful in allowing you to find out what work has been done on a particular organism.
- Using the *Science Citation Index* (SCI): this is a very valuable source of new references, because it lets you see who has cited a given paper; in effect, SCI allows you to move *forward* through the literature from an existing reference. The Index is published regularly during the year and issues are collated annually. Some libraries have copies on CD-ROM: this allows rapid access and output of selected information.

For specialized information

You may need to consult reference works and bibliographies. For instance, ecological research invariably requires good maps. These may be found in your institute's library, but may also be kept in the geography department. Each library will have a section of reference books. Three worth noting are:

- The *Handbook of Chemistry and Physics* (Lide, 1990): the Chemical Rubber Company's publication (affectionately known as the 'Rubber Bible') giving all manner of physical constants, radioisotope half-lives, etc.
- *The Merck Index* (Budavari *et al.*, 1989), which gives useful information about organic chemicals, e.g. solubility, whether poisonous, etc.
- The *Geigy Scientific Tables* (Diem and Lentner, 1970), which gives a wide range of information centred on biochemistry, e.g. buffer formulae, properties of constituents of living matter.

White (1991) provides an excellent overview of all types of sources of scientific information, especially for media other than books.

Obtaining copies of research papers

Photocopying

Assuming your library takes the relevant journal, the simplest method of obtaining a copy of a paper is to photocopy it. In the UK, the Copyright, Designs and Patents Act (1988) allows the Copyright Licensing Agency to license institutions so that students and researchers may take photocopies for personal research purposes — no more than a single article per journal issue, one chapter of a book, or extracts to a total of 10% of a book. This permission may depend on the country of publication and the publisher.

Making an Inter-library Loan

If your library does not take the journal or book of interest, it may be possible for them to borrow it from a nearby institute or from the British Library Document Supply Centre (BLDSC). For papers, the BLDSC may supply a photocopy or the journal itself. You will have to fill in a form giving full bibliographic details of the paper and where it was cited and there may be a charge for the service.

Making a reprint request

Here, you send off a reprint request card to a designated author of the paper and ask for a copy to be sent to you. It is sensible to keep a note of which papers you have requested, perhaps by photocopying each batch of cards you send, because not all requests will receive a reply. The reprint request system is wasteful: almost certainly the total cost of sending a request card, printing a reprint and posting it on will greatly exceed your costs if you simply photocopy the article of interest. Think carefully before making a reprint request to a country where postage and other costs may eat into the author's research funds.

Filing and indexing reprints and photocopied articles

In the course of writing a literature survey or carrying out a research project, you are likely to accumulate a number of references to the literature on your topic. Regardless of whether you have a copy of the article of interest, there will come a time when your database needs to be organized so that you can (a) rapidly find details about any article or the hard copy itself and (b) collate information about reprints for the bibliography of a report, thesis or paper.

Physical storage of papers

Copies of papers can simply be filed in alphabetical order. The priority rule is: first author name, subsequent author names, date. Papers can either be stored in filing boxes (choose foolscap size to allow for large format reprints) or inserted into the leaves of a vertical (drawer) file. With this method, papers are easy to locate, but as the collection grows, the system becomes increasingly unwieldy: the papers will need to be reorganized and refiled as the collection becomes too big for the space originally allocated.

Alternatively, an alphabetical listing of papers can be kept on an index card system and the papers filed by an 'accession number' which you allocate when they are received. To find a paper, you look it up in the index card system, note its accession number, and go to the appropriate file box to find it. As the collection expands, new filing space is only required at one 'end' rather than 'in between'. A further advantage is that the accession numbers can be used to construct a simple cross-referencing system.

Key point
Make sure you write down *all* the information of potential relevance for bibliographies on your cards.

Card index systems

Index cards (Fig. 16.1) are a useful adjunct to any filing system. Firstly, you may not have a copy to file yet may still wish the reference information to be recorded somewhere for later use. Secondly, a selected pile of cards can be handed to a typist for typing out a reference list in a specified format. Thirdly, the cards can be used to help when organising a review (see p. 274). Fourthly, the back of each card can be used to record key points and comments on the paper. The priority rule for storage in card boxes is again first author name, subsequent author name(s), date.

Setting up a computerized index system — this may take a lot of effort: you should only consider using one if the time invested will prove worthwhile.

Computerized index systems

If facilities are available, the information written out on a card can be entered into a computerized database (e.g. 'Archivist' published by IRL Press, Oxford). These systems allow you to find a 'lost' paper from any word in the title, any author's name or keywords and abstract words if present. You can also cross-reference subject details so that you can be sure you have all relevant papers at your fingertips when writing and the cross-referencing categories don't need to be decided at the time of filing. You can also make the program print out details of the database. Thus, you can create a 'hard copy' card system if desired, or a word processor file of specified format ready to go into a project report or thesis.

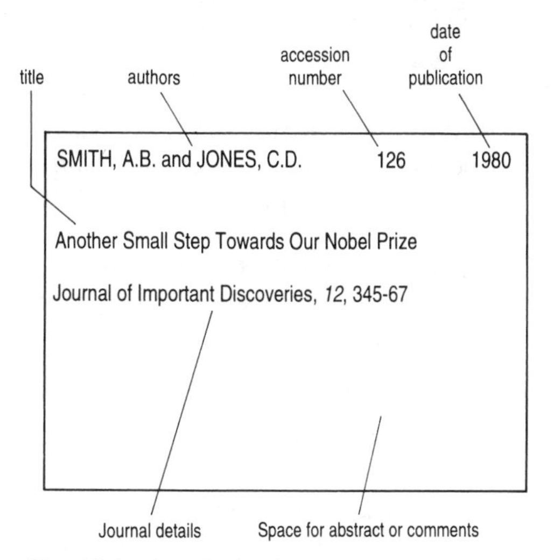

Fig. 16.1 A typical reference card.

Making citations in text

There are two main ways of citing articles and creating a Bibliography (also referred to as 'References' or 'Literature Cited').

Key point
The main advantages of the Harvard system are that the reader might recognize the paper being referred to and that it is easily expanded if extra references are added.

The Harvard system

For each citation, the author name(s) and the date of publication are given at the relevant point in the text. The Bibliography is organized alphabetically and by date of publication for papers with the same authors. Formats normally adopted are, for example, 'Smith and Jones (1983) stated that . . .' or 'it has been shown that . . . (Smith and Jones, 1983)'. Lists of references within parentheses are separated by semi-colons, e.g. (Smith and Jones, 1983; Jones and Smith, 1985), normally in order of date of publication. To avoid repetition within the same paragraph, a formula such as 'the investigations of Smith and Jones indicated that' could be used following an initial citation of the paper. Where there are more than two authors it is usual to write 'et al.' (or *et al.* if an italic font is available); this stands for the Latin *et alia* meaning 'and others'. If citing more than one paper with the same authors, put, for example, 'Smith and Jones (1987; 1990)' and if papers by a given set of authors appeared in the same year, letter them (e.g. Smith and Jones, 1989a; 1989b).

Choosing a citation system — you might adopt the Vancouver system when listing many authors and dates as the Harvard system would interrupt the flow of the text or make it lengthy; however, adding or taking away references from the list is tedious.

The numerical or Vancouver system

Papers are cited via a superscript or bracketed reference number inserted at the appropriate point. Normal format would be, for example: 'DNA sequences[4,5] have shown that . . .' or 'Jones [55,82] has claimed that . . .'. Repeated citations use the number from the first citation. In the true numerical method (e.g. as in *Nature*), numbers are allocated by order of citation in the text, but in the alpha-numerical method (e.g. the *Annual Review* series), the references are first ordered alphabetically in the Bibliography, then numbered, and it is this number which is used in the text.

How to list your citations in a bibliography

Whichever citation method is used in the text, comprehensive details are required for the Bibliography so that the reader has enough information to find the reference easily. Citations should be listed in alphabetical order with the priority: first author, subsequent author(s), date. Unfortunately, in terms of punctuation and layout, there are almost as many ways of citing papers as there are journals! Your department or institute may specify an exact format for project work, but if not, the following list gives typical arrangements of the most common sorts of reference.

Ensuring a consistent approach — decide on a style and stick to it. If you do not pay attention to the details of citation your marks may suffer.

Paper in journal: Smith, A.B., Jones, C.D. and Professor, A. (1990) The most important piece of research ever carried out. Journal of Negative Results, 333; 123–5.

Book: Smith, A.B. (1990) The Most Interesting Book Ever. Megadosh Publishing Corp., Big City. ISBN 0-123-45678-9.

Chapter in edited book: Jones, C.D. and Smith, A.B. (1990) Novel earthshattering work from our laboratory. In: Research Compendium 1980 (ed. A. Professor), pp. 123–456. Bigbucks Press, Booktown.

Thesis: Smith, A.B. (1980) Modesty is my middle name. PhD thesis, University of Life, Fulchester.

Citing journal and book titles — underlining used here specifies italics in print: use an italic font if available to you.

Care is required over the following details of citation:

- Authors and editors: give details of *all* authors and editors, even if given as *et al.* in the text.
- Abbreviations for journals: there are standard abbreviations for the titles of journals (consult library staff for lists), but it is probably a good idea to give the whole title if possible.
- Books: the edition should always be specified as contents may change between editions. Add, for example, '(5th edition)' after the title of the book. You may be asked to give the International Standard Book Number (ISBN), a unique reference number for each book published.
- Unsigned articles, e.g. newspaper articles without a by-line and quotes from unattributed instruction manuals — refer to the author(s) in text and bibliography as 'Anon.'.
- Unread articles: you may be forced to refer to a paper via another without having seen it. If possible, refer to another authority who has cited the paper, e.g. '... Jones (1980), cited in Smith (1990), claimed that ...'. Alternatively, you could denote such references in the Bibliography by an asterisk and add a short note to explain at the start of the reference list.
- Personal communications: information received in a letter, seminar or conversation can be referred to in the text as, for example, '... (Smith, pers. comm.)'. These citations are not generally listed in the bibliography of papers though in a thesis you could give a list of personal communicants and their addresses.

Obtaining specimens

17 Collecting animals and plants

If you are required to collect material you may need to use a formal sampling procedure (see Chapter 13) or simple qualitative collecting. Your choice of equipment will depend upon your objectives. Some of the main reasons for collecting include:

- Collecting for subsequent laboratory experimentation or observation: this requires collection of living material while causing minimum stress and damage to the specimens.
- Making estimations of population and community parameters: this can be destructive (requiring killing of specimens) or non-destructive, depending upon the objectives of the study and any requirements for subsequent laboratory work, e.g. sorting and identification.
- Collecting for museum-type collections of preserved specimens: here, the main objective is to obtain undamaged and representative specimens, usually in a preserved form (see Chapter 18).
- Collecting for subsequent chemical/biochemical analysis: this may require formal sampling procedures to be used. This objective requires care in both the method of collection and subsequent storage to avoid inducing chemical changes which are artefacts of the collection and storage processes. Deep freezing is usually the preferred storage method when subsequent chemical analysis is likely.

The main 'rules' for collecting are:

- Be aware of any legislation relating to the species or habitats you are intending to use. Obtain any permits required and obey any regulations strictly.
- Collect only enough for your purposes.
- Treat animals with respect at all times: do not cause unnecessary stress or suffering. There are formal rules for the handling of many vertebrate species but the same attitude should be taken towards all living organisms.
- Minimize damage and stress during collection and transport: stressed organisms are of little use for realistic experimentation.
- Be aware of the limitations and bias of your collecting equipment: this is particularly important for formal sampling procedures where collecting devices almost always have such problems. This may require specific testing for your particular usage.
- Keep good records of collection details.

Equipment for collecting

There is an immense variety of collecting/sampling devices available. In general, the collection of remote and/or animal samples presents the greatest problems. The more remote the operator is from the point of sampling, the more difficult is the evaluation of the quality of the sample in terms of its representativeness. Animal collection can be difficult because of factors such as mobility and complications introduced when allowing for e.g. avoidance behaviour.

Collecting formal samples

Here the objective is to obtain specimens that both qualitatively and quantitatively represent some well-defined habitat(s). Some of the more obvious practical considerations are:

Definition
Edge effect — in this context, any phenomenon associated with the sampling procedure at the edge of a sampling device.

- Are there 'edge effects' associated with the sampling method? Because of such effects you may need to adopt a special protocol, e.g. to determine whether or not a specimen falling on the boundary is included or excluded from the sample (p. 58).
- Will the sample be uniform? For example, remote sampling of marine sediments of different texture using grabs often results in 'bites' of different depth being taken.
- Does the method sample all components of the biota equally well? This may not be important as long as it adequately samples those components in which you are interested.
- How accurately can the location of the sample be defined, especially with regard to other samples intended as replicates? This is particularly important in remote sampling.
- Is the size of the sampling unit adequate for the size and distribution of the object, species or communities being sampled? (see Chapter 13). In remote sampling, this can often be a problem and frequently the method used is a compromise.
- Can inter-species interactions affect the integrity of the sample after collection but before processing? Factors such as predation can be prevented by using an appropriate chemical fixative immediately upon collection (see Chapter 18). Freezing is a poor option here since, upon thawing, most animal and plant tissues tend to disintegrate and the specimens are in poor condition for subsequent identification, etc.

General collecting strategies

For qualitative collecting, where a representative sample of the population or habitat is not the objective, the effectiveness of equipment may be of less significance. Here, suitability for capturing living and undamaged specimens is the principal criterion in choosing your equipment. For mobile animals this often involves nets and traps of various kinds, combined with narcotizing agents such as smoke for insects. If specimens can be killed before capture, then spray insecticides can be a useful aid. Some useful general points about all collections are:

Recording information — the following details should be recorded upon collection: date, time, location (Grid Ref.), habitat details, collecting technique, preservation technique.

- Think ahead and be prepared. Your collecting equipment must be appropriate for the task!
- Keep good records of collection details: this is particularly important if the collection is for museum or herbarium purposes.
- Keep collected plants in humid atmospheres to ensure good condition. The vasculum originally designed for this purpose has largely been replaced by the polythene bag in most circumstances except where mechanical damage is likely.
- Keep all living animals in conditions as similar as possible to the environment from which they were collected. Aquatic specimens are usually particularly temperature-sensitive and should be kept in a Thermos flask to prevent rapid temperature changes.
- Rigid containers are better than plastic bags for most purposes since they help to prevent damage to the specimens. Plastic containers should be chosen in preference to glass for most purposes. Make sure that the container seals properly to avoid the loss of specimens and to prevent loss of water.

18 Fixing and preserving animals and plants

Take care when fixing specimens — remember anything that will fix your specimen is also capable of fixing you! Be very careful in your handling of these materials and obey all safety precautions.

Fixation is a chemical process which stops autolysis and stabilizes the protein components of tissues so that during subsequent processing, the tissues retain as fully as possible the form they had in life. Preservation allows material to be stored indefinitely by destroying any bacteria, fungi, etc. which could degrade the specimen. Preservation and fixation used to be synonymous in that most of the commonly used preservatives also had a fixative action. This has changed with the introduction of 'phenoxetols' as good preservatives but since they do not arrest autolysis, their use must be preceded by treatment with a true fixative: for that reason, they are called post-fixation preservatives.

Your choice of fixative will depend upon the material to be fixed and the purpose for which it is being fixed. Some of the most common fixatives (Table 18.1) may be used alone although more often they are used in mixtures, the object being to combine the virtues of the various ingredients. Fixation is a critical process when used for histological or cytological work and you should consult specialist texts such as Kiernan (1990).

Table 18.1 Some of the most widely used fixatives/preservatives and their properties

Substance	Fixation	Usage	Notes
Formaldehyde	+	4% v/v	Comes as 40% v/v solution (= formalin): normal dilution 1 + 9. Make up with sea water for marine specimens; buffer for calcareous specimens. Use in a fume cupboard — health hazard.
Ethanol	+	70% v/v aqueous	Highly volatile; inflammable; containers must be well sealed to prevent evaporation; causes shrinkage and decolorization as well as loss of lipids
Acetic acid	+	In mixtures	Pungent vapour
Picric acid	+	In mixtures	Risk of explosion; detonates readily on contact with some metals. Not recommended for routine student use (COSHH Regulations)
Mercuric chloride	+	In mixtures	Extremely poisonous; corrosive to metal implements; tissue will contain mercuric salt deposits
Osmium tetroxide	+	1% v/v in buffer or vapour	Both fluid and vapour highly toxic; use in fume cupboard only. Excellent for cytological detail but is expensive and can only be used for very small specimens due to poor penetration speed. Vapour good for protozoa
Propylene phenoxetol	–	1–2% v/v aqueous	Relatively expensive but innocuous and effective preservative: needs pre-fixation stage
Glutaraldehyde	+	2–4% v/v in buffer	Must be used cold

Narcotization

Narcotization (relaxation) is usually advisable for animals since many are highly contractile and assume grossly distorted postures if placed straight into fixative; it is also more humane to narcotize before killing. Failure to narcotize may result in contortion, rupture of the body wall or evisceration and a reduction in the scientific value of the specimen. The need for narcotization varies with the type of organism and the objectives of the study: some of the most widely used narcotizing agents are given in Table 18.2.

Table 18.2 Narcotizing agents and their characteristics

Agent	Usage	Notes
Cold (chilling)	Cold-blooded animals	Effective form of relaxing many animals such as tropical and sub-tropical invertebrates
Heat (warming)	Slow heat	Works for some animals. Start from ambient but keep time period as short as possible
Magnesium sulphate or Magnesium chloride	7% w/v in water 20% w/v in water	Quite effective for many invertebrates but beware of osmotic problems if made up in sea water: keep exposure times fairly short (1–2 h)
Menthol crystals	Float on water	Slow but effective for many aquatic animals
Chloral hydrate	1% w/v in sea water	General narcotizing agent
Ethanol	10% v/v dropwise	Slow and rather tedious process for all but very small specimens
MS-222 (Tricaine)	Use as 0.05% w/v aqueous solution	Good for marine and freshwater fish: very rapid effect (15 s–1 min)
Chloretone	0.1–0.5% w/v in water	General narcotizing agent
Ethyl acetate	Vapour	Effective for most insects (kills as well); highly volatile
Ether	Vapour	Effective for vertebrates
Chloroform	Vapour	Effective for vertebrates

Selecting your fixative/preservative

The main factors you must consider in selecting the preservative/fixative suitable for your materials are:

- Speed of penetration, which determines the size of object that can be fixed. Some fixatives penetrate very slowly (e.g. osmium salts) and only very small pieces of tissue can be fixed. Others such as formaldehyde penetrate very rapidly and allow fixation of relatively large, whole animal specimens. It may be necessary to inject fixatives into the body cavities of large specimens to ensure adequate fixation.
- Shrinkage: fixatives such as ethanol and mercuric chloride cause tissues to shrink, thus distorting them significantly. The addition of glacial acetic acid is frequently used to reduce this problem.
- Hardening: ethanol is particularly liable to cause tissues to harden on prolonged exposure to concentrations above 70% v/v. For whole specimens, the addition of 3% v/v glycerol to 70% v/v alcohol reduces this, making

it a useful medium for specimen storage (not to be used for histological studies). The time in hardening solutions must be carefully controlled for histological material.

- Decolorization: for whole specimens, the loss of colour may be detrimental and solvents such as ethanol must be used with care since they readily remove many pigments.
- Decalcification: acidic fixatives such as unbuffered formalin and those containing picric acid and acetic acid readily dissolve calcareous structures. For histological preparations this may be desirable, but for whole specimens it is highly undesirable. The acidic properties of formaldehyde can be overcome by neutralizing with calcium salts.
- Osmotic problems: distortion of tissues by osmotic movements of water can be rapid and serious. Some fixatives are best made up in sea water if they are to be used for marine specimens to avoid osmotic swelling of tissues. Others such as osmium salts and glutaraldehyde usually need to be made up in a buffer solution isotonic with the tissues.
- Other chemical reactions: fixatives containing mercuric chloride usually result in the deposition of mercuric precipitates in the tissues. For histological preparations, these must be removed by thorough washing and if necessary by post-treating with iodized alcohol. There is no good reason for using mercuric chloride fixatives other than for histological work and the benefits must be weighed against the dangers and difficulties. Note also that you must not use metal implements with such fixatives as they are extremely corrosive — use only wooden or glass implements.

Preservation

Wet preservation

This is used mainly for animals and is usually preceded by narcotization. The process of fixation is then comparatively straightforward: sometimes it is necessary to arrange the body and appendages of the animal using tapes, elastic bands, etc. before fixation begins. The solutions most commonly used for wet preservation have been either:

- a 5–10% v/v aqueous solution of formalin (= 40% v/v aqueous formaldehyde), neutralized with calcium carbonate or some other agent to prevent any acidity in the solution resulting in the slow dissolution of calcareous structures; or
- 50–70% v/v ethanol.

Formaldehyde is cheap and non-inflammable but tends to stiffen and harden tissues on prolonged exposure: safety regulations have made it particularly problematical to use because of its noxious vapour. Ethanol is inflammable, highly volatile and tends to cause shrinkage and decolorization in soft-bodied animals: pass them slowly through a graded series of concentrations to minimize this problem. Industrial methylated spirit is quite suitable for preservation if ethanol is unavailable.

Working with formaldehyde — always use formaldehyde in a fume cupboard as it is not only an excellent fixative but a significant health hazard.

A 1–2% v/v aqueous solution of propylene phenoxetol has been widely used for vertebrate and invertebrate preservation provided it is preceded by the use of a fixative: it preserves the natural colour well and leaves the material pliable. It is comparatively expensive but is non-inflammable and non-volatile.

Wet preservation of plant material is relatively straightforward and has the advantage of preserving the form of the specimen providing it is done carefully. Herbaceous plants should be preserved in a solution with an alcohol base, while for succulents a formaldehyde base is better. Algae, fungi, ferns, lichens and seed plants can be stored wet in formalin acetic acid (FAA) solution: this is made from formalin, acetic acid and ethanol in the ratio 85:10:5. To retain the green coloration of ferns and seed plants, add a crystal of copper sulphate to the solution and incubate for 3–10 days, then transfer to ordinary FAA solution. Colours of flowers cannot easily be preserved using wet preservatives.

For marine specimens, make up a stock of 1 part propylene glycol with 1 part of formalin. Use this in the ratio of 1 part stock to 8 parts sea water for fixation and 1 part stock to 9 parts sea water for subsequent preservation. This markedly reduces the formaldehyde concentration without loss of the fixative/preservative effect and has been shown to be satisfactory even for specimens with calcareous components.

Dry preservation

This is used mainly for vertebrate taxidermy, and for arthropod and plant material, although the development of freeze-drying has made this procedure applicable to almost any type of small or medium sized organism. Arthropods and vascular plants, with their hard exoskeletons and vascular tissue respectively, are most easily dealt with since the skeletons prevent collapse and loss of form upon drying. Similarly, some types of fungal fruiting bodies such as 'bracket fungi' can be preserved in a dry state.

Using a plant press — be sure to label each specimen carefully in a way that ensures that the label cannot be lost during paper changes.

Flowering plants, or parts of a plant, mosses and liverworts, should be dried in a plant press between sheets of absorbent paper. Interpose a sheet of muslin or greaseproof paper between the paper and the specimen to prevent the specimen adhering to the paper. Great care must be exercised in the cleaning and arrangement of the plant so that essential features are not obscured. Change the drying sheets daily until dry. Drying by heating the press is particularly good for preserving colour: use a low temperature oven or place on a radiator.

Algae are usually dried in a press after floating them onto the sheet of paper which is immersed in sea water: this allows careful arrangement of the often delicate algal fronds. Once suitably arranged, the process is essentially the same as for flowering plants although regular changes of the absorbent paper layers are necessary and the drying procedure usually takes several days. Fungi are best preserved by drying without any attempt at pressing.

Storage

Careful maintenance of specimens after the fixation/preservation processes is very important. Specimens may be stored for long periods provided they are protected from pests such as mites: this may require the use of chemicals such as naphthalene in air-tight containers. Seeds are frequently stored after air-drying only but they have a finite period of viability.

The storage conditions for all dried materials are critical; these include moderate to low temperatures and very low humidity. Storage of wet material should be in appropriate, vapour-tight vessels: even then the containers will need checking and topping up over time, particularly when volatile preservatives are used. Labelling must be comprehensive and contain information on the fixation and preservation processes as well as ecological and taxonomic details.

19 Collecting and isolating microbes

Microbes can be studied by taking samples for analysis, usually in one of the following ways:

- by direct examination of individual cells of a particular microbe, e.g. using fluorescence microscopy (p. 112);
- by isolating/purifying a particular species or related individuals of a taxonomic group, e.g. the faecal indicator bacterium *Escherichia coli* in sea water;
- by studying microbial processes, rather than individual microbes, either *in situ*, or in the laboratory.

The sampling process

The main factors to be considered when deciding on a suitable sampling procedure are discussed in Chapter 13.

Sampling techniques include the use of swabs, sellotape strips and agar contact methods for sampling surfaces, bottles for aquatic habitats, plastic bags and corers for soils and sediments. A wide range of complex apparatus is available for accurately sampling water or soil at particular depths. An important feature of all techniques is that the sampling apparatus must be sterile.

The sampling method must minimize the chance of contamination with microbes from other sources, especially the exterior of the sampling apparatus and the operator. For example, if you are sampling an aquatic habitat, stand downstream of the sampling site. Strict aseptic technique must be used throughout the sampling process (p. 120), e.g. using a portable Bunsen burner.

Process the sample as quickly as possible, to minimize any changes in microbiological status. As a general guideline, many procedures require samples to be analysed within 6 h of collection. Changes in aeration, pH and water content may occur after collection. Some microbes are more susceptible to such effects, e.g. anaerobic bacteria may not survive if the sample is exposed to air.

Sub-sampling — to minimize the effects of changes in temperature, aeration and water status during transportation, a primary sample may be returned to the laboratory, where the working sample (sub-sample) is then taken (e.g. from the centre of a large block of soil).

Soil and water samples are often kept cool (at 0–5°C) during transport to the laboratory. In contrast, microbes adapted to grow in association with warm-blooded animals may be damaged by low temperatures. An alternative approach is to keep the sample near the ambient sampling temperature using a thermostatic vessel (e.g. Thermos flask).

Isolation techniques

Several different approaches may be used to obtain microbes in pure culture. The choice of method will depend upon the microbe to be isolated: some organisms are relatively easy to isolate, while others require more involved procedures.

Separation methods

Most isolation procedures involve some form of separation to obtain individual microbial cells. The most common approach is to use an agar medium for primary isolation, with streak dilution, spread plating or pour plating to produce single colonies, each derived from an individual microorganism (p. 123). It is often necessary to dilute samples before isolation, so that a small number

Obtaining a pure culture — if a single colony from a primary isolation medium is used to prepare a streak dilution plate and all the colonies on the second plate appear identical, then a pure culture has been established. Otherwise, you cannot assume that your culture is pure.

of individual microbial cells can be transferred to the growth medium. Strict decimal dilutions or doubling dilutions (p. 19) of a known amount of sample are required for quantitative work.

If your aim is to isolate a particular microbe, perhaps for further investigation, you will need to sub-culture individual colonies from the primary isolation plate to establish a pure culture, also known as an axenic culture. Pure cultures of most microbes can be maintained indefinitely, using sterile technique and microbial culture methods (Chapter 27).

Other separation techniques include:

- Dilution to extinction. This involves diluting the sample to such an extent that only one or two microbes are present per millilitre: small volumes of this dilution are then transferred to a liquid growth medium. After incubation, most of the tubes will show no growth, but some tubes may show growth, having been inoculated with a single microbial cell at the outset. This should give a pure culture. However, this method is wasteful in terms of time and resources.
- Micromanipulation. It may be possible to separate the microbe from contaminants using a micropipette and dissecting microscope (p. 136). The microbe can then be transferred to an appropriate growth medium, to give a pure culture. However, this is rarely an easy task for the novice.
- Sonication/homogenization. Useful for separating individual microbial cells from each other and from inert particles, prior to isolation. However, some decrease in viability is likely. Sonicators must be used thoughtfully — minimize damage by using short treatment times, cooling the sample after each treatment.
- Filtration. This can be useful where the number of microbes is low. Samples can be passed through a sterile cellulose ester filter (pore size 0.2 μm), which is then incubated on the surface of an appropriate solidified medium. Sieving (p. 22) and filtration techniques are often used in soil microbiology to subdivide a sample on the basis of particle size.
- Motility. Phototactic microbes (including photosynthetic flagellates and motile cyanobacteria) will move towards a light source, while heterotrophic flagellate bacteria will move through a filter of appropriate pore size into a nutrient solution, or away from unfavourable conditions.

Selective and enrichment methods

Selective methods are based on the use of conditions that will permit the growth of a particular group of microbes while inhibiting others. Laboratory incubation under selective conditions will allow the particular microbes to be isolated in pure culture. Enrichment techniques encourage the growth of certain bacteria, usually by providing additional nutrients in the growth medium. The difference between selective and enrichment techniques is that the former use growth conditions unfavourable for competitors while the latter provide improved growth conditions for the chosen microbes.

Selective and enrichment techniques can be considered together, since they both enhance the growth of a particular microbe when compared with its competitors. Methods based on specific physical conditions include:

Definitions
Psychrophile — a microbe with an optimum temperature for growth of <20°C (Lit. 'cold-loving').
Psychrotroph — a microbe with an optimum temperature for growth of ≥20°C, but capable of growing at lower temperature, typically 0–5°C (Lit. 'cold-feeding').

- Temperature. Psychrophilic and psychrotrophic bacteria can be isolated by incubating the growth medium at 4°C. Short-term heat treatment of samples can be used to select for endospore-forming bacteria, e.g. 70–80°C for 5–10 min, prior to isolation.
- Atmosphere. Most eukaryotic microbes are obligate aerobes, requiring an

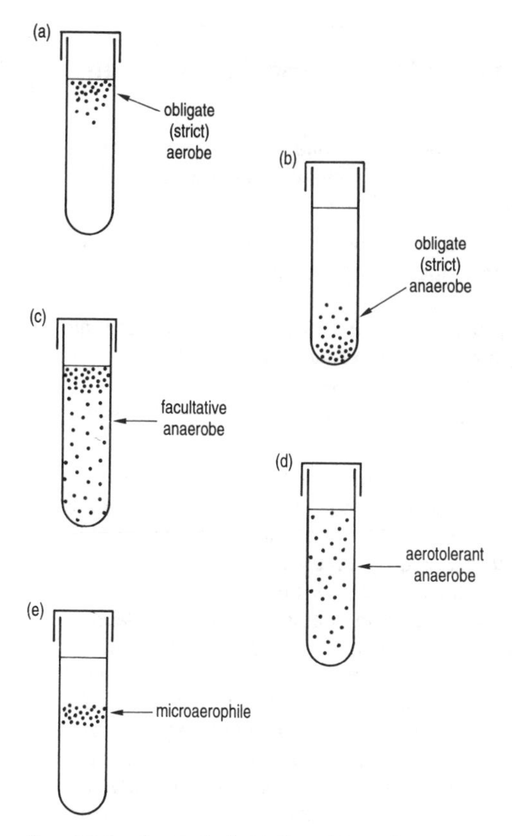

Fig. 19.1 Agar shake tubes. Bacteria are suspended in molten agar at 45–50°C and allowed to cool. The growth pattern after incubation reflects the atmospheric (oxygen) requirements of the bacterium.

Table 19.1 Selective agents in bacteriological media

Substance	Selective for
Azide salts	*Enterococcus* spp.
Bile salts	Intestinal bacteria
Brilliant green	Gram-negative bacteria
Crystal violet	*Streptococcus* spp.
Gentian violet	Gram-negative bacteria
Lauryl sulphate	Gram-negative bacteria
Methyl violet	*Vibrio* spp.
Malachite green	*Mycobacterium*
Polymyxin	*Bacillus cereus*
Sodium selenite	*Salmonella* spp.
Sodium chloride	Halotolerant bacteria *Staphylococcus aureus*
Sodium tetrathionate	*Salmonella* spp.
Trypan blue	*Streptococcus* spp.

adequate supply of oxygen to grow. Bacteria vary in their responses to oxygen: obligate anaerobes are the most demanding, growing only under anaerobic conditions (e.g. in an anaerobic cabinet). The oxygen requirements of a bacterium can be determined using the agar shake tube method as part of the isolation procedure (Fig. 19.1). Some pathogenic bacteria grow best in an atmosphere with a reduced oxygen status and increased CO_2 concentration: such carboxyphilic bacteria (capnophiles) are isolated and grown in a special incubator where the gas composition can be adjusted.

- Centrifugation. This can be used to separate buoyant microbes from their non-buoyant counterparts — on centrifugation, such organisms will collect at the surface while the remaining microbes will sediment. Alternatively, density gradient methods may be used (p. 190). Centrifugation can be combined with repeated washing, to separate microbes from their contaminants.
- Ultraviolet irradiation. Some microbes are tolerant of UV treatment and can be selected by exposing samples to UV light. However, the survivors may show a greater rate of mutation, due to the mutagenic effects of UV radiation.
- Illumination. Samples may be enriched for cyanobacteria and microalgae by incubation under a suitable light regime. For dilute samples, where the number of photosynthetic microbes is too low to give the sample any visible green coloration, there is a risk of photoinhibition and loss of viability if the irradiance is too high. Such samples should be shaded during the initial stages of growth.

Chemical methods form the mainstay of bacteriological isolation techniques and various media have been developed for the isolation of specific groups of bacteria (Table 19.1). The chemicals involved can be subdivided into the following groups:

- Selectively toxic substances: for example, salt-tolerant, Gram-positive cocci can be grown in a medium containing 7.5% w/v NaCl, which prevents the growth of most common heterotrophic bacteria. Several media include dyes as selective agents, particularly against Gram-positive bacteria.
- Antibiotics: for example, the use of antibacterial antibiotics (penicillin, streptomycin, chloramphenicol) in media designed to isolate fungi, or the use of antifungal agents (cycloheximide, nystatin) in bacterial media. Some antibacterial antibiotics show a narrow spectrum of toxicity and these can be incorporated into selective isolation media for bacteria tolerant to these agents.
- Nutrients which encourage the growth of certain microbes: including the addition of a particular carbon source, or specific inorganic nutrients.
- Substances that affect the pH of the medium: for example, the use of alkaline peptone water at pH 8.6 for the isolation of *Vibrio* spp.

Note that sub-cultures from primary isolation media should be grown in a non-selective medium, to confirm the purity of the isolate.

Many of the selective and enrichment media used in bacteriology are able to distinguish between different types of bacteria: such media are termed differential media or diagnostic media and they are often used in the preliminary stages of an identification procedure. Box 19.1 gives details of the constituents of MacConkey medium, a selective, differential medium used in clinical microbiology (e.g. for isolation of certain pathogenic faecal bacteria).

Box 19.1 Differential media for bacterial isolation: an example

MacConkey agar is both a selective and a differential medium, useful for the isolation and identification of intestinal Gram-negative bacteria. Each component in the medium has a particular role:

- Crystal violet: selectively inhibits the growth of Gram-positive bacteria.
- Bile salts: toxic to most microbes apart from those growing in the intestinal tract.
- Peptone: (a meat digest) provides a rich source of complex organic nutrients, to support the growth of non-exacting bacteria.
- Lactose: present as an additional specific carbon source.
- Neutral red: a pH indicator dye, to show the decrease in pH which accompanies the breakdown of lactose.

Any intestinal Gram-negative bacterium capable of fermenting lactose will grow on MacConkey agar to produce large purple–red colonies, the red coloration being due to the neutral red indicator under acidic conditions while the purple coloration, often accompanied by a metallic sheen, is due to the precipitation of bile salts and crystal violet at low pH.

In contrast, enteric Gram-negative bacteria unable to metabolize lactose will give colonies with no obvious pigmentation. This differential medium has been particularly useful in medical microbiology, since many enteric bacteria are unable to ferment lactose (e.g. *Salmonella, Shigella*) while others metabolize this carbohydrate (e.g. *Escherichia coli, Klebsiella* spp.). Colonial morphology on such a medium can give an experienced bacteriologist important clues to the identity of an organism, e.g. capsulate *Klebsiella* spp. characteristically produce large, mucoid colonies with a weak pink coloration, due to the fermentation of lactose, while *E. coli* produces smaller, flattened pink colonies with a metallic sheen.

Further details on methods can be found in Gerhardt (1981). Note that isolation procedures often combine several of the techniques described above. For instance, a protocol for isolating food-poisoning bacteria from a foodstuff might involve:

1. homogenization of a known amount of sample in a suitable diluent;
2. serial decimal dilution;
3. separation procedures using spread or pour plates to quantify the number of bacteria of a particular type in the foodstuff and provide a viable count (p. 130);
4. selective/enrichment procedures, e.g. specific media/temperatures/atmospheric conditions, depending on the bacteria to be isolated;
5. confirmation of identity: any organism growing on a primary isolation medium would require subculture and further tests, to confirm the preliminary identification (Chapter 22).

Identifying organisms

20 Naming and classifying organisms

Definitions
Systematics — the study of the diversity of living organisms and of the evolutionary relationships between them.
Classification (*taxonomy*) — the study of the theory and methods of organization of taxa and therefore, a part of systematics.
Taxon (plural *taxa*) — an assemblage of organisms sharing some basic features.
Nomenclature — the allocation of names to taxa.
Identification — the placing of organisms into taxa (see Chapters 21 and 22).

The use of scientific names is fundamental to all aspects of biological science since it aims to provide a system of identification which is precise, fixed and of universal application. Without such a system, comparative studies would be impossible.

There are two bases for such classification:

- Phenetic taxonomy, which involves grouping on the basis of phenotypic similarity, frequently using complex statistical techniques to obtain objective measures of similarity. The characters used have been largely morphological and anatomical, but biochemical, cytological and other characters are increasingly used.
- Phyletic taxonomy, which involves grouping on the basis of presumed evolutionary, and therefore genetic, relationships.

These two systems usually are fairly similar in outcome, since closely related organisms are usually fairly similar to each other and because judgements of evolutionary relationships are usually themselves based upon similarities. The situation is made more complex by phenomena such as convergent and divergent evolution; phyletic classifications are also liable to subjective bias.

The hierarchical system

The classification of organisms into kingdoms is somewhat arbitrary and no scheme has yet been universally adopted; however, it is usual today to recognize six kingdoms if the viruses are included:

- Virales. The viruses: very simple in structure, obligate parasites of prokaryotic or eukaryotic cells.
- Monera. The bacteria and cyanobacteria, prokaryotic: non-nucleate and lacking mitochondria, plastids, endoplasmic reticulum, Golgi apparatus.
- Protista. Includes algae, protozoans and slime moulds: eukaryotic, mainly unicellular and often autotrophs. A heterogenous group.
- Plantae. The plants: eukaryotic, walled, mainly multicellular and usually photoautotrophs.
- Fungi. Eukaryotic, usually syncytial, walled and saprotrophic.
- Animalia. The animals: eukaryotic, multicellular and heterotrophs.

Six other levels of taxa are generally accepted: phylum, class, order, family, genus and species, although in botany and microbiology division is used instead of phylum. The levels of taxa above genus are rather subjective and may vary among authorities, but the use of genus and species are governed by strict, internationally agreed conventions called Codes of Nomenclature. There are three such Codes, the Botanical, Zoological and Bacteriological Codes, which operate on similar but not identical principles.

The basis of classification

The basic unit of classification is the species, which represents a group of recognizably similar individuals, clearly distinct from other such groups. No

Table 20.1 Example of taxonomies for a plant, an animal and a bacterium

Common name	English oak	Honey-bee	Pseudomonas
Kingdom	Plantae	Animalia	Monera
Phylum/Division	Anthophyta	Arthropoda	Gracilicutes
Class	Dicotyledonae	Insecta	Scotobacteria
Order	Fagales	Hymenoptera	Pseudomonadales
Family	Fagaceae	Apidae	Pseudomonadaceae
Genus	*Quercus*	*Apis*	*Pseudomonas*
Species	*Q. robur*	*A. mellifera*	*P. aeruginosa*

simple definition of a species is possible, but there are two generally used definitions:

- A group of organisms capable of interbreeding and producing fertile offspring — this, however, excludes all asexual, parthenogenetic and apomictic forms!
- A group of organisms showing a close similarity in phenotypic characteristics — this would include morphological, anatomical, biochemical, ecological and life history characters.

When species are compared, groups of species may show a number of features in common; they are then arranged into larger groupings known as genera (singular genus). This process can be repeated at each taxonomic stage to form a hierarchical system of classification whose different levels are known as taxonomic ranks. The number of levels in this system is arbitrary and based upon practical experience — the seven levels normally used have been found sufficient to accommodate the majority of the variation observed in nature.

When a generic or specific name is changed as a result of further study, the former name becomes a synonym; you should always try to use the latest name. Where a generic name has been changed recently, the old name is occasionally given in parentheses to allow easy reference to the extensive use of the old name. Thus when the cockle *Cardium edule* was renamed *Cerastoderma edule*, it was commonly referred to in textbooks as *Cerastoderma (Cardium) edule*: this is strictly not correct practice but can be helpful for non-specialists. There are taxonomic reference works available for each discipline or sub-discipline, such as the *Flora Europea* (for plants) and the *Plymouth Marine Fauna* (for British marine animals) which can provide the current versions of a name and often its synonyms.

Nomenclature in practice

The scientific name of an organism is effectively a symbol or cipher which removes the need for repeated use of descriptions. It normally comprises two words and is, therefore, called a binomial term. The name of the genus is followed by a second term which identifies the species, e.g. *Quercus robur, Apis mellifera*, or *Pseudomonas aeruginosa* (Table 20.1). Common names are often interesting, but totally unsatisfactory for use in biological nomenclature because of the lack of consistency in their use: the Codes of Nomenclature were established to prevent any ambiguities.

The Codes require that all scientific names are either Latin or treated as Latin, written in the Latin alphabet and subject to the rules of Latin grammar. Consequently, you must be very precise in your use of such names. In some cases, the Codes stipulate a standardized ending for the names of all taxa of a given taxonomic rank, e.g. names of all animal families must end in -idae while plant, fungal and bacterial families end in -aceae. When used in a formal scientific context, you should follow the specific name by the authority on which that name is based, i.e. the name of the person describing that species and the date of the description.

Some basic rules for the writing of taxonomic names are:

- Names of the seven levels of taxa should take lower case initial letters, e.g. class Mollusca or kingdom Fungi.
- The Latin forms of all taxon names except the specific name take initial capital letters, e.g. 'the Arthropoda . . .' but anglicized versions do not, e.g. 'the arthropods . . .'.
- The names of the higher taxa are all plural, hence 'the Mollusca are . . .' while the singular of the anglicized version is used for a single member of that taxon, hence 'a mollusc is . . .'.
- The binomial system gives each species two terms, the first being the generic name and the second the specific name, which must never be used by itself. The generic and species names are identified from the rest of the text either:
  (a) by being underlined (when handwritten or typed), e.g. Patella vulgata; or
  (b) by being set in italics (in print or on an appropriate word processor), e.g. *Patella vulgata*.
- The generic name is singular and always takes an initial capital letter. If you use the generic name this implies that the point being made is a generic characteristic unless the specific name is present. Write the generic name in full when first used in a text, e.g. *Patella vulgata*, but subsequent references can be abbreviated to its initial letter, e.g. *P. vulgata*, unless this will result in confusion with another genus also being considered.
- The abbreviation 'sp.' should be used in place of the specific name if a single unspecified species of a genus is being referred to, e.g. *Patella* sp.; it is not underlined or italicized. If more than one unspecified species is meant, then the correct form is 'spp.', e.g. *Patella* spp.
- Common names should not normally be written with a capital letter, e.g. limpet.
- Each species name should be followed by the authority: on first usage in formal reports and in titles, the name or names of the person(s) to whom that name is attributed and the date of that description should be quoted. These names may sometimes be abbreviated, e.g. L. for Linnaeus and the standard abbreviations must be used. If the species was first described under its current generic name, the authority's name, often in abbreviated form is added, e.g. *Quercus robur* L. If, however, the species were first described under a different genus, the name of the author of the original description is presented in parentheses, e.g. *Quercus petraea* (Mattuschkal) Liebl. Note that in zoology, the date the description was published is also included, e.g. *Ischnochiton kermadecensis* Iredale, 1914. The use of authorities should be confined to formal papers, final year project reports, etc.; they would not normally be used in practical reports, short assignments or examinations.

Writing taxonomic names — always underline or italicize generic and species names to avoid confusion: thus bacillus is a descriptive term for rod-shaped bacteria, while *Bacillus* is a generic name.

Taxa below the rank of species

Some use is made of taxa below the rank of species. Within zoology, this is confined to the term subspecies, so the names of subspecies have three components, e.g. *Mus musculus domesticus*, no rank name being necessary.

In bacteriology, the use of subspecies is again acceptable although a word indicating rank is usually inserted, e.g. *Bacillus subtilis* subsp. *niger*. The Bacteriological Code considers ranks from the level of subspecies up to, and including, class: the use of the term variety is discouraged, as the term is synonymous with subspecies. However, other terms are in widespread use for taxa below the species level, especially in medical microbiology and plant pathology, when a particular strain of bacterium has been identified (p. 104). Subspecies identification is often referred to as typing and the following terms apply:

- biovar, or biotype: sub-divided according to biochemical characteristics;
- serovar or serotype: sub-divided by serological methods, using antibodies (see Chapter 32);
- pathovar: sub-divided according to pathogenicity (ability to cause disease);
- phagovar or phage type: sub-divided according to susceptibility to particular viruses.

The term 'strain' is now more widely used, particularly in the context of the practice of lodging microbiological strains with culture collections. Many microorganisms are now referred to by their generic and specific names followed by a culture collection reference number, e.g. *Bacillus subtilis* NCTC 10400, where NCTC stands for the National Collection of Type Cultures and 10400 is the reference number of that strain in the collection.

In botany, several categories below the rank of species are recognized and a term of rank is used before the name, e.g. *Salix repens* var. *fusca*: the term var. is short for the Latin word *varietas* and is subordinate to the term subspecies in the Botanical Code. The term cultivar (cv.) is an important modern term frequently used in experimental work and refers to cultivated varieties of plants.

The special case of viruses

The classification and nomenclature of viruses is less advanced than for cellular organisms and the current nomenclature has been arrived at on a piecemeal, *ad hoc* basis. Recently, the International Committee for Virus Taxonomy proposed a unified classification system, dividing viruses into fifty families on the basis of:

- host preference;
- nucleic acid type (i.e. DNA or RNA);
- whether the nucleic acid is single or double stranded;
- the presence or absence of a surrounding envelope.

Virus family names end in -viridae and genus names in -virus. (Note that these names are *not* latinized and the genus-species binomial is not now approved). However, this system has not yet been adopted universally and many viruses are still referred to by their trivial names or by code-names, e.g. the bacterial viruses ϕX174, T4, etc. Many of the names used reflect the diseases caused by the virus. Often, a three-letter abbreviation is used, e.g. HIV (for human immunodeficiency virus), TMV (for tobacco mosaic virus).

Example
The virus that causes tobacco mosaic disease belongs to the genus tobamovirus and can be referred to as tobacco mosaic tobamovirus, tobacco mosaic virus or TMV.

21 Identifying plants and animals

The normal way to identify plants or animals is to use identification guides. These consist of two parts:

1. Written and pictorial descriptions of organisms, which you compare with your unknown specimen to aid in its identification. Good descriptions direct you to the crucial diagnostic features for the relevant taxon, explain the range of variability found and point out biological and ecological characteristics of importance.
2. Keys, which help you find the likely description for your specimen rapidly and simply. Most keys are arranged to present you with a series of choices, usually dichotomous (dividing in two). The paired statements of each 'couplet' are framed to be contrasting and mutually exclusive. Each choice you make narrows down the possibilities for your specimen until you find the appropriate description.

The authors of identification guides assume that you have a live or preserved specimen to hand and the means to observe it closely and measure it. The terminology in guides is designed to combine precision with brevity. You need to know enough of the vocabulary to understand the choices presented to you, but all identification guides provide both a glossary and a list of abbreviations to help with this. The best identification guides are those which lead you in the simplest way to a correct identification. If you need to choose one for your area of interest, think about the following questions:

- What degree of prior knowledge is assumed? Some guides are written for novices, while others assume an expert's command of terminology. If tempted to go for the former type, consider whether it will always be suitable for your needs.
- What is the scope of the guide? Guides may be restricted in the taxa they consider or in the geographical region that is covered; this will suit you if your interests are similarly narrow. However, if your interests are wide, the relevant guide may be so large as to be unwieldy in the field.
- How well is the guide illustrated? Good quality illustrations enhance the ease of use of a guide — features can be shown pictorially that might involve an off-putting specialized vocabulary to describe. Accurately coloured illustrations can be helpful, but note that colour can be a variable character: look for good line diagrams that highlight the critical diagnostic points.
- Is the guide divided into parts? Good guides are arranged in short parts dealing with different levels of the taxonomic hierarchy. This speeds up identification by allowing you to skip initial material when you have a fair idea of the specimen's identity.
- How old is the guide? Taxonomists frequently change the names of taxa and update their classification. An up-to-date identification requires an up-to-date guide!
- Do you like the style of the key? As discussed below, there are several ways in which a key can be presented, one of which may suit you more than the others. If you can't actually test out a key yourself on real specimens, the next best is to ask for the opinion of someone who has used it.

Types of keys

Bracketed keys

Here, numbered pairs of adjacent lines in the key present you with a choice and either 'send' you to a new couplet or provide the tentative identification (Box 21.1).

Box 21.1 Example of a bracketed key

Part of key to ragworts native to the British Isles (modified after Stace, 1991):

1. Ligules $<$ 8 mm or 0; capitula cylindrical, about $2\times$ as long as wide 2
   Ligules $>$ 8 mm; capitula bell-shaped in flower, about $1.5\times$ as long as wide 3
2. Ligules usually 0; achenes $\leq$ 2.5 mm *Senecio vulgaris*
   Ligules usually present; achenes $>$ 3 mm *Senecio cambrensis*

If there are no ligules on your specimen or they are less than 8 mm in length, you should proceed to choice 2, but if they are present and greater than 8 mm in length, you should proceed to choice 3.

In this case, the choice at 2 is sufficient to pin down the species; sometimes quite early in a key a distinctive characteristic may allow the specimen to be 'identified' to species level, while for the other options the specimen's identity remains open. Note the use of more than one comparison in each couplet to provide confirmation.

Key point
Bracketed keys have the advantage that they keep the couplets close together for ready comparison, but indented keys show the relationships between taxa more clearly and allow you to back-track more easily if an error has been made.

Indented keys

In this method, the pairs of choices are indented and given the same number. They are separated by other choices further down the sequence. Having made a choice, you look at the next couplet below which will be one indent level further in. When a choice is sufficiently distinctive, the tentative identity of your specimen will be given (Box 21.2).

Box 21.2 Example of an indented key

Part of key for common species of true bumblebee in the British Isles (modified after Prys-Jones and Corbet, 1987):

1. Thorax with black area(s)
    2. Thorax all black
        3. Pollen baskets with red hairs *Bombus rudarius*
        3. Pollen baskets with black hairs *Bombus lapidarius*
    2. Thorax black with yellow or brown patches
        4. Tail white, buff or brown
            5. Scutellum black *Bombus terrestris*
            5. Scutellum yellow *Bombus hortorum*
        4. Tail red or orange *Bombus pratorum*
1. Thorax without any black *Bombus pascuorum*

If your bumblebee has a black thorax with yellow patches, proceed to choice 4; if its tail is brown, carry on to choice 5, etc.

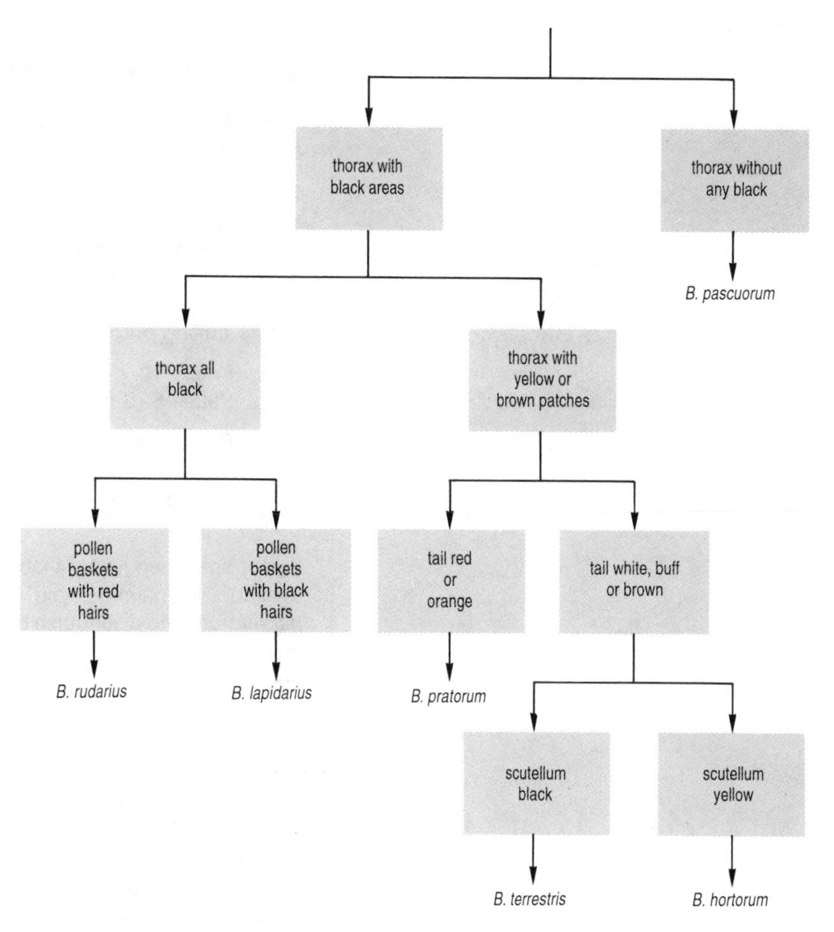

Fig. 21.1 A key for British bumblebees (*Bombus* spp.) laid out in the form of a flowchart.

Flowchart keys

In this form of key, the choices are laid out in the form of a flowchart (Fig. 21.1), which allows easier cross-checking of options but is only feasible where there is a small number of choices. To use this type of key, follow the arrows after making each choice in the sequence; this will lead you onto another choice and eventually to the tentative identification.

Multi-access keys

These allow you to choose the characters used in the key according to the state of your specimen (Box 21.3). They are useful in situations where:

- important characters are difficult to observe;
- characters are likely to be misinterpreted;
- a single character would be unreliable in isolation;
- a part is missing or seems abnormal.

Another way of presenting a multi-access key is in the form of a table. For instance, the taxa to be distinguished could make up the rows of the table and relevant characteristics the columns (see Table 22.1). Like the flowchart, this type of key is limited to a small number of choices.

Box 21.3 Example of a multi-access key

Part of key to species of willowherbs native to the British Isles (modified after Stace, 1991):

Stigma 4-lobed	A
Stigma club-shaped	B
Seeds minutely uniformly papillose	C
Seeds with longitudinal papillose ridges	D
Stems erect or erect at apex	E
Stems trailing on ground	F

ACE	Petals 10–16 mm, purplish-pink	*Epilobium hirsutum*
	Petals 5–9 mm, paler	*Epilobium parviflorum*
BCE	Petioles 4–15 mm, plant perennating by rosettes	*Epilobium roseum*
	Petioles $\leq$ 4 mm, plant perennating by stolons ending in tight bud	*Epilobium palustre*

If your specimen had a 4-lobed stigma and erect stems, but no seeds were available to examine, you could 'identify' it as either *E. hirsutum* or *E. parviflorum* and distinguish between these choices on the basis of petal size and colour.

Advice for using keys

- Note down the route taken (i.e. the numbering system for the decision tree) as you go: this makes it easier to trace back your path through the key.
- At each step, read the full description for both choices before arriving at a decision about which one to take.
- Never guess if you do not know the precise meaning of the terms used — consult the key's glossary and list of abbreviations. Where measurements are required, use a ruler — do not guess sizes.
- If features are very small, use an appropriate lens to inspect them clearly.
- If the key is a multi-part one, look carefully at the descriptions for higher levels of taxa before progressing to the species key: this not only acts as a check that you are correct up to this stage, but may also provide definitions of useful terminology.
- If both of a pair of choices seem reasonable, try out each route — one will usually prove to be unsuitable at a later stage.
- Be aware of problems likely to confound your best attempts to identify the specimen, such as the existence of sexual dimorphism or polymorphism, juvenile and adult phases (e.g. gametophyte and sporophyte phases for certain plants), local forms, non-native taxa, etc. A good guide will point out these problems where they occur.

Key point
Take special care not to collect, disturb or destroy rare plants and animals in the course of your observations.

Comparing specimens with descriptions

When you arrive at the end of a key's path, *do not simply accept this as a reliable identification of your specimen*. Compare your specimen with the full description of the species. If the specimen doesn't fit the description properly, follow the instructions outlined below:

- Compare the specimen with neighbouring descriptions: in a well-organized guide those of similar species will be together.

- Go back along the path of the key and re-examine each decision you have made. Try going down the alternative route for any that might have been questionable.
- Check to see whether you inadvertently went down the wrong pathway even though you made the correct diagnosis.
- Bear in mind the possibility that your specimen is not typical. A good key will use characteristics that are constant, but biological variation will often throw up an oddity to confuse you. Try to obtain another specimen, preferably not genetically related to the original.
- Consider the possibility that it could be outside its normal geographical range or even new to science!

The ultimate check on an identification is a comparison of the specimen with an authentically named specimen in a museum or herbarium.

22 Identifying microbes

Key point
Identification of bacteria is often based on a combination of:
- **growth characteristics,**
- **microscopic examination,**
- **physiological and biochemical characterization,**
- **where necessary, immunological tests.**

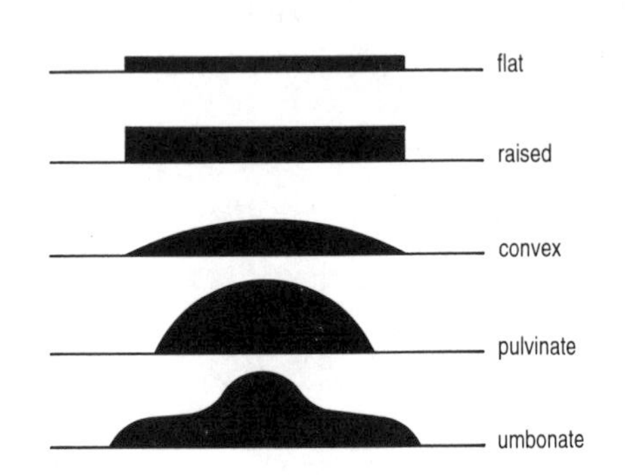

Fig. 22.1 Colony elevation (cross-sectional profile).

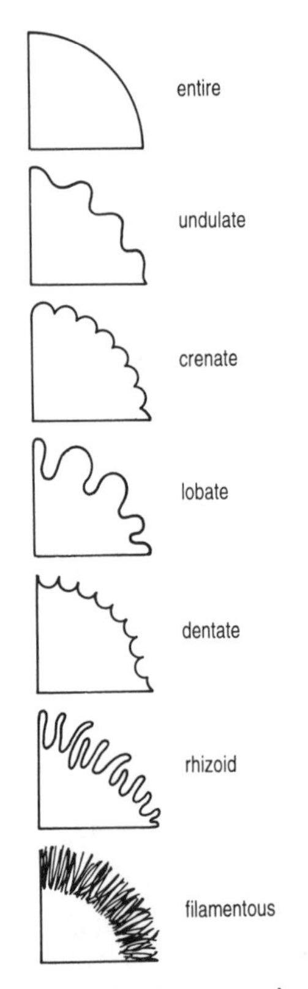

Fig. 22.2 Types of colony margin (edge).

Most of the methods described in this chapter were developed for the identification of bacteria, and bacterial examples are used to illustrate the principles involved. While the basic techniques are equally applicable to other types of microbe, the identification systems for some protozoa, fungi and algae rely predominantly on microscopic appearance. Identification of viruses requires electron microscopy or immunological techniques.

Direct observation

The visual appearance of individual colonies on the surface of a solidified medium may provide useful information. Filamentous fungi usually grow as large, spreading colonies with a matt appearance and are identified by microscopy, using the morphological characteristics of their reproductive structures. Yeasts produce smaller, glistening colonies. Identification usually involves microscopy, combined with physiological and biochemical tests similar to those used for bacteria. Bacteria typically produce smooth, glistening colonies, varying in diameter from <1 mm to >1 cm. Actinomycete colonies are often <1 cm, with a shrivelled, powdery surface.

Colony characteristics

The characteristics of a bacterial colony on a particular medium include:

- Size: some bacteria produce punctiform colonies, with a diameter of less than 1 mm, while motile bacteria may spread over the entire plate. You should measure the diameter of a typical colony, well spaced from any others, as colony size is affected by competition for nutrients.
- Form: colonies may be circular, irregular, lenticular (spindle-shaped) or filamentous.
- Elevation: various terms are used to describe the depth and cross-sectional appearance of a bacterial colony (Fig. 22.1).
- Margin: the edge of a colony may be distinctive (Fig. 22.2).
- Consistency: colonies may be viscous (or mucoid), butyrous (of similar consistency to butter) or friable (dry and granular).
- Colour: some bacteria produce characteristic pigments. A few pigments are fluorescent under UV light.
- Optical properties: colonies may be translucent or opaque.
- Odour: some actinomycetes and cyanobacteria produce earthy odours, while certain bacteria and yeasts produce fruity or 'off' odours. However, odour is not a reliable characteristic in bacterial identification. Never smell mould cultures because of the risk of inhaling large numbers of spores.
- Haemolytic reactions on blood agar: many pathogenic bacteria produce characteristic zones of haemolysis. Alpha haemolysis is a partial breakdown of the haemoglobin from the erythrocytes, producing a green zone around the colony, while beta haemolysis is the complete destruction of haemoglobin, producing a clear zone.

Microscopic examination

Bacteria are usually observed using an oil immersion objective at a total magnification of $\times 1000$ (p. 136).

Hanging drop technique — place a drop of bacterial suspension on a coverslip and invert over a cavity slide so that the drop does not make contact with the slide: motile aerobes are observed at the edge of the droplet, where oxygen is most abundant.

Assessing motility — if you have not seen bacterial motility before, it is worth comparing your unknown bacterium to a positive and a negative control.

Motility

Wet mounts can be prepared by placing a small drop of bacterial suspension on a clean, degreased slide, adding a coverslip and examining the film without delay. For aerobes, areas near air bubbles or by the edge of the coverslip will give best results, while anaerobes will show greatest motility in the centre of the preparation, with rapid loss of motility due to oxygen toxicity.

Prepare wet mounts using young cultures in exponential growth in a liquid medium (p. 126). It is best to work with cultures grown at 20 or 25°C, since those grown at 37°C may not be actively motile on cooling to room temperature. It is essential to distinguish between the following:

- True motility, due to the presence of flagella: bacteria dart around the field of view, changing direction in zigzag, tumbling movements.
- Brownian motion: non-motile bacteria show a localized, vibratory, random motion, due to bombardment of bacterial cells by molecules in the solution.
- Passive motion, due to currents within the suspension: all cells will be swept in the same direction at a similar rate of movement.
- Gliding motility: a slower, intermittent movement, parallel to the longitudinal axis of the cell, requiring contact with a solid surface.

Cell shape

Bacteria are subdivided into the following groups:

- Cocci (singular, coccus): spherical, or almost spherical, cells, sometimes growing in pairs (diplococci), chains or clumps.
- Rods: straight, cylindrical cells with flattened, tapered or rounded ends, termed bacilli. Short rods are sometimes called cocco-bacilli.
- Curved rods: the curvature varies according to the organism, from short curved rods, sometimes tapered at one end, to spiral shapes.
- Branched filaments: characteristic of actinomycete bacteria.

Many bacteria are pleomorphic, varying in shape according to growth conditions/culture age; thus other characteristics are required for identification.

Gram staining

This is the most important differential staining technique in bacteriology (Box 22.1 gives details). It enables us to divide bacteria into two distinct groups, Gram-positive and Gram-negative, according to a particular staining procedure (the technique is given a capital letter, since it is named after its originator). The basis of the staining reaction is the different structure of the cell walls of Gram-positive and Gram-negative bacteria. Heat fixation of air-dried bacteria causes some shrinkage, but cells retain their shape: to measure cell dimensions use a chemical fixative (p. 81).

Gram staining should be carried out using light smears of young, active cultures, since older cultures may give variable results. In particular, certain Gram-positive bacteria may stain Gram-negative if older cultures are used. This Gram-variability is due to autolytic changes in the cell wall of Gram-positive bacteria. If a pure culture of an unknown bacterium gives both Gram-positive and Gram-negative cells, identical in size and shape, it can be regarded as a Gram-positive organism. Spores are often visible as unstained areas within older vegetative cells of *Bacillus* and *Clostridium*. Other stains can be used to demonstrate spores, capsules or flagella (p. 117).

Box 22.1 Preparation of a heat-fixed, Gram-stained smear

Preparation of a heat-fixed smear

The following procedure will provide you with a thin film of bacteria on a microscope slide, for staining.

1. **Take a clean microscope slide and pass it through a Bunsen flame twice**, to ensure it is free of grease. Allow to cool.
2. **Using a sterile inoculating loop, place a single drop of water in the centre of the slide and then mix in a small amount of sample** from a single bacterial colony with the drop, until the suspension is slightly turbid. Smear the suspension over the central area of the slide, to form a thin film. For liquid cultures, use a single drop of culture fluid, spread in a similar manner.
3. **Allow to air-dry at room temperature**, or high above a Bunsen flame: air drying must proceed gently, or the cells will shrink and become distorted.
4. **Fix the air-dried film by passage through a Bunsen flame**. Using a slide holder or forceps, pass the slide, film side up, rapidly through the hottest part of the flame (just above the blue cone). The temperature of the slide should be just too hot for comfort on the back of your hand: note that you must not overheat the slide or you may burn yourself (you will also ruin the preparation).
5. **Allow to cool**: the smear is now ready for staining.

Gram-staining procedure

The version given below is a modification of the Hucker method, since acetone is used to decolorize the smear. Note that some of the staining solutions used are flammable, especially the acetone decolorizing solvent: you must make sure that all Bunsens are turned off during staining. The procedure should be carried out with the slides suspended over a sink, using a staining rack.

1. **Flood a heat-fixed smear with 2% w/v crystal violet in 20% v/v ethanol:water** and leave for 1 min.
2. **Pour off the crystal violet and rinse briefly with tap water. Flood with Gram's iodine** (2 g KI and 1 g I_2 in 300 ml water) for 1 min.
3. **Rinse briefly with tap water** and leave the tap running gently.
4. **Tilt the slide and decolorize with acetone** for 2–3 s: acetone should be added dropwise to the slide until no colour appears in the effluent. This step is critical, since acetone is a powerful decolorizing solvent and must not be left in contact with the slide for too long.
5. **Immediately immerse the smear in a gentle stream of tap water**, to remove the acetone.
6. **Pour off the water and counterstain for 10–15 s using 2.5% w/v safranin** in 95% v/v ethanol:water.
7. **Pour off the counterstain, rinse briefly with tap water, then dry the smear** by blotting gently with absorbent paper: all traces of water must be removed before the stained smear is examined microscopically.
8. **Place a small drop of immersion oil on the stained smear: examine directly** (without a coverslip) using an oil-immersion objective (p. 00).

Gram-positive bacteria retain the crystal violet (primary stain) and appear purple while Gram-negative bacteria are decolorized by acetone and counterstained by the safranin, appearing pink or red when viewed microscopically.

Other decolorizing solvents are sometimes used, including ethanol:water, ethyl ether:acetone and acetone:alcohol mixtures. The time of decolorization must be adjusted, depending upon the strength of the solvents used, e.g. 95% v/v ethanol:water is less powerful than acetone, requiring around 30 s to decolorize a smear.

Basic laboratory tests

At least two simple biochemical tests are usually performed:

Oxidase test

This identifies cytochrome *c* oxidase, an enzyme found in obligate aerobic bacteria. Soak a small piece of filter paper in a fresh solution of 1% (w/v) N-N-N′-N′-tetramethyl-p-phenylenediamine dihydrochloride on a microscope slide. Rub a small amount from the surface of a young, active colony onto the filter paper using a glass rod or *plastic* loop: a purple-blue colour within 10 s is a positive result.

Catalase test

This identifies catalase, an enzyme found in obligate aerobes and in most facultative anaerobes, which catalyses the breakdown of hydrogen peroxide into water and oxygen. Transfer a small sample of your unknown bacterium onto a coverslip using a disposable plastic loop or glass rod. Invert onto a drop of hydrogen peroxide: the appearance of bubbles within 30 s is a positive reaction. This method minimizes the dangers from aerosols formed when gas bubbles burst.

The oxidase and catalase tests effectively allow us to sub-divide bacteria on the basis of their oxygen requirements, without using agar shake cultures (p. 87) and overnight incubation, since, for the most part:

- obligate aerobes will be oxidase and catalase positive;
- facultative anaerobes will be oxidase negative and catalase positive;
- microaerophilic bacteria, aerotolerant anaerobes and strict anaerobes will be oxidase and catalase negative — the latter group will grow only under anaerobic conditions (p. 87).

Once you have reached this stage (colony characteristics, motility, shape, Gram reaction, oxidase and catalase status) it may be possible to make a tentative identification, at least for certain Gram-positive bacteria, at the generic level. To identify Gram-negative bacteria, particularly the oxidase-negative, catalase-positive rods, further tests are required.

Identification tables: further laboratory tests

Bacteria are asexual organisms and strains of the same species may give different results for individual biochemical/physiological tests. This variation is allowed for in identification tables (multi-access keys, p. 97), based on the results of a large number of tests. Identification tables are often used for particular sub-groups of bacteria, after Gram staining and basic laboratory tests have been performed: an example is shown in Table 22.1.

A large number of specific biochemical and physiological tests are used in bacterial identification including:

- Carbohydrate utilization tests. Some bacteria can use a particular carbohydrate as a carbon and energy source. Acidic end-products can be identified using a pH indicator dye (p. 30) while CO_2 is detected in liquid culture using an inverted small test tube (Durham tube). Aerobic breakdown

Table 22.1 Identification table (tabular multi-access key) for selected motile, oxidase-negative, catalase-positive, Gram-negative rods

	Biochemical test								
Bacterium	1	2	3	4	5	6	7	8	9
Escherichia coli	v	+	−	+	−	v	v	+	−
Proteus mirabilis	−	−	v	−	+	−	+	−	+
Morganella morganii	−	−	−	−	+	−	+	+	−
Vibrio parahaemolyticus	−	+	v	−	−	+	+	+	−
Salmonella spp.	−	+	v	−	−	+	+	−	+

Key to biochemical tests:
1. sucrose utilization,
2. mannitol utilization,
3. citrate utilization,
4. *o*-nitrophenol-β-D-galactoside hydrolysis,
5. urease activity,
6. lysine decarboxylase activity,
7. ornithine decarboxylase activity,
8. indole production,
9. H_2S production.

Key to symbols:
+, >90% of strains tested positive,
−, < 10% of strains tested positive,
v, 10–90% of strains tested positive.

is termed oxidation while anaerobic breakdown is known as fermentation. Identification tables usually incorporate tests for several different carbohydrates e.g. Table 22.1.

- Enzyme tests. Most of these incorporate a substance which changes colour if the enzyme is present, e.g. a pH indicator, or a chromogenic substrate.
- Tests for specific end-products of metabolism, e.g. the production of indole from tryptophan, or H_2S from sulphur-containing amino acids.

Identification kits

Some biochemical tests are now supplied in kit form, e.g. the API® 20E system incorporates 20 tests within a sterile plastic strip. After inoculation and overnight incubation, the results of the tests are converted into a 7-digit code, for comparison with known bacteria using either a reference book (the Analytical Profile Index), or a computer program. While kit identification systems save time and labour, they are more expensive and less flexible than conventional biochemical tests.

Immunological tests

Tests used in diagnostic microbiology include:

- Agglutination tests: based on the reaction between specific antibodies and a particular bacterium (p. 153). These tests are particularly useful for subdividing biochemically similar bacteria.
- Fluorescent antibody tests: the reaction between a labelled antibody and a particular bacterium can be visualized using UV microscopy. The direct fluorescent antibody test uses fluorescein isothiocyanate as the label.
- Enzyme-linked immunoassay tests using antibodies labelled with a particular enzyme, e.g. the double antibody sandwich ELISA test (p. 157).

While such tests can give specific and accurate confirmation of the identity of a bacterium under controlled laboratory conditions, they are too expensive and time-consuming for routine identification purposes.

Typing methods

The identification of bacteria at subspecies level is known as typing: this is usually done in a specialist laboratory, e.g. as part of an epidemiological study to establish the source of an infection. Various methods are used:

- Antigen typing or serotyping is based on immunological tests.
- Phage typing is based on the susceptibility of different strains to certain bacterial viruses (phages).
- Biotyping is based on biochemical differences between different strains.
- Bacteriocin typing: bacteriocins are proteins released by bacteria which inhibit the growth of other members of the same species.

Manipulating specimens

23 The purpose and practice of dissection

Dissection in zoology involves the display or removal of parts of any dead animal while vivisection is an operation on a living animal. Dissection is usually associated with teaching animal structure but the skills involved are used widely within the life sciences for:

- investigating anatomy and morphology;
- making physiological preparations of nerves, muscles and other organs;
- investigating parasites in various body organs;
- removing specific body organs/tissues for chemical analysis;
- investigating reproductive status;
- removing organs/tissues for histological/histochemical investigation;
- manipulating living material as in grafting processes.

Key points
Primary objectives of dissection:
- **Personal exploration of animal structure and function.**
- **Development of manipulative skills.**
- **Production of reports based upon personal observation.**

There is considerable debate about the use of animals for dissection but this problem should be distinguished from that of vivisection. It is part of a complex and emotive issue but one viewpoint is that dissection must be used for effective zoology teaching: experience of dissection at an *appropriate* stage of the curriculum is both enlightening and teaches an essential technical skill. However, if dissection is definitely not for you, it is best to discover this as early as possible in your career!

The ground-rules of dissection

There are some important rules to be considered if dissection is to be an acceptable procedure in zoological teaching.

Humane treatment

The use of any animal, whatever its level of organization, for experimentation or dissection must be a considered act with due regard for humane treatment and killing. Remember there are very specific regulations for the use of vertebrate species and some higher invertebrates: check with your supervisor.

You have a particular responsibility to turn up and make careful and maximal use of a dissection specimen.

Maximum benefit

Any animal should be used for as many investigations as possible.

Preparation

Prepare for the exercise by ensuring that practical schedules and relevant texts are consulted before attempting a dissection. If you have to miss a practical, inform the organizer in advance to prevent animals being killed or prepared unnecessarily.

Types of dissection

Dissection can be carried out at three levels of sophistication, related directly to the size of the organism or structure being investigated. Gross dissection of large organisms requires equipment very different from that used for 'normal' dissection in that large knives replace scalpels, etc. Normal-scale dissection involves creatures from a dog down to an earthworm, and requires equipment typical of commercial dissection kits/instruments. Fine-scale dissection for the removal or display of organs, glands, etc., from small animals usually requires only mounted needles, fractured glass edges for cutting and a dissecting microscope or magnifier. Fine-scale dissection requires extensive practice and the preparation of material (relaxation and fixation (p. 82)) is much more critical than for larger specimens.

Equipment

Table 23.1 lists basic equipment required for normal dissection: some comments on use are given below. Commercial dissection kits often contain inappropriate components and you should buy your instruments individually from specialist suppliers if possible. Equipment for fine-scale dissections can be assembled easily to your own specifications.

Table 23.1 List of basic equipment recommended for dissection

Quantity	Description
1	All-metal scalpel, stainless steel, 45 mm blade
1 each	Swann-Morton scalpel handles, sizes 3 and 4
1 each	Swann-Morton blades, packets of nos. 10, 11, 12, 15, 22 and 24
2	Dissecting needles, straight, stainless steel
2	Blunt seekers, stainless steel: metal handles
1	Fine forceps, sharp points, stainless steel, 112 mm length
1	Coarse forceps, blunt points, stainless steel, 112 mm length
1	Coarse scissors, open shanks, straight points, stainless steel, 150 mm length
1	Fine scissors, open shanks, straight fine points, stainless steel, 110 mm length
1	Section lifter

Scalpels

There are two basic types of scalpels, fixed blade and replaceable blade. Buy at least one 45-mm blade length solid forged scalpel, preferably made of stainless steel: use this for coarser cutting procedures as it can be resharpened using an oil-stone. The Swann–Morton scalpel comprises a handle and a disposable blade: blades come in a variety of shapes and sizes, curved-edge ones being the most useful. Blades fit only specific handle sizes so make sure that they match.

Forceps

Buy at least one each of coarse and fine stainless steel forceps. The latter are very delicate and you should check that the points meet precisely before purchasing. Look after them very carefully as the points are easily damaged: use only on soft tissues. Use large and small forceps for general purposes but keep fine ones for delicate work only.

Dissecting scissors

Buy two pairs of stainless steel, pointed scissors, one medium-large for coarse work and cutting small bones and one fine pair for delicate work. Points and edges are easily damaged and require professional sharpening when repair is needed so use carefully. A pair of bone cutters is optional but do *not* use scissors to cut large bones.

Dissecting (mounted) needles
Buy at least two with metal handles and protect the points; use them for dissecting membranes in areas where damage is acceptable. Needles have many non-dissection functions associated with other fine manipulative techniques.

Dissecting seekers
These have blunt points and should have metal handles. Use for breaking down membranes holding delicate organs together. Again, they have multiple uses.

Recommended accessory equipment
A teat pipette is valuable for washing delicate organs/tissues. A camel-hair brush is also useful for removing material from delicate structures. Pins and awls are usually provided by your department and are important for fixing the dissection specimen to the board (awls) or wax dish (pins): small pins are invaluable for pinning organs aside for display purposes. For measurement, and as an aid to drawing, a pair of dividers is very useful, as is a small steel ruler.

Tips for better dissection

The basic sequence of steps in a typical dissection is shown in Box 23.1 and Fig. 23.1. Specific instructions will normally be given in your practical schedule. However, the following tips should help you develop your dissection technique:

- Use a blunt seeker and controlled tension on the tissue to remove the connective tissue that binds tissues and organs together: use scalpels sparingly.
- Insert pins/awls obliquely so that they do not interfere with further dissection. In segmented animals such as earthworms, fine pins can be used to mark the position of specific segments.
- Dissect most invertebrates under water: the water buoys up the tissues/organs and assists dissection. Change the water if it becomes clouded but do not allow flowing water to run directly onto the specimen as it will damage delicate structures.
- Dissect vertebrates and larger invertebrates in air on a dissection board, using cotton wool swabs to remove excess blood and other body fluids.
- Keep tissues under tension while dissecting, but avoid damaging them with forceps/fingers.
- Cut away from the organs when using scalpels or scissors and keep scissor points away from structures you are attempting to free. This is essential when opening the body cavities of small animals.
- Use appropriate sized equipment; e.g. do not attempt to use large scissors to open an earthworm. Never use delicate instruments for coarse work.
- *Never* remove anything until you know what you are removing.
- Dissect along structures and not across them, particularly for tubular structures such as nerves and blood vessels.
- Use fresh material whenever possible. Fixatives make tissues more brittle and inelastic: alcohol storage tends to harden skin, muscle and connective tissue.
- Keep your instruments clean and sharp: you can't dissect with blunt or dirty equipment. Dry instruments after washing and wipe with an oily cloth. Cut nothing but tissues with scalpels and scissors. Do not sharpen pencils with scalpels or stick mounted needles into the bench or the dissecting board.

- Be hygienic! Wear rubber gloves if you have cuts or lesions on your hands. Wash hands and equipment thoroughly when finished. Dispose of animal remains carefully as instructed.

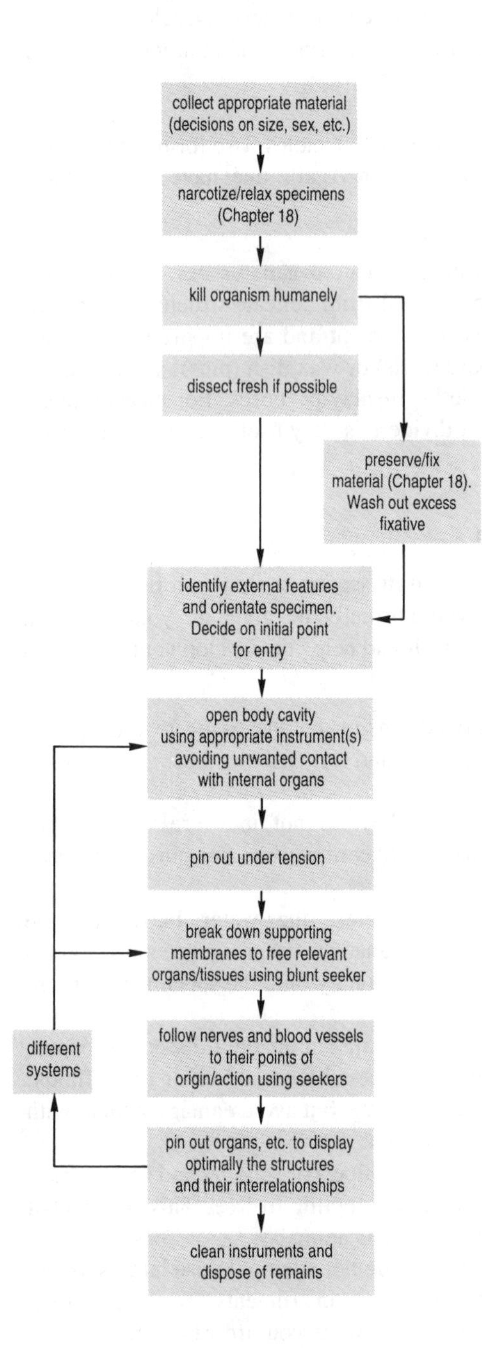

Fig. 23.1 Flowchart for dissection.

Box 23.1 Basic stages of an animal dissection

The basic sequence of steps is outlined in Fig. 23.1.

1. **The animal should have been killed as humanely and as recently as possible**: this will probably be done for you by the class supervisor or technician according to the rules appropriate to the type of animal. The method of killing should be chosen carefully to keep the specimen relaxed. If preserved material is used, wash out excess fixative thoroughly before dissecting.
2. **Orientate the specimen carefully**; determine the dorsal/ventral, anterior/posterior or other oral/aboral axes (see Fig. 29.1) and work out the correct orientation of the specimen for dissection. Invertebrates are usually dissected from the dorsal surface (the nerve cord being ventral in position) and vertebrates from the ventral surface (the nerve cord being dorsal in position): however, special objectives may require a different orientation.
3. **Open the body cavity carefully**. This is usually done using forceps to lift the skin away from underlying organs while using a fine pair of scissors to make an initial opening; the scissors are then used to extend this opening in antero-posterior and lateral directions until the skin flaps can be pinned back. Pin out by placing tension on the skin flaps, breaking down any restricting membranes using a seeker.
4. **Subsequent dissection procedure depends upon your objectives**. To display the system being investigated as clearly and neatly as possible requires:
   (a) identification of organs, blood vessels and nerves initially visible;
   (b) separation of these structures from the membranes which hold them in place, best achieved using a blunt instrument such as a seeker to avoid damage;
   (c) removal or displacement of organs which obscure parts of the system you wish to display. Displacement using pins to hold the organ in position is the preferred option when possible;
   (d) if your objective is to display a blood system or a nervous system, you must follow individual vessels/nerves from their point of origin to their destination organ; do this using a blunt seeker to remove covering membranes by working carefully along the structure. Do not pull sideways during this procedure as this often results in breakages; such tissues are usually much stronger in the direction of their length than when pulled laterally!
   (e) tidy up loose pieces and wash away residual blood, etc.; use more pins to finalize the display of the system and then make notes and drawings as necessary (see e.g. Fig. 15.7).

24 Introduction to microscopy

Many features of interest in biology are too small to be seen by the naked eye and can only be observed with a microscope. All microscopes consist of a coordinated system of lenses arranged so that a magnified image of a specimen is seen by the viewer (Fig. 24.1). The main differences are the wavelengths of electromagnetic radiation used to produce the image, the nature and arrangement of the lens systems and the methods used to view the image.

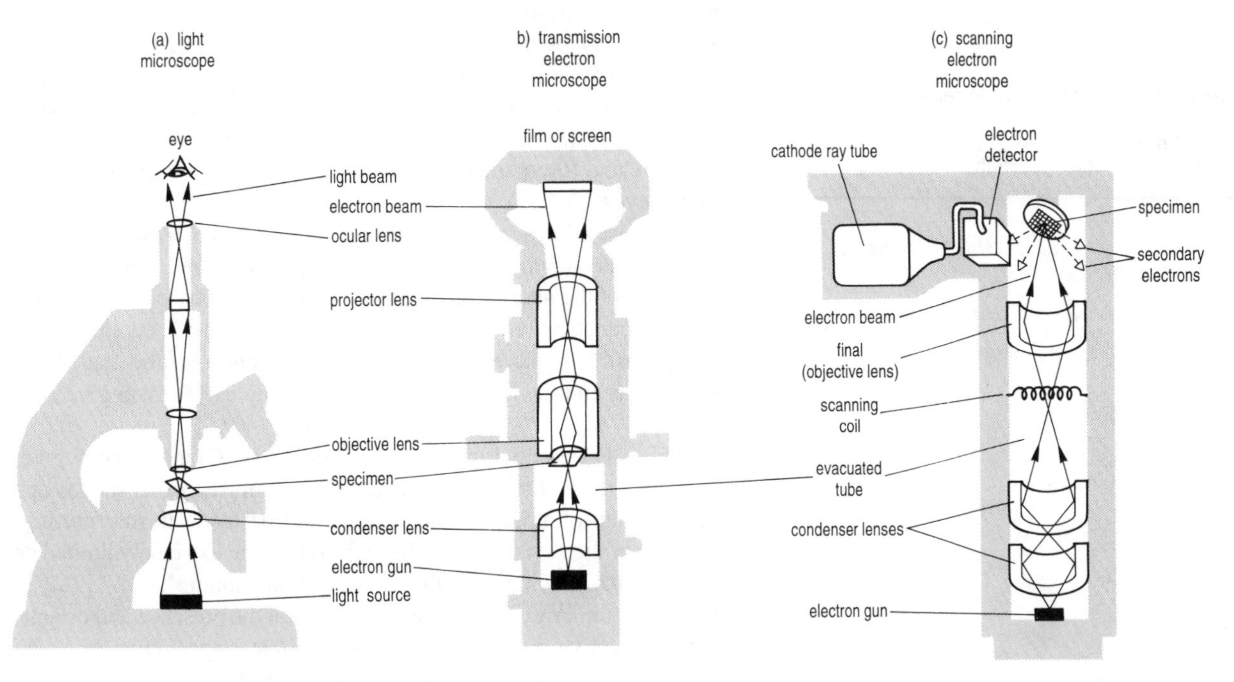

Fig. 24.1 Simplified diagrams of light and electron microscopes. Note that the electron microscopes are drawn upside-down to aid comparison with the light microscope.

Microscopes allow objects to be viewed with increased resolution and contrast. Resolution is the ability to distinguish between two points on the specimen — the better the resolution, the 'sharper' the image. Resolution is affected by lens design and inversely related to the wavelength of radiation used. Contrast is the difference in intensity perceived between different parts of an image. This can be enhanced (a) by the use of stains, and (b) by adjusting microscope settings, usually at the expense of resolution.

The three main forms of microscopy used in biology are light microscopy, transmission electron microscopy (TEM) and scanning electron microscopy (SEM). Their main properties are compared in Table 24.1.

Light microscopy

Two forms of the standard light microscope, the binocular (compound) microscope and the dissecting microscope, are described in detail in Chapter 28. These are the instruments most likely to be used in your practical work. In more advanced project work, you may be able to use one or more of the

Table 24.1 Comparison of microscope types. Resolution is that obtained by a skilled user. LM, light microscope; SEM, scanning electron microscope; TEM, transmission electron microscope

	Type of microscope		
Property	LM	SEM	TEM
Resolution	0.2 μm	10 nm	1 nm
Depth of focus	Low	High	Medium
Field of view	Good	Good	Limited
Specimen preparation (ease)	Easy	Easy	Skilled
Specimen preparation (speed)	Rapid	Quite rapid	Slow
Relative cost of instrument	Low	High	High

following more sophisticated variants of light microscopy to improve image quality:

- Dark field illumination involves a special condenser which causes reflected and diffracted light from the specimen to be seen against a dark background. The method is particularly useful for near-transparent specimens and for delicate structures like flagella. Care must be taken with the thickness of slides used, air bubbles and dust must be avoided and immersion oil must be used between the dark field condenser and the underside of the slide.
- Ultraviolet microscopy utilizes short-wavelength UV light to increase resolution. Fluorescence microscopy uses radiation at UV wavelengths to make certain naturally fluorescent substances (e.g. chlorophyll) emit light of visible wavelengths which may aid identification or localization. The method is especially useful when specimens have been stained with fluorescent dyes which bind to specific cell components. Special light sources, lenses and mountants are required for UV and fluorescence microscopy and filters must be used to eliminate the possibility of damage to users' eyes.
- Phase contrast microscopy is useful for increasing contrast when viewing transparent specimens. It is superior to dark field microscopy because a better image of the interior of specimens is obtained. Phase contrast operates by causing constructive and destructive interference effects in the image, which are visible as increased contrast. Adjustments must be made for each objective lens and the microscope must be set up carefully to give optimal results.
- Nomarski or Differential Interference Contrast (DIC) microscopy gives an image with a three-dimensional quality. However, the relief seen is optical rather than morphological, and care should be taken in interpreting the result. One of the advantages of the technique is the extremely limited depth of focus which results: this allows 'optical sectioning'.
- Polarized light microscopy can be used to reveal the presence and orientation of optically active components within biological specimens (e.g. starch grains, cellulose fibres), showing them up brightly against a dark background.

Electron microscopes

Electron microscopes offer an image resolution about 200 times better than light microscopes (Table 24.1) because they utilize radiation of shorter wavelength in the form of an electron beam. The electrons are produced by a tungsten filament operating in a vacuum and are focused by electro-magnets. TEM and SEM differ in the way in which the electron beam interacts with the specimen: in TEM, the beam passes through the specimen (Fig. 24.1b), while in SEM the beam is scanned across the specimen and is reflected from the surface (Fig. 24.1c). In both cases, the beam must fall on a fluorescent screen before the image can be seen. Permanent images ('micrographs') are produced after focusing the beam on photographic film.

You are unlikely to use either type of electron microscope as part of undergraduate practical work because of the time required for specimen preparation and the need for detailed training before these complex machines can be operated correctly. However, electron microscopy is extremely important in biology and you will probably be shown electron micrographs, with one or more of the following objectives:

- to demonstrate cell ultrastructure (TEM);
- to show surface features of organisms (SEM);
- to investigate changes in the number, size, shape and condition of cells and organelles (TEM);
- to carry out quantitative studies of cell and organelle disposition (TEM).

Aspects of the interpretation of electron micrographs are dealt with in Chapter 29.

Preparative procedures

Without careful preparation of the material being studied, the biological structures viewed with any microscope can be rendered meaningless. Figure 24.2 summarizes the processes involved for the main types of microscopy discussed above. The processes involved in preparing material for light microscopy are outlined in Chapter 25.

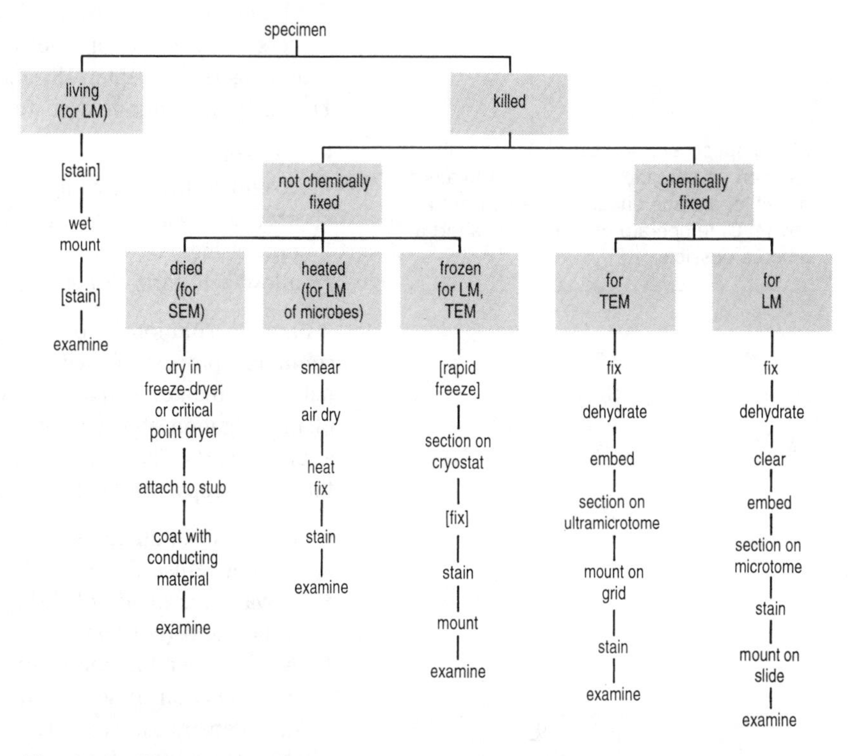

Fig. 24.2 Flowchart of procedures necessary to prepare specimens for different forms of microscopy. Steps enclosed in brackets are optional. LM, light microscope; SEM, scanning electronic microscope; TEM, transmission electron microscope.

25 Preparing specimens for light microscopy

Preparative techniques are crucial to successful microscopical investigation because the chemical and physical processes involved have the potential for making the material difficult to work with and for producing artefacts. The basic steps (outlined in Fig. 24.2) are similar in most cases, but the exact details (e.g. timing, chemicals used and their concentrations) differ according to the material being examined and the purpose of the investigation. It is usually best to follow a recipe that has worked in the past for your material (see, for example, Grimstone and Skaer, 1972).

Chemical fixation

The main purpose of fixation is to preserve material in a lifelike manner. The process of fixation for microscopy is much more critical than for whole specimens (see Chapter 18) and only small pieces of tissue should be used. The fixation solutions used for microscopy are intended to:

Fixing specimens — tissues must be fixed as soon as possible after death. If this is not possible, storage should be carried out at low (4°C) temperatures and for as short a time as possible.

- penetrate rapidly to prevent post-mortem changes in the cells;
- coagulate the cell contents into insoluble substances;
- protect tissues against shrinkage and distortion during subsequent processing;
- allow cell parts to become selectively and clearly visible when stained.

Fixative solutions are usually mixtures of chemicals selected for their combined properties (see Chapter 18). Your choice of fixative from the numerous recipes available in reference texts will depend upon both the type of investigation and the nature of the material. Poor fixation can produce artefacts, particularly where coagulant fixatives are used. When using a fixative for microscopy, observe the following points:

- Use fresh solutions: some of the fluids are unstable and do not keep well. Do not re-use fixative.
- Always use plenty of fixing fluid compared with the volume of material to be fixed (not less than a 10:1 fixative:sample volume ratio).
- Avoid under-fixation or over-fixation: in general the optimum time will be a function of several factors including:
  (a) Penetration capacity of the fixative.
  (b) The size of the piece of tissue: this should always be small and have as large a surface:volume ratio as possible.
  (c) The type of tissue to be fixed: uniform tissues fix more quickly than complex tissues, where one component may form a barrier to others. The presence of chitin usually means a slow rate of penetration. Tissues filled with air can limit penetration and make submergence of the specimen difficult; this can be overcome by fixation in a partial vacuum.
  (d) The temperature: increased temperature results in increased penetration rate, but also tends to make tissue brittle.
- Wash the specimen thoroughly after fixation: residues of fixative can interfere with subsequent processes. The washing may be in water or another appropriate solution.

Decalcifying specimens — this may be necessary if calcareous structures remain after fixation: this is usually done using a 5–10% solution of EDTA followed by thorough washing.

Dehydration and clearing

A high water content in tissues will usually hinder subsequent processing so they must be dehydrated. This is done with an organic solvent using a series of solutions graded from pure water to pure solvent. Ethanol/water mixtures are often used in histology. Dehydration must be carried out carefully, using prescribed time schedules to avoid distortion and hardening — never be tempted to rush, because incomplete dehydration will do more than anything else to ruin a preparation. Protect delicate specimens by mixing solutions in the container rather than by transferring the specimen as this is when damage will occur.

The chemicals used for dehydration are not usually soluble in the embedding medium and tissues must therefore be infiltrated with an intermediate fluid, miscible with the waxes and resins normally used: such fluids make the tissues transparent and are termed clearing agents. The most widely used ones are hydrocarbons such as xylene; many are volatile, pose significant health risks, and tend to harden tissues rapidly. Clearing oils such as clove oil and cedarwood oil are safer, but slower in action. Terpineol is useful since it does not require such complete dehydration and can be used straight from 90% v/v ethanol. Note that every trace of oil must be removed during the infiltration (embedding) stage, or specimens will not embed properly.

Embedding and sectioning

Embedding involves infiltrating the specimen with a solid medium which will support it when sectioned. The specimen is either passed through a series of gradually increasing concentrations of the embedding material (e.g. wax or epoxy resin) dissolved in the dehydrating or clearing agent, or placed straight into pure embedding agent. Several changes of embedding material are made to remove the last traces of solvent, then the embedding agent is solidified by cooling or by a polymerization treatment. Precise protocols can be found in specialist books (e.g. Kiernan, 1990). The specimen must be orientated carefully during the solidification process to aid section cutting in known planes.

The aim of sectioning is to provide a thin slice through the tissues suitable for observing cellular details at maximum resolution. After trimming, blocks of embedded material are sectioned using a microtome. Section-cutting techniques are complex and require time for familiarization. Sections are attached to slides after flattening the sections by a flotation procedure carried out either directly on the glass slide or in a water bath. The sections are attached to the slide by coating the latter with a very thin layer of albumen before drying them down onto this sticky surface.

For some procedures, specimens are frozen in isopentane, freon or dry-ice/ethanol mixtures and then maintained below −20°C until sectioned. Sectioning is performed in the cold chamber (cryostat) where there is a microtome; sections are removed directly to glass slides. If storage is necessary, slides are placed over a desiccant in a refrigerator, but this is possible only for short periods.

Hand-cut sections can be made through certain relatively stiff materials, notably stems and roots of herbaceous plants:

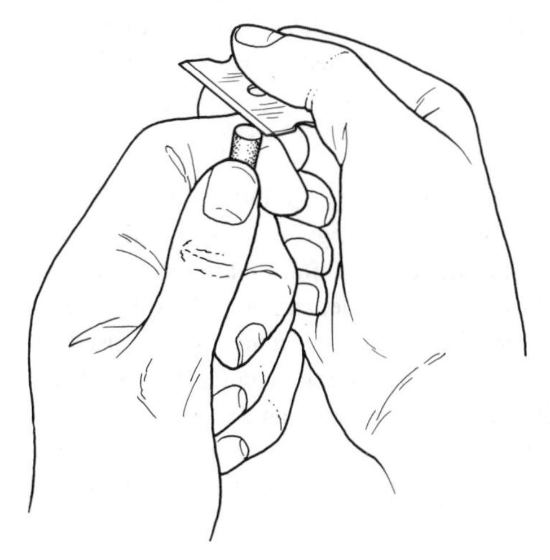

Fig. 25.1 Preparation of a hand-cut section.

1. Grasp the object firmly between your index finger and thumb.
2. Brace your elbows against your ribs to steady your hands.
3. Rest the side of a fresh and sharp razor blade on the index finger holding the specimen (Fig 25.1).

4. Pull the blade towards your body so that you cut the specimen with a slicing action.
5. Repeat this action quickly, pushing the object slowly upwards with your thumb.
6. Float off the cut sections onto a watch glass of water.

This is a relatively tricky procedure, requiring practice. You may find it helps to lubricate the razor and object by wetting them before cutting. Thin material such as leaves is best sectioned when supported between the halves of a longitudinally split cylinder of pith or fresh carrot. Always cut lots of sections, because only a small proportion will be thin enough to use — the best ones are often wedge-shaped, tapering off to thin edges.

Staining

The purpose of staining in microscopy is to:

- add contrast to the image;
- identify chemical components of interest;
- locate particular tissues, cells or organelles.

This is achieved in different ways for different types of microscopy. In standard light microscopy, contrast is achieved by staining the structure of interest with a coloured dye; in UV microscopy, contrast is obtained using fluorescent stains. Physico-chemical properties of the stain cause it to attach to certain structures preferentially or be taken up across cell membranes.

Stains for light microscopy are categorized according to the charge on the dye molecule. Stains like haematoxylin, whose coloured part is a cation (i.e. basic dyes), stain acidic, anionic substances like nucleic acids: such structures are termed basophilic. Stains like eosin, whose coloured part is an anion (i.e. acid dyes), stain basic, cationic substances: such structures are termed acidophilic. Acid dyes tend to stain all tissue components, especially at low pH, and are much used as counterstains. Staining is progressive if it results in some structures taking up the dye preferentially. Staining is regressive if it involves initial over-staining followed by decolorization (differentiation) of those structures which do not bind the dye tightly (e.g. Gram staining, p. 102).

Certain 'vital' stains (e.g. neutral red) are used to determine cell viability or the pH of cell compartments such as plant vacuoles. 'Mortal' stains (e.g. Evans' blue) are excluded from living cells but diffuse into dead ones and are used to assay cell mortality.

Definitions
Metachromic stains — have the capacity to stain different structures different colours.
Orthochromic stains — never change colour whatever they stain.
Mordants — chemicals (salts and hydroxides of divalent and trivalent metals) that increase the efficiency of stains usually by forming complexes with the stain.
Counterstains — stains that apply a background colour to contrast with stained structures.
Negative staining — where the background is stained rather than the structure of interest.

Stains and staining procedures

There is a huge range of stains for light microscopy but the features of those used commonly for cytology in botany, microbiology and zoology are given in Tables 25.1–25.3. Consult appropriate texts for full details of (a) how to make up stains and (b) the protocol to use. Results depend on technique: follow the recommended procedures carefully. There are many stains used in histochemistry for identifying various classes of macromolecules such as DNA, RNA, proteins, lipids, carbohydrates (chitin, cellulose, starch, callose, pectins, glycogen) and heteropolymers (e.g. lipopolysaccharides, peptidoglycans, proteoglycans). Consult specialist texts for methods (e.g. Grimstone and Skaer, 1972, Lillie, 1977).

Table 25.1 A selection of stains for light microscopy of sections of plant tissues

Stain	What it stains	Comments
Chlorazol black	Cell walls: black Nuclei: black, yellow or green Suberin: amber	The solvent used (70% ethanol in water or water alone) affects colours developed
Neutral Red	Living cells: pink (pH < 7)	A vital stain used to determine cell viability or to visualize plant protoplasts in plasmolysis experiments; best used at neutral external pH
Phloroglucinol/HCl	Lignified cell walls: red	Care is required because the acid may damage microscope lenses
Ruthenium red	Pectins: red	Shows up the middle lamella
Safranin + Fast green	Nuclei, chromosomes, cuticle and lignin: red Other components: green	Stain in safranin first, then counterstain with fast green (light green will substitute). A differentiation step is required
Toluidine blue	Lignified cell walls: blue Cellulose cell walls: purple	Best to apply dilute and allow progressive staining to occur

Table 25.2 A selection of stains for light microscopy of microorganisms (bacteria and fungi)

Stain	What it stains	Comments
Giemsa	Bacterial chromosome: purple Bacterial cytoplasm: colourless	Also used in zoology to stain protozoa
Gram	Gram-positive bacteria: violet/purple Gram-negative bacteria: red/pink Yeasts: violet/purple	See p. 102 for procedure
Gray	Bacterial flagella: red	Uses toxic chemicals: mercuric chloride and formaldehyde. Leifson's stain is an alternative
Lactophenol cotton blue	Fungal cytoplasm: blue (hyphal wall unstained)	Shrinkage may occur
Nigrosin or India ink	Background: grey–black	Negative stains for visualisation of capsules: requires a very thin film
Proca–Kayser	Viable bacteria: purple Dead bacteria: pink/red Viable endospores: pink Dead endospores: blue	Recent method involves acridine orange and UV microscopy
Shaeffer and Fulton	Bacterial endospores: green Vegetative cells: pink/red	Malachite green is primary stain, heated for 5 min. Counterstained with safranin
Ziehl–Nielsen	Actinomycetes: red Bacterial endospores: red Other microbes: blue	Requires heat treatment of fuchsin primary stain, decolorization with ethanol–HCl and a methylene blue counterstain (acid-fast structures remain red)

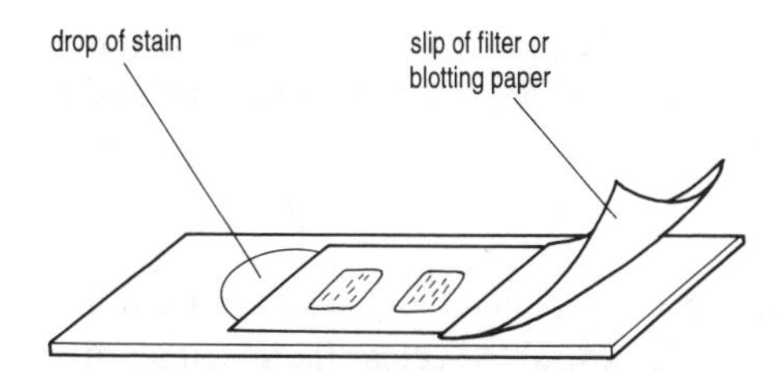

Fig. 25.2 How to irrigate a section with stain by drawing it through with a piece of filter paper.

Transferring sections — a camel-hair paint brush is useful for transferring sections between staining solutions and onto a slide.

Stains for light microscopy are normally applied by one of four methods:

- Floating the sections on the stain.
- Applying the stain to a smear fixed onto a slide, e.g. when staining bacteria and blood.
- Drawing the stain through under a coverslip as shown in Fig. 25.2.
- Immersing slides with sections attached into a staining trough. This is best for bulk staining.

Most stains need to act in an aqueous medium, so sections that have been embedded in wax must be rehydrated before staining. The wax is dissolved, e.g. in Histo-clear® for 1–3 min, the Histo-clear® replaced by 100% ethanol (1 min) then 70% (v/v) ethanol:water for 1 min, finally transferring to distilled water. Fresh sections of plants or heat-fixed smears of microbes generally require no pretreatment before staining if they are not to be retained after examination.

Table 25.3 A selection of stains for light microscopy of sections of animal cells

Stain	What it stains	Comments
Azure A/eosin B	Nuclei, RNA: blue Basophilic cells: blue–violet Most other cells: pale blue Muscle cells: pink Necrosing cells: pink Cartilage matrix: red–violet Bone: pink Red blood cells: orange–red Mucins: green–blue/blue–violet	Used in pathology — shows up bacteria as blue; must be fresh; care required over pH: Mann's methyl blue/eosin gives similar results
Chlorazol black	Chitin: greenish-black Nuclei: black, yellow or green Glycogen: pink or red	Solvent (70% v/v ethanol in water or water alone) affects colours formed
Iron haematoxylin	Nuclei, chromosomes and red blood cells: black Other structures: grey or blue–black	Good for resolving fine detail; iron alum used as mordant before haematoxylin to differentiate
Mallory	Nuclei: red Nucleoli: yellow Collagen, mucus: blue Red blood cells: yellow Cytoplasm: pink or yellow	Simple, one-stage stain; fades within a year; not to be used with osmium-containing fixatives. Heidenhain's azan gives similar results but does not fade. Cason's one-step Mallory is a rapidly applied stain which is particularly good for connective tissue
Masson's trichrome	Collagen, mucus: green Cytoplasm: orange or pink	Used as a counterstain after, for example, iron haematoxylin which will have stained nuclei black. Not to be used after osmium fixation
Mayer's (haemalum and eosin; 'H&E')	Nuclei: blue/purple Cytoplasm: pink	Alum used as mordant for haematoxylin; eosin is the counterstain. To show up collagen, use van Gieson's stain as counterstain

Mounting sections

Wet mounts

These are used for observing fresh specimens. The following steps are involved:

1. Isolate the specimen.
2. Place the specimen in a small droplet of the relevant fluid (fresh water, sea water, etc.) on a microscope slide.
3. Gently lower a coverslip onto the droplet, using forceps or two needles and avoiding bubbles.
4. Remove any excess water on or around the coverslip with absorbent paper.

Using coverslips — all wet specimens for microscopic examination must be covered with a coverslip to protect the objective lens from water, oil, stains and dirt.

Entire specimens can be examined under the light microscope providing they are small enough to be mounted on a glass slide. They may be mounted in cavity slides or using ring mounts.

Temporary mounts

These essentially involve wet mounting in a mountant with a short useful life, e.g. for identification purposes. It may be desirable to clear the specimen first and a dual purpose substance such as lactophenol, which will clear from 70% (v/v) ethanol, is recommended.

Using mountants — most mountants require that all water is removed from the section by transfer through increasing concentration of ethanol until 100% ethanol is reached.

Permanent mounts

These protect sections during examination and allow storage without deterioration. A permanent mount involves sealing your section under a coverslip in a mountant. The mountants used are clear resins dissolved in a slowly evaporating solvent. A good mountant has a similar refractive index

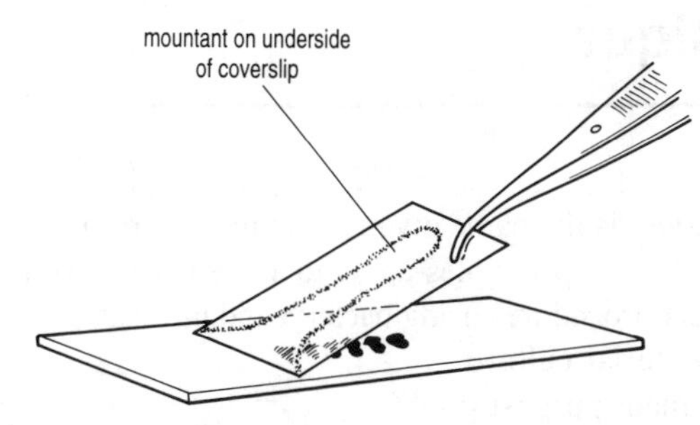

Fig. 25.3 How to lower a coverslip painted with mountant on its underside onto sections on a slide.

to the tissue being mounted, remains clear through time, is chemically inert and will harden quickly. Natural resins like Canada balsam take a long time to dry, are variable in quality and tend to colour-up and crack in time. The newer synthetic resins and plastics such as DPX® are superior: they dry quickly, are available in a range of refractive indices, and do not yellow with age. For tissue components soluble in organic solvents, aqueous mounting media based on, for example, gelatine or glycerol should be used. Find out which mountant is recommended for your particular sections.

The recommended procedure when mounting sections on a slide is:

1. Apply a little mountant to a coverslip of appropriate size.
2. Turn the coverslip over and place on its edge to one side of the sections as in Fig. 25.3.
3. Lower the coverslip slowly down onto the sections so as to displace all the air and sandwich the sections between the slide and the coverslip.
4. Press firmly from the centre outwards to distribute the mounting medium evenly.
5. Allow the solvent to evaporate — best results come from slow drying when time allows, but many synthetic mountants will tolerate brief heating when speed is essential.

Squash preparations

These may be required for any type of mount. The smallest specimens can be squashed after mounting by applying gentle pressure on the coverslip with your forceps. Larger specimens can be squashed between two slides before fixing and mounting — this ensures higher pressures are applied evenly.

26 Sterile technique

Sterile technique (aseptic technique) is the name given to the procedures used in cell culture. While the same general principles apply to all cell types, you are most likely to learn the correct procedures using bacteria and the examples given in this section refer to bacterial culture.

Sterile technique serves two main purposes:

- to prevent accidental contamination of laboratory cultures due to microbes from external sources, e.g. skin, clothing or the surrounding environment;
- to prevent microbial contamination of laboratory workers, in this instance you and your fellow students.

All microbial cell cultures should be treated as if they contained potentially harmful organisms, for the following reasons:

- You may accidentally isolate a harmful microbe as a contaminant when culturing a relatively harmless strain.
- Some individuals are more susceptible to infection and disease than others — not everyone exposed to a particular microbe will become ill.
- Laboratory culture involves purifying and growing large numbers of microbial cells — this represents a greater risk than small numbers of the original microbe.
- A microbe may change its characteristics, perhaps as a result of gene exchange or mutation.

Key point
Sterile technique forms an important part of safety procedures, and must be followed whenever cell cultures are handled in the laboratory.

Sterilization procedures

Given the ubiquity of microbes, it is clear that the only way to achieve a sterile state is by destruction or removal of all microbes. Several methods can be used to achieve this objective:

Heat treatment

This is the most widespread form of sterilization, employed in several basic laboratory procedures, including the following:

- Red heat sterilization. Achieved by heating metal inoculating loops, forceps, needles, etc., in a Bunsen flame. This is a simple and effective form of sterilization as no microbe will survive even a brief exposure to a naked flame. Flame sterilization using alcohol is used for glass rods and similar items (Fig. 26.2).
- Dry heat sterilization. Here, a hot air oven is used at a temperature of at least 160°C for a minimum of 2 h. This method is used for the routine sterilization of laboratory glassware. Dry heat procedures are relatively inefficient and of little value for items requiring repeated sterilization during use.
- Moist heat sterilization. This is the method of choice for many laboratory items, including most fluids, apart from heat-sensitive media. It is also used to decontaminate liquid media and glassware after use. The laboratory autoclave is used for these purposes. Typically, most items will be sterile after 15 min at 121°C, although large items may require a longer period. The rapid killing action results from the latent heat of condensation of the pressurized steam, released on contact with cool materials in the autoclave.

Using glass pipettes — Pasteur and volumetric pipettes should be plugged with cotton wool at the top prior to autoclaving to prevent subsequent contamination.

Radiation

Many disposable plastic items used in microbiology and cell biology are sterilized by exposure to UV or ionizing radiation. They are supplied commercially in sterile packages, ready for use. Ultraviolet radiation has limited use in the laboratory, while ionizing radiation requires industrial facilities and cannot be operated on a laboratory scale.

Filtration

Heat-labile solutions (e.g. complex macromolecules, including proteins, antibiotics, serum) are particularly suited to this form of sterilization. The filters come in a variety of shapes, sizes and materials, usually with a pore size of either 0.2 μm or 0.45 μm. The filtration apparatus and associated equipment is usually sterilized by autoclaving, or by dry heat. Passage of liquid through a sterile filter of pore size 0.2 μm into a sterile vessel is usually sufficient to remove bacteria but not viruses, so filtered liquids are not necessarily virus-free.

Chemical agents

These are known as disinfectants, or biocides, and are most often used for the disposal of contaminated items following laboratory use, e.g. glass slides and pipettes. They are also used to treat spillages. The term 'disinfection' implies destruction of disease-causing bacterial cells, although spores and viruses may not always be destroyed. Remember that disinfectants require time to exert their killing effect — any spillage should be treated by covering with an appropriate disinfectant and then leaving for at least 10 min before mopping up.

Use of laboratory equipment

Media

Cells may be cultured in either a liquid medium (broth), or a solidified medium (p. 125). The gelling agent used in most solid media is agar, a complex polysaccharide extracted from red algae that produces a stiff transparent gel when used at 1–2% (w/v). Agar is used because it is relatively resistant to degradation by most bacteria and because of its rheological properties — an agar medium melts at 98°C, remaining solid at all temperatures used for routine laboratory culture. However, once melted, it will not solidify until the temperature falls to approximately 44°C. This means that heat-sensitive constituents (e.g. vitamins, blood, etc.) can be added aseptically to media after autoclaving. Even the cells themselves can be added to molten agar (p. 124).

Plastic disposable loops — these are used in many research laboratories: pre-sterilized and suitable for single use, they avoid the hazards of naked flames and the risk of aerosol formation during heating. Discard into a disinfectant solution after use.

Inoculating loops

The initial isolation and subsequent transfer of cells between containers can be achieved using a sterile inoculating loop. Most teaching laboratories use nichrome wire loops in a metal handle. A wire loop can be repeatedly sterilized by heating the wire, loop downwards and almost vertical, in the hottest part of a Bunsen flame until the whole wire becomes red hot. Then the loop is removed from the flame to minimize heat transfer to the handle. Once the loop has been allowed to cool for 8–10 s (without touching any other object), it is ready for use.

When re-sterilizing a contaminated loop in a Bunsen flame after use, do not heat the loop too rapidly, as the sample may spatter, creating an aerosol: it is better to soak the loop for a few minutes in disinfectant than to risk heating a fully charged (contaminated) inoculating loop.

Containers

There is a risk of contamination whenever a sterile bottle, flask or test tube is opened. One method that reduces the chance of airborne contamination is quickly to pass the open mouth of the glass vessel through a flame. This destroys any microbes on the outer surfaces nearest to the mouth of the vessel. In addition, by heating the air within the neck of the vessel, an outwardly-directed air flow is established, reducing the likelihood of microbial contamination from the air.

It is general practice to flame the mouth of each vessel immediately after opening and then repeat the procedure just before replacing the top. Caps, lids and cotton wool plugs must not be placed on the bench during flaming and sampling: they should be removed and held using the smallest finger of one hand, to minimize the risk of contamination. This also leaves the remaining fingers free to carry out other experimental manipulations. After suitable practice, it is possible to remove the tops from two tubes at one time, flaming each tube and transferring material from one to the other while holding one top in each hand.

Laminar flow cabinets

These are designed to prevent airborne contamination, e.g. when preparing media or subculturing microbes or tissue cultures. Sterile air is produced by passage through a high efficiency particulate air (HEPA) filter: this is then directed over the working area, either horizontally (towards the operator) or downwards. The operator handles specimens, media, etc., through an opening at the front of the cabinet. Standard laminar flow cabinets do *not* protect the worker from contamination and must not be used with pathogenic microbes: special microbiological safety cabinets are used in clinical laboratories for work with ACDP hazard group 3 and 4 microbes (Table 26.1).

Microbiological hazards

The most obvious risks when handling microbial cultures are those due to ingestion or entry via a cut in the skin — all cuts should be covered with a plaster or disposable plastic gloves. A less obvious source of hazard is the formation of aerosols of liquid droplets from microbial suspensions, with the risk of inhalation, or surface contamination of other objects. The following steps will minimize the risk of aerosol formation:

- Use stoppered tubes when shaking, centrifuging or mixing microbial suspensions.
- Pour solutions gently, keeping the difference in height to a minimum.
- Discharge pipettes onto the side of the container.

Other general rules which apply in all microbiology laboratories include:

- Take care with sharp instruments, including needles and glass Pasteur pipettes.
- Do not pour cultures down the sink — waste cell suspensions should be autoclaved.
- Put other contaminated items (e.g. slides, pipettes) into disinfectant after use.
- Wipe down your bench with disinfectant when practical work is complete.
- Always wash your hands before leaving the laboratory.

Table 26.1 Classification of microbes on the basis of hazard. The following categories are recommended by the UK Advisory Committee on Dangerous Pathogens (ACDP)

Hazard group	Comments
1	Unlikely to cause human disease
2	May cause disease: possible hazard to laboratory workers, minimal hazard to community
3	May cause severe disease: may be a serious hazard to laboratory workers, may spread to community
4	Causes severe disease: is a serious hazard to laboratory workers, high risk to community

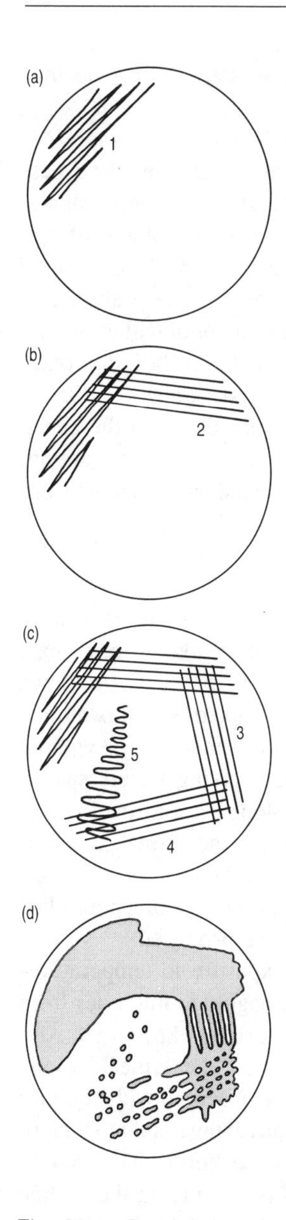

Fig. 26.1 Preparation of a streak plate for single colonies. (a) Using a sterile metal loop, take a small sample of the material to be streaked. Distribute the sample over a small sector of the plate (area 1), then flame the loop and allow to cool (approximately 8–10 s). (b) Make several small streaks from the initial sector into the adjacent sector (area 2), taking care not to allow the streaks to overlap. Flame the loop and allow to cool. (c) Repeat the procedure for areas 3 and 4, re-sterilizing the loop between each step. Finally, make a single, long streak, as shown for area 5. (d) The expected result after incubation at the appropriate temperature (e.g. 37 °C for 24 h): each step should have diluted the inoculum, giving individual colonies within one or more sectors on the plate. Further sub-culture of an individual colony should give a pure (clonal) culture.

Plating methods

Many culture methods make use of a solidified medium within a Petri plate. A variety of techniques can be used to transfer and distribute the organisms prior to incubation. The three most important procedures are described below.

Streak dilution plate

Streaking a plate for single colonies is one of the most important basic skills of sterile technique, since it is used in the initial isolation of a cell culture and in maintaining stock cultures, where a streak dilution plate with single colonies all of the same type confirms the purity of the strain. A sterile inoculating loop is used to streak the organisms over the surface of the medium, thereby diluting the sample (Fig. 26.1). The aim is to achieve single colonies at some point on the plate: ideally, such colonies are derived from single cells (e.g. in the case of unicellular bacteria, animal and plant cell lines) or from groups of cells of the same species (in filamentous or colonial forms). Single colonies, containing cells of a single species and derived from a single parental cell, form the basis of all pure culture methods (p. 125).

Note the following:

- Keep the lid of the Petri plate as close to the base as possible to reduce the risk of aerial contamination.
- Allow the loop to glide over the surface of the medium. Hold the handle at the balance point (near the centre) and use light, sweeping movements, as the agar surface is easily damaged and torn.
- Work quickly, but carefully. Do not breathe directly onto the exposed agar surface and replace the lid as soon as possible.

Spread plate

This method is used with cells in suspension, either in a liquid growth medium or in an appropriate sterile diluent. It is one method of quantifying the number

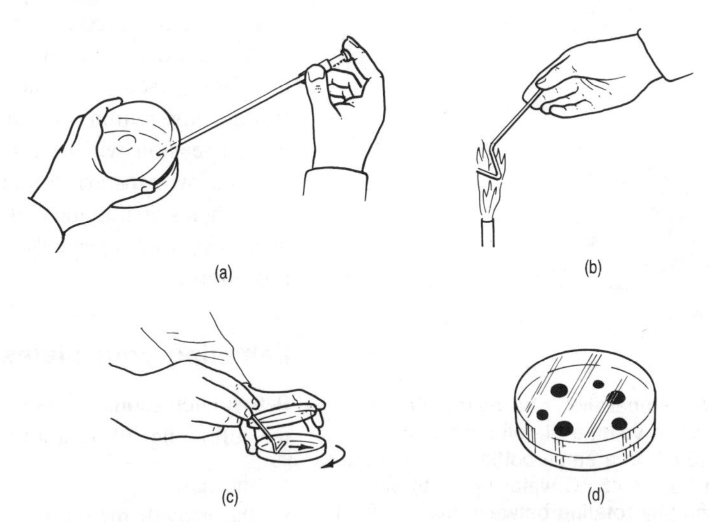

Fig. 26.2 Preparation of a spread plate. (a) Transfer a small volume of cell suspension (0.05–0.5 ml) to the surface of a solidified medium in a Petri plate. (b) Flame sterilize a glass spreader and allow to cool (8–10 s). (c) Distribute the liquid over the surface of the plate using the sterile spreader. Make sure of an even coverage by rotating the plate as you spread: allow the liquid to be absorbed into the agar medium. Incubate under suitable conditions. (d) After incubation, the microbial colonies should be distributed over the surface of the plate.

of viable cells (or colony forming units) in a sample, after appropriate decimal dilution (p. 19).

An L-shaped glass spreader is sterilized by dipping the end of the spreader in a beaker containing a small amount of 70% v/v alcohol, allowing the excess to drain from the spreader and then igniting the remainder in a Bunsen flame. After cooling, the spreader is used to distribute a known volume of cell suspension across the plate (Fig. 26.2). *There is a significant fire risk associated with this technique*, so take care not to ignite the alcohol in the beaker, e.g. by returning an overheated glass rod to the beaker. The alcohol will burn with a pale blue flame that may be difficult to see, but will readily ignite other materials (e.g. a laboratory coat). Another source of risk comes from small droplets of flaming alcohol shed by an overloaded spreader onto the bench and this is why you *must* drain excess alcohol from the spreader *before* flaming. Some laboratories now provide plastic disposable spreaders for student use to avoid the risk of fire.

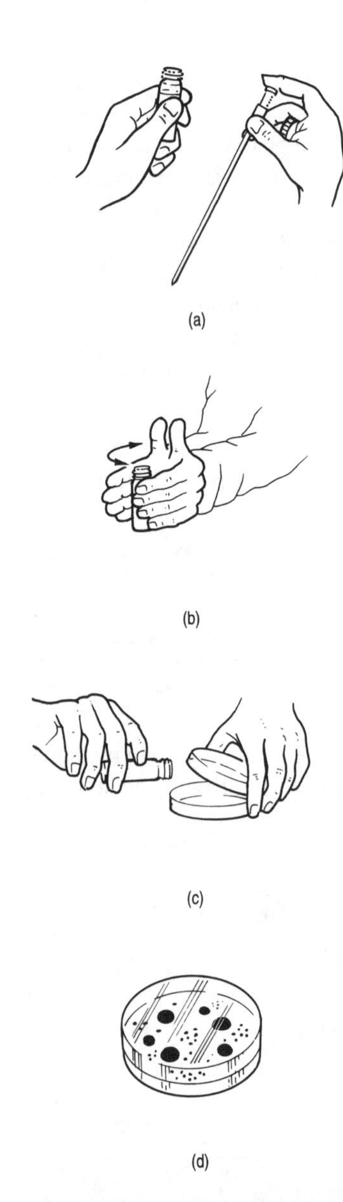

Fig. 26.3 Preparation of a pour plate. (a) Add a known volume of cell suspension (0.05–1.0 ml) to a small bottle of molten agar medium from a 45 °C water bath. (b) Mix thoroughly, by rotating between the palms of the hands: do not shake or this will cause frothing of the medium. (c) Pour the mixture into an empty, sterile Petri plate and allow to set. Incubate under suitable conditions. (d) After incubation, the microbial colonies will be distributed throughout the medium: any cells deposited at the surface will give larger, spreading colonies.

Pour plate

This procedure also uses cells in suspension, but requires molten agar medium, usually in screw-capped bottles containing sufficient medium to prepare a single Petri plate (i.e. 15–20 ml), maintained in a water bath at 45–50°C. A known volume of cell suspension (typically 0.05–1.0 ml) is mixed with this molten agar, distributing the cells throughout the medium. This is then poured without delay into an empty sterile Petri plate and incubated, giving widely spaced colonies (Fig. 26.3). Furthermore, as most of the colonies are formed within the medium, they are far smaller than those of the surface streak method, allowing higher cell numbers to be counted (e.g. up to 1 000 colonies per plate): some workers pour a thin layer of molten agar onto the surface of a pour plate after it has set, to ensure that no surface colonies are produced.

Most bacteria and fungi are not killed by brief exposure to temperatures of 45–50°C, though the procedure may be more damaging to microbes from low temperature conditions, e.g. psychrophilic bacteria (p. 86).

One disadvantage of the pour plate method is that the typical colony morphology seen in surface grown cultures will not be observed with colonies that develop within the agar medium. A further disadvantage is that some of the suspension will be left behind in the screw-capped bottle. This can be avoided by transferring the suspension to the Petri plate, adding the molten agar, then swirling the plate to mix the two liquids. However, even when the plate is swirled repeatedly, the liquids are not mixed as evenly as in the former procedure.

Labelling your plates

Petri plates should always be labelled on the base, rather than the lid, with the following information:

- the date,
- the growth medium,
- your name or initials,
- brief details of the experimental treatment.

Restrict your labelling to the outermost region of the plate, to avoid problems when counting colonies, assessing growth, etc. After labelling, Petri plates usually are incubated upside-down in a temperature-controlled incubator (often at 37°C) for an appropriate period (usually 24–72 h).

27 Cell culture

Microbial, animal and plant cell culture methods are based on the same general principles, requiring:

- a pure (axenic) culture, perhaps isolated as part of an earlier procedure, or from a culture collection;
- a suitable nutrient medium to provide the necessary components for growth. This medium must be sterilized before use (p. 120);
- satisfactory growth conditions including temperature, pH, atmospheric requirements, ionic and osmotic conditions;
- sterile technique to maintain the culture in pure form.

Precise details are often organism-specific and you should consult appropriate manuals (e.g. Freshney, 1983, for animal cells; Dodds and Robert, 1982, for plant cells; and Collins *et al.*, 1989, for microbes).

Choice of medium

Heterotrophic organisms

Animal cells, fungi and many bacteria require appropriate organic compounds as sources of carbon and energy. Non-exacting bacteria can utilize a wide range of compounds and they are often grown in media containing complex natural substances (including meat extract, yeast extract, soil, blood). Animal cells have more stringent growth requirements and are cultured in media containing serum, or in complex synthetic media, to provide the necessary amino acids, growth factors, attachment factors and vitamins.

Photoautotrophic organisms

Photosynthetic bacteria, cyanobacteria and algae are grown in a mineral medium containing inorganic ions including chelated iron, with a light source and CO_2 supply. Plant cells require additional vitamins and hormones.

Growth on solidified media

Harvesting bacteria from a plate — colonies can be harvested using a sterile loop, providing large numbers of cells without the need for centrifugation. The cells are relatively free from components of the growth medium; this is useful if the medium contains substances which interfere with subsequent procedures.

Subculturing — when subculturing microbes from a colony on an agar medium, take your sample from the growing edge, so that viable cells are transferred.

Many organisms can be cultured on agar-based media (p. 121). An important benefit is that an individual cell inoculated onto the surface will develop to form a visible colony: this is the basis of most microbial isolation and purification methods, including the streak dilution, spread plate and pour plate procedures (p. 123).

Plant tissue cultures can be established by growing sterile tissue slices on an appropriate medium, to give a callus of undifferentiated cells. For many plants, such cultures can be established from a broad range of tissue types. Callus may be induced to differentiate, forming organs on a solidified medium containing appropriate plant hormones: in many cases, these cultures will develop to form plants. Anther cultures will give rise to haploid cells and haploid plants — often useful for experimental genetics and breeding.

Animal tissues can be digested enzymically (e.g. using trypsin) or mechanically (using gentle homogenization) to produce single cells, termed a primary cell line. These are often grown as an adherent monolayer on the surface of a glass or plastic culture vessel, rather than an agar-based medium.

Several types of culture vessel are used:

- Petri plates (Petri dishes): usually the pre-sterilized, disposable plastic type, providing a large surface area for growth.
- Glass bottles or test tubes: these provide sufficient depth of agar medium for prolonged growth of bacterial and fungal cultures, avoiding problems of dehydration and salt crystallization. Inoculate aerobes on the surface and anaerobes by stabbing down the centre, into the base (stab culture).
- Flat-sided bottles: these are used for animal cell culture, to provide an increased surface area for attachment and alter growth of cells as a surface monolayer. Usually plastic and disposable.

The dynamics of growth are usually studied in liquid culture, apart from certain rapidly growing filamentous fungi, where increases in colony diameter can be measured accurately using Vernier calipers (p. 20).

Growth in liquid media

Many cells, apart from primary cultures of animal cells, can be grown as a homogeneous unicellular suspension in a suitable liquid medium, where growth is usually considered in terms of cell number (population growth). Most liquid culture systems need agitation, to ensure adequate mixing and to keep the cells in suspension. An Erlenmeyer flask (100–2 000 ml capacity) can be used to grow a batch culture on an orbital shaker, operating at 20–250 cycles per minute. For aerobic organisms, the surface area of such a culture should be as large as possible: restrict the volume of medium to not more than 20% of the flask volume. Larger cultures may need to be gassed with sterile air and mixed using a magnetic stirrer. The simplest method of air sterilization is filtration, using glass wool, non-absorbent cotton wool or a commercial filter unit of appropriate pore size (usually 0.2 μm). Air is introduced via a sparger (a glass tube with many small holes, so that small bubbles are produced) near the bottom of the culture vessel. More complex fermenters have baffles and paddles to further improve mixing and gas exchange.

Protoplasts are bacterial, fungal or plant cells whose walls have been removed, usually enzymically. Partial digestion yields sphaeroplasts (protoplasts with adherent wall fragments). Protoplasts and sphaeroplasts are spherical and osmotically sensitive, bursting in media with a low solute concentration. They can be used for gene transfer and in physiological studies, using techniques developed for single-celled microbes. On prolonged incubation in a suitable medium, protoplasts and sphaeroplasts may regenerate their cell walls, to grow as single-celled cultures. Sometimes, protoplasts of different species can be fused to produce novel 'hybrid' organisms.

Cultivation methods may be sub-divided under two broad headings:

Batch culture

This is the most common system for routine liquid culture. Cells are inoculated into a sterile vessel containing a fixed amount of growth medium: your choice of vessel will depend upon the volume of culture required. Larger-scale vessels (e.g. 1 litre and above) are often called fermenters or bioreactors, particularly in biotechnology. Growth within the vessel usually follows a predictable S-shaped (sinusoidal) curve (Fig. 27.1), divided into four components:

1. Lag phase: the initial period when no increase in cell number is seen. The larger the inoculum of active cells the shorter the lag phase will be, provided the cells are transferred from similar growth conditions.

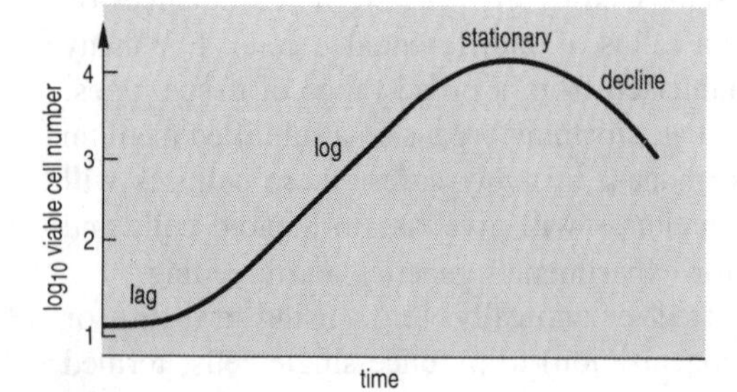

Fig. 27.1 Population growth curve for cells in batch culture (liquid medium).

2. Log phase, or exponential phase: where cells are growing at their maximum rate. This may be quantified by the growth rate constant, or specific growth rate (μ), where:

$$\mu = \frac{2.303\ (\log N_x - \log N_0)}{(t_x - t_0)} \qquad [27.1]$$

where N_0 is the initial number of cells at time t_0 and N_x is the number of cells at time t_x. For times specified in hours, μ is expressed as h^{-1}.
Prokaryotes grow by binary fission while eukaryotes grow by mitotic cell division; in both cases each cell divides to give two identical offspring. Consequently, the doubling time or generation time (g, or T_2) is:

$$g = \frac{0.301\ (t_x - t_0)}{\log N_x - \log N_0} \qquad [27.2]$$

Cells grow at different rates, with doubling times ranging from under 20 min for some bacteria to 24 h or more for animal and plant cells. Exponential phase cells are often used in laboratory experiments, since growth and metabolism are nearly uniform.
3. Stationary phase: growth decreases as nutrients are depleted and waste products accumulate. Any increase in cell number is offset by death.
4. Decline phase, or death phase: this is the result of prolonged starvation and toxicity, unless the cells are sub-cultured. Like growth, death often shows an exponential relationship with time, which can be characterized by a rate constant (death rate), equivalent to that used to express growth or, more often, as the decimal reduction time (d, or T_{90}), the time required to reduce the population by 90%:

$$d = \frac{t_x - t_0}{\log N_0 - \log N_x} \qquad [27.3]$$

Some cells undergo rapid autolysis at the end of the stationary period while others show a slower decline.

Batch culture methods are used to maintain stocks of particular organisms; cells are sub-cultured onto fresh medium before they enter the decline phase. However, primary cultures of animal cells have a finite life unless transformed to give a continuous cell line, capable of indefinite growth.

Continuous culture

This is a method of maintaining cells in exponential growth for an extended period by continuously adding fresh growth medium to a culture vessel of fixed capacity. The new medium replaces nutrients and displaces some of the culture, diluting the remaining cells and allowing further growth.

After inoculating the vessel, the culture is allowed to grow for a short time as a batch culture, until a suitable population size is reached. Then medium is pumped into the vessel: the system is usually set up so that any increase in cell number due to growth will be offset by an equivalent loss due to dilution, i.e. the cell number within the vessel is maintained at a steady state. The cells will be growing at a particular rate (μ), counterbalanced by dilution at an equivalent rate (D), where:

$$D = \frac{\text{flow rate}}{\text{vessel volume}} \qquad [27.4]$$

In a chemostat, the growth rate is limited by the availability of some nutrient in the inflowing medium, usually either carbon or nitrogen (see Fig. 15.2). In a turbidostat, the input of medium is controlled by the turbidity of the culture, measured using a photocell. A turbidostat is more complex than a chemostat, with additional equipment and controls.

To determine the specific growth rate (μ) of a continuous culture:

1. Measure the flow of medium through the vessel over a known time interval (e.g. connect a sterile measuring cylinder or similar volumetric device to the outlet), to calculate the flow rate.
2. Divide the flow rate by the vessel volume (eqn [27.4]) to give the dilution rate (D).
3. This equals the specific growth rate, since $D = \mu$ at steady state.
4. If you want to know the doubling time (g), calculate using the relationship:

$$g = \frac{0.693}{\mu} \quad [27.5]$$

(Note that eqn [27.5] also applies to exponential phase cells in batch culture and is useful for interconverting g and μ.)

Continuous culture systems are more complex to set up than batch cultures. They are prone to contamination, having additional vessels for fresh medium and waste culture: strict aseptic technique is necessary when the medium reservoirs are replaced, and during sampling and harvesting. However, they offer several advantages over batch cultures, including the following:

- The physiological state of the cells is better, since actively growing cells at the same stage of growth are provided over an extended time period. This is useful for biochemical and physiological studies.
- Monitoring and control can be automated and computerized.
- Modelling can be achieved for biotechnology and fermentation technology.

Measuring growth in cell cultures

Most methods of measuring growth are based on cell number.

Key point
Growth may be measured as:
- **cell number,**
- **mass (biomass),**
- **turbidity or absorbance,**
- **a cellular component (e.g. protein, ATP).**

Direct microscopic counts

One of the simplest methods is to count the cells in a known volume of medium using a microscope and a counting chamber or haemocytometer (Box 27.1). While this gives a rapid assessment of the total cell number, it does not discriminate between living and dead cells. It is also time-consuming as a large number of cells must be counted for accurate measurement. It may be difficult to distinguish individual cells, e.g. for cells growing as clumps.

Electronic counters

These instruments can be used to give a direct (total) count of a suspension of microbial cells. The Coulter® counter detects particles due to change in electrical resistance when they pass through a small aperture in a glass tube. It gives a rapid count based on a larger number of cells than direct microscopy. It is well suited for repeat measurements or large sample numbers and can be linked to a microcomputer for data processing. A major limitation of electronic counters is the lack of discrimination between living cells, dead cells, cell clumps and inanimate particles (e.g. dust). In addition, the instrument must be set up and calibrated by trained personnel.

Box 27.1 How to use a counting chamber or haemocytometer

A counting chamber is a specially designed slide containing a chamber of known depth with a grid etched onto its lower surface. When a flat coverslip is placed over the chamber, the depth is uniform. Use as follows:

1. **Place the special coverslip over the chamber**. Press the edges firmly, to ensure that the coverslip makes contact with the surface of the slide, but take care that you do not break the slide or coverslip by using too much force. When correctly positioned, you should be able to see interference rings (Newton's rings) at the edge of the coverslip.
2. **Add a small amount of your cell suspension to fill the central space above the grid**. Place on the microscope stage and allow the cells to settle (2–3 min).
3. **Examine the grid microscopically**, using the ×10 objective lens first, since the counting chamber is far thicker than a standard microscope slide. Then switch to the ×40 objective: take care not to scratch the surface of the objective lens, as the special coverslip is thicker than a normal coverslip. For a dense culture, the small squares are used, while the larger squares are used for dilute suspensions. You may need to dilute your suspension if it contains more than thirty cells per small square.
4. **Count the number of cells in several squares**: at least 600 cells should be counted for accurate measurements. Include those cells that cross the upper and right-hand boundaries, but not those that cross the lower or left-hand rulings. A hand tally may be used to aid counting. Motile cells must be immobilized prior to counting (e.g. by killing with glutaraldehyde).
5. **Divide the total number of cells (*C*) by the number of squares counted (*S*)**, to give the mean cell count per square.
6. **Determine the volume (in ml) of liquid corresponding to a single square (*V*)**, e.g. a Petroff–Hausser chamber has small squares of linear dimension 0.2 mm, giving an area of 0.04 mm^2; since the depth of the chamber is 0.02 mm, the volume is $0.04 \times 0.02 = 0.000\,8$ mm^3; as there are 1 000 mm^3 in 1 ml, the volume of a small square is 8×10^{-7} ml; similarly, the volume of a large square (equal to 25 small squares) is 2×10^{-5} ml. Note that other types of counting chamber will have different volumes: check the manufacturer's instructions.
7. **Calculate the cell number per ml by dividing the mean cell count per square by the volume of a single square**.
8. **Remember to take account of any dilution of your original suspension** in your final calculation by multiplying by the reciprocal of the dilution (*M*), e.g. if you counted a one in twenty dilution of your sample, multiply by twenty, or if you diluted to 10^{-5}, multiply by 10^5.

The complete equation for calculating the total microscopic count is:

$$\text{Total cell count (per ml)} = (C / S / V) \times M \quad [27.6]$$

e.g. if the mean cell count for a hundred-fold dilution of a cell suspension, counted using a Petroff–Hausser chamber, was 12.4 cells in ten small squares, the total count would be

$$(12.4 / 10 / 8 \times 10^{-7}) \times 10^2 = 1.55 \times 10^8 \text{ ml}^{-1}.$$

Definition
c.f.u. — colony-forming unit: a cell or group of cells giving rise to a single colony on a solidified medium.

Viable counts

A variety of techniques have been developed for determining the number of viable bacterial cells:

- Spread or pour plate methods. The most widespread approach is to transfer a suitable amount of the sample to an agar medium, incubate under appropriate conditions and then count the resulting colonies (Box 27.2).
- Membrane filtration. For bacterial samples where the expected cell number is lower than 100 c.f.u. ml^{-1}, pass the sample through a sterile filter (pore size 0.2 μm). The filter is then incubated on a suitable medium until colonies are produced, giving a viable count by dividing the mean colony count per filter by the volume of sample filtered.
- Most probable number (MPN). A bacteriological technique. The sample is diluted and known volumes are transferred to several tubes of liquid medium (typically, five tubes at three volumes), chosen so that there is a low probability of the smallest volumes containing a viable bacterial cell. After incubation, the number of tubes showing bacterial growth (turbidity) is compared to tabulated values to give the most probable number.

Box 27.2 How to make a viable count of bacteria using an agar medium

1. **Prepare serial decimal dilutions of the sample in a sterile diluent (p. 19).** The most widely used diluents are 0.1% w/v peptone water or 0.9% w/v NaCl, buffered at pH 7.3. Take care that you mix each dilution before making the next one. For soil, food, or other solid samples, make the initial decimal dilution by taking 1 g of sample and making this up to 10 ml using a suitable diluent. Gentle shaking or homogenization may be required for organisms growing in clumps. The number of decimal dilutions required for a particular sample will be governed by your expected viable count: dilute until the expected number of viable cells is around 100–1 000 ml^{-1}.
2. **Transfer an appropriate volume (0.05–0.5 ml) of the lowest dilution to an agar plate** using either the spread plate method or the pour plate procedure (p. 123). At least two, and preferably more, replicate plates should be prepared for each sample. You may also wish to prepare plates for more than one dilution, if you are unsure of the expected number of viable cells.
3. **Incubate under suitable conditions for 24–72 h, then count the number of colonies on each replicate plate at the most appropriate dilution.** The most accurate results will be obtained for plates containing 30–300 colonies. Mark the base of the plate with a spirit-based pen each time you count a colony. Determine the mean colony count per plate at this dilution (C).
4. **Calculate the colony count per ml of that particular dilution** by dividing by the volume (in ml) of liquid transferred to each plate (V).
5. **Now calculate the viable count per ml of the original sample** by multiplying by the reciprocal of the dilution: this is the multiplication factor (M); e.g. for a dilution of 10^{-3}, the multiplication factor would be 10^3. For soil, food or other solid samples, the viable count should be expressed per g of sample.

The complete equation for calculating the viable count is:

$$\text{Viable count per ml (or per g)} = (C / V) \times M \qquad [27.7]$$

e.g. for a sample with a mean colony count of 5.5 colonies per plate for a volume of 0.05 ml at a dilution of 10^{-7}, the viable count would be:

$$(5.5 / 0.05) \times 10^7 = 1.1 \times 10^9 \text{ c.f.u. ml}^{-1}$$

Strictly speaking, the count should be reported as colony forming units (c.f.u.) per ml, rather than as cells per ml, since a colony may be the product of more than one cell, particularly in filamentous microbes or in organisms with a tendency to aggregate.

- Vital/mortal staining. Some stains can be used to distinguish between living and dead cells. The direct epifluorescence technique (DEFT) uses acridine orange and UV epi-fluorescence microscopy to separate living and dead bacteria, while neutral red is a vital stain used for plant cells (p. 117).

The principal advantage of viable counting procedures is that dead cells will not be counted. However, for the cultural techniques, the incubation conditions and media used may not allow growth of all cells, underestimating the viable count. Further problems are caused by cell clumping and dilution errors. In addition, such methods require sterile apparatus and media and the incubation period is lengthy before results are obtained.

Observing specimens

28 Setting up and using a light microscope

The light microscope is probably the most important instrument used in biology practicals and its correct use is one of the basic and essential skills of biology. A standard undergraduate binocular microscope (Fig. 28.1) consists of three main types of optical unit: eyepiece, objective and condenser. These are attached to a stand which holds the specimen on a stage. A monocular microscope is constructed similarly but has one eyepiece lens rather than two.

Setting up a binocular light microscope

Before using any microscope, familiarize yourself with its component parts. Never assume that the previous person to use your microscope has left it set up correctly: apart from differences in users' eyes, the microscope needs to be properly set up for each lens combination used.

The procedures outlined below are simplified to allow you to set up microscopes like those of the Olympus CH series (Fig. 28.1) in minimum time. For monocular microscopes, disregard instructions for adjusting eyepiece lenses in (5).

1. Place the microscope at a convenient position on the bench. Adjust your seating so that you are comfortable operating the focus and stage controls. Unwind the power cable, plug in and switch on after first ensuring that the lamp setting is at a minimum. Adjust the lamp setting to about two-thirds of the maximum.
2. Select a low-power (e.g. ×10) objective. Make sure that the lens clicks home.
3. Set the eyepiece lenses to your interpupillary distance; this can usually be read off a scale on the turret. You should now see a single circular field of vision. If you do not, try adjusting a small amount in either direction.
4. Put a prepared slide on the stage. Examine it first against a light source and note the position, colour and rough size of the specimen. Place the slide on the stage (coverslip up!) and, viewing from the side, position it with the stage adjustment controls so that the specimen is illuminated.
5. Focus the image of the specimen using first the coarse and then the fine focusing controls. The image will be reversed and upside-down compared to that seen by viewing the slide directly.
   (a) If both eyepiece lenses are adjustable, set your interpupillary distance on the scale on each lens. Close your left eye, look through the right eyepiece with your right eye and focus the image with the normal controls. Now close your right eye, look through the left eyepiece with your left eye and focus the image by rotating the eyepiece holder. Take a note of the setting for future use.
   (b) If only the left eyepiece is adjustable, close your left eye, look with the right eye through the static right eyepiece and focus the image with the normal controls. Now close your right eye, look through the left eyepiece with your left eye and focus the image by rotating the eyepiece holder. Take a note of the setting for future use.

Using binocular eyepieces — if you do not know your interpupillary distance, ask someone to measure it with a ruler. You should stare at a fixed point in the distance while the measurement is taken. Take a note of the value for future use.

Problems with spectacles — those who wear glasses can remove them for viewing as microscope adjustments will accommodate most deficiencies in eyesight (except astigmatism). This is more comfortable and stops the spectacle lenses being scratched by the eyepiece holders. However, it creates difficulties in focusing when drawing diagrams.

Adjusting a microscope with a field–iris diaphragm — adjust this *before* the condenser–iris diaphragm: close it until its image appears in view, if necessary focusing its image with the condenser controls and centring it. Now open it so the whole field is just illuminated.

High-power objectives — never remove a slide while a high power objective lens (i.e. ×40 or ×100) is in position. Always turn back to the ×10 first. Having done this, lower the stage and remove the slide.

6. Close the condenser–iris diaphragm (aperture–iris diaphragm), then open it to a position such that further opening has no effect on the brightness of the image (the 'threshold of darkening'). The edge of the diaphragm should not be in view. Turn down the lamp if it is too bright.
7. Focus the condenser. Place an opaque pointed object (the tip of a mounted needle or a sharp pencil point) on the centre of the light source. Adjust the condenser setting until both the specimen and needle tip/pencil point are in focus together. Check that the condenser–iris diaphragm is just outside the field of view.
8. For higher magnifications, swing in the relevant objective (e.g. ×40), carefully checking that there is space for it. Adjust the focus using the fine control only. If the object you wish to view is in the centre of the field with the ×10 objective, it should remain in view (magnified, of course) with the ×40. Adjust the condenser–iris diaphragm and condenser as before — the correct setting for each lens will be different.
9. When you have finished using the microscope, remove the last slide and clean the stage if necessary. Turn down the lamp setting to its minimum, then switch off. Clean the eyepiece lenses with lens tissue. Check that the objectives are clean. Unplug the microscope from the mains and wind the cable round the stand and under the stage. Replace the dust cover.

If you have problems in obtaining a satisfactory image, refer to Box 28.1; if this doesn't help, refer the problem to the class supervisor.

Box 28.1 Problems in light microscopy and possible solutions

- No image; very dark image; image dark and illuminated irregularly
  - Microscope not switched on (check plug and base)
  - Illumination control at low setting or off
  - Objective nosepiece not clicked into place over a lens
  - Diaphragm closed down too much or off centre
  - Lamp failure
- Image visible and focused but pale and indistinct
  - Diaphragm needs to be closed down further
  - Condenser requires adjustment
- Image blurred and cannot be focused
  - Dirty objective
  - Dirty slide
  - Slide in upside-down
  - Slide not completely flat on stage
  - Eyepiece lenses not set up properly for user's eyes
  - Fine focus at end of travel
- Dust and dirt in field of view
  - Eyepiece lenses dirty
  - Slide dirty
  - Dirt on lamp glass or upper condenser lens

Procedure for observing transparent specimens

Some stained preparations and all colourless objects are difficult to see when the microscope is adjusted as above. Contrast can be improved by closing down the condenser–iris diaphragm. Note that when you do this, diffraction haloes appear round the edges of objects. These obscure the image of the true structure of the specimen and may result in loss of resolution. Nevertheless, an image with increased contrast may be easier to interpret.

Fig. 28.1 Diagram of the Olympus binocular microscope model CH.

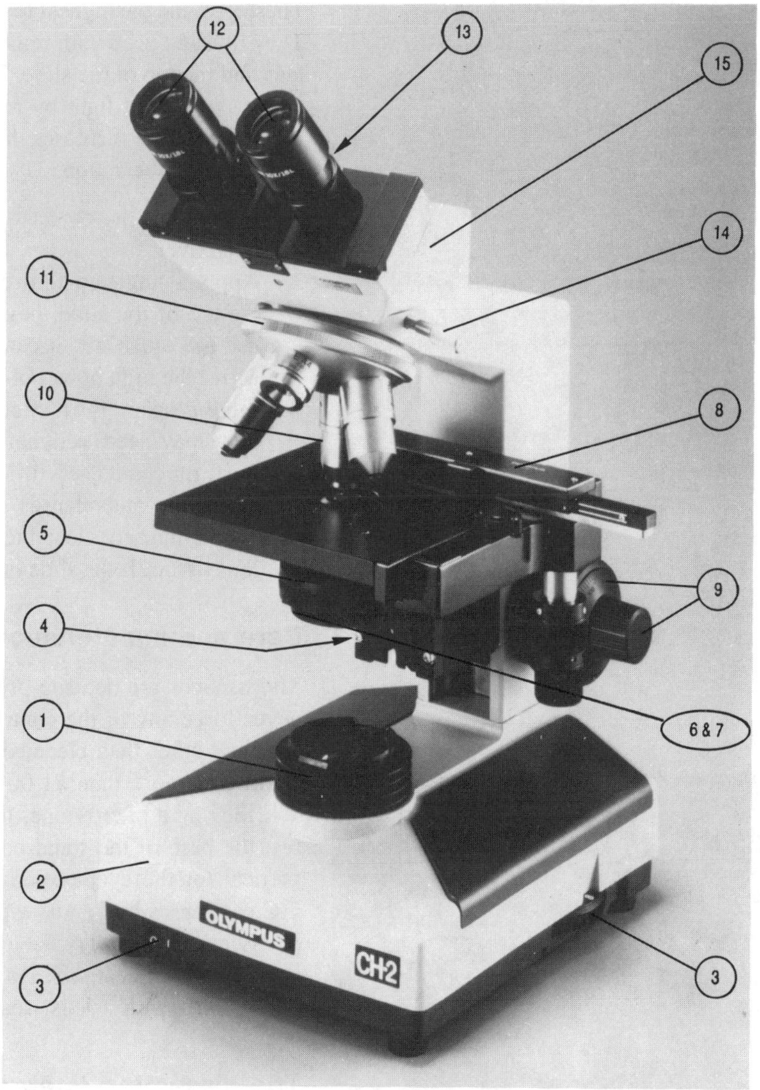

- The lamp (1) in the base of the stand (2) supplies light; its brightness is controlled by an on–off switch and voltage control (3). Never use maximum voltage or the life of the bulb will be reduced — a setting two-thirds to three-quarters of maximum should be adequate for most specimens. A field-iris diaphragm may be fitted close to the lamp to control the area of illumination (not present on this model).
- The condenser control (4) focuses light from the condenser lens system (5) onto the specimen and projects the specimen's image onto the front lens of the objective. Correctly used, it ensures optimal resolution.
- The condenser–iris diaphragm (6) controls the amount of light entering and leaving the condenser; its aperture can be adjusted using the condenser–iris diaphragm lever (7). Use this to reduce glare and enhance image contrast by cutting down the amount of stray light reaching the objective lens.
- The specimen (normally mounted on a slide) is fitted to a mechanical stage or slide holder (8) using a spring mechanism. Two controls allow you to move the slide in *x* and *y* planes. Vernier scales (see p. 20) on the slide holder can be used to return to the same place on a slide. The fine and coarse focus controls (9) adjust the height of the stage relative to the lens systems. Take care when adjusting the focus controls to avoid hitting the lenses with the stage or slide.
- The objective lens (10) supplies the initial magnified image; it is the most important component of any microscope because its qualities determine resolution, depth of field and optical aberrations. The objective lenses are attached to a revolving nosepiece (11). Take care not to jam the longer lenses onto the stage or slide as you rotate the nosepiece. You should feel a distinct click as each lens is moved into position. The magnification of each objective is written on its side; a normal complement would be $\times 4$, $\times 10$, $\times 40$ and $\times 100$ (oil immersion).
- The eyepiece lens (12) is used to further magnify the image from the objective and to put it in a form and position suitable for viewing. Its magnification is written on the holder (normally $\times 10$). By twisting the holder for one or both of the eyepiece lenses you can adjust their relative heights to take account of optical differences between your eyes. The interpupillary distance scale (13) and adjustment knob allow compensation to be made for differences in the distance between users' pupils.
- The turret clamping screw (14) allows the eyepiece turret (15) to be rotated so a demonstrator can view your specimen without exchanging position with you. If loosened too much, the turret can come off, so take care and always re-tighten after use.

Procedure for oil immersion objectives

These provide the highest resolution of which the light microscope is capable. They must be used with immersion oil filling the space between the objective lens and the top of the slide. The oil has the same refractive index as the glass lenses, so loss of light by reflection and refraction at the glass/air interface is reduced. This increases the resolution, brightness and clarity of the image and reduces aberration. Use oil immersion objective(s) as follows:

1. Check that the object of interest is in the field of view using the ×40 objective.
2. Apply a single small droplet of immersion oil to the illuminated spot on the top of the slide, having first swung the ×40 objective away. Never use too much oil: it can run off the slide and mess up the microscope.
3. Move the high power objective into position carefully, checking first that there is space for it. Focus on the specimen using the fine control only. You may need a higher lamp setting.
4. Perform condenser–iris diaphragm and condenser focusing adjustments as for the other lenses.
5. When finished, clean the oil immersion lens by gently wiping it with clean lens tissue. If the slide is a prepared one, wipe the oil off with lens tissue.

Care and maintenance of your microscope

Microscopes are delicate precision instruments. Handle them with care and never force any of the controls. Never touch any of the glass surfaces with anything other than clean, dry lens tissue. Bear in mind that a replacement could cost more than £1 000!

If moving a microscope, hold the stand above the stage with one hand and rest the base of the stand on your other hand. Always keep the microscope vertical (or the eyepieces may fall out). Put the microscope down gently.

Clean lenses by gently wiping with clean, dry, lens tissue. Use each piece of tissue once only. Try not to touch lenses with your fingers as oily fingerprints are difficult to clean off. Do not allow any solvent (including water) to come into contact with a lens; sea water is particularly damaging.

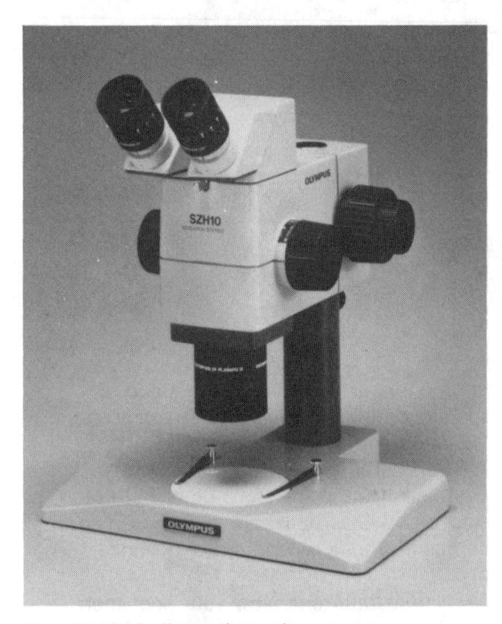

Fig. 28.2 A dissecting microscope.

Measuring specimens using a dissecting microscope — because of the low magnification, sizes can generally be estimated by comparison with a ruler placed alongside the specimen, but if more accurate measurements are required, eyepiece graticules can be used.

The dissecting (stereoscopic) microscope

The dissecting microscope (Fig. 28.2) is a form of stereoscopic microscope used for observations at low total magnification (×4 to ×50) where a large working distance between objectives and stage is required, perhaps because the specimen is not flat or dissecting instruments are to be used. A stereoscopic microscope essentially consists of two separate lens systems, one for each eye. Some instruments incorporate zoom objectives. The eyepiece–objective combinations are inclined at about 15° to each other and the brain resolves the compound image in three dimensions as it does for normal vision. The image is right side up and not reversed, which is ideal for dissections. Specimens are often viewed in a fresh state and need not be placed on a slide — they might be in a Petri dish or on a white tile. Illumination can be from above or below the specimen, as desired.

Most of the instructions for the binocular microscope given above apply equally well to dissecting microscopes, although the latter do not normally have adjustable condensers or diaphragms. With stereoscopic microscopes, make specially sure to adjust the eyepiece lenses to suit your eyes so that you can take full advantage of the stereoscopic effect.

29 Interpreting microscope images

Microscope images, whether viewed directly or as photomicrographs, need to be interpreted with extreme care. This is not only because of the potentially damaging procedures involved in specimen preparation, but also because scale, section orientation, and staining combine to create images that can be misinterpreted by the human mind.

Establishing scale and measuring objects

Working out the length of objects on photomicrographs

On a micrograph, the scale will usually be provided as a magnification factor (e.g. $\times 500$) or in the form of a bar of defined length (e.g. 100 nm). If you need to estimate the dimensions of an object in the micrograph, follow the steps below:

1. Measure the object as it appears on the micrograph with a ruler or set of Vernier calipers (see p. 20). It may be difficult to decide exactly where the boundary of a structure lies; rather than a discrete boundary, you may be dealing with shades of grey. It is essential to be consistent!
2. If the scale is given as a bar, measure the bar too and find the object's size by proportion. For example, if the object measures 32 mm and a bar representing 100 nm is 20 mm long, then the size of the object is $32/20 \times 100$ nm $= 160$ nm.
3. If the scale is given as a magnification factor, divide your measurement by this number to obtain the object's size, taking care to enter the correct units in your calculator. For example, if the object measures 32 mm and the print magnification is stated as $\times 200\,000$, then its size is 32×10^{-3} m$/200\,000 = 1.6 \times 10^{-7}$ m $= 160$ nm.

Spurious accuracy — avoid putting too many significant figures in any estimates of dimensions: there may be quite large errors in estimating print magnifications which could make the implied accuracy meaningless (p. 39).

Adding linear scales to drawings

The magnification of a light microscope image is calculated by multiplying the objective magnification by the eyepiece magnification. However, the magnification of the image bears no certain relation to the magnification of any drawing of the image — you may equally well choose to draw the same image 10 mm or 10 cm long. For this reason, *it is essential to add a scale to all your diagrams*. You can provide either a bar giving the estimated size of an object of interest, or a bar of defined length (e.g. 100 μm).

The simplest method of estimating linear dimensions is to compare the size of the image to the diameter of the field of view. You can make a rough estimate of the field diameter by focusing on the millimetre scale of a transparent ruler using the lowest power objective. Estimate the diameter of this field directly, then use the information to work out the field diameters at the higher powers pro rata. For example, if the field at an overall magnification of $\times 40$ is 4 mm, at an overall magnification of $\times 100$ it will be: $40/100 \times 4$ mm $= 1.6$ mm (1 600 μm).

Using an eyepiece graticule — choose the eyepiece lens corresponding to your stronger eye (usually the same side as the hand you write with) and check that you have made the correct adjustments to the eyepiece lenses as detailed on p. 133.

Greater accuracy can be obtained if an eyepiece micrometer (graticule) is used. This carries a fine scale and fits inside an eyepiece lens. The eyepiece micrometer is calibrated using a stage micrometer, basically a slide with a fine scale on it. Say you find at $\times 400$ overall magnification that 1 stage micrometer unit of 0.1 mm $= 39$ eyepiece micrometer units. Each eyepiece micrometer unit = 0.1/39 mm = 2.56 μm. The eyepiece scale can now be used to measure objects: in the above example, the scale reading is multiplied by 2.56 to give a value in micrometres. Alternatively, you could put a 100 μm bar on your diagram corresponding to the length of 39 eyepiece micrometer units on your specimen.

Estimating the area and area density of objects

You can calculate the area of the field at each magnification using a field diameter estimate obtained as above (obtain the area from πr^2, but don't forget that radius r = diameter/2). The units you use to measure the diameter will be squared; thus, a field diameter of 0.45 mm converts to a field area of 0.159 mm^2. Use the field area to estimate the cross-sectional area of a specimen by proportion (e.g. an object occupying one-third of the above field would have a cross-sectional area of 0.159/3 = 0.053 mm^2). If several objects appear in a field, you can express their frequency on an area basis by dividing the number seen by the field area (e.g seven objects in the above field would be present at a density of 7/0.159 = 44 objects mm^{-2}). It is nearly always best to take an average from several fields of view before calculating.

Volume–density estimates

The density of particles in a liquid suspension can be *roughly* estimated as follows: place a drop of well mixed suspension onto a slide with a Pasteur pipette. You can estimate the drop volume by weighing drops of water and assuming 1 mg = 1 mm^3. Alternatively, use a pipettor (see p. 14). Quickly cover the drop with a coverslip (see p. 119), avoiding air bubbles — a 22 × 22 mm coverslip should allow a 30 mm^3 drop to spread out completely. Now count the numbers of the object of interest within each of several fields and calculate a mean value. The volume of each field will be:

$$\text{(drop volume} \times \text{field area)/coverslip area} \qquad [29.1]$$

and the density will be:

$$\text{(mean no. of objects per field)/field volume} \qquad [29.2]$$

Where accurate estimates of volume–density are required, you should use a haemocytometer (see p. 129).

For a field area of 0.159 mm^2, a droplet volume of 30 mm^3 and a coverslip of area 484 mm^2, the field volume would be 30 × (0.159/484) = 0.01 mm^3, so if you saw 25 objects per field, their density would be 25/0.01 = 2 500 mm^{-3}.

Determining the position and orientation of the section within the specimen

When interpreting microscope images, you need to determine and take into account what position and orientation the section was taken from. The terminology used for sections taken in different planes is shown in Figs 29.1 and 29.2.

Always remember that a section is a two-dimensional slice through a three-dimensional structure. Thus, a circular profile in two dimensions could arise from sectioning numerous three-dimensional shapes (Fig. 29.3); you cannot determine which shape without further information (e.g. from serial sections or sections taken in another plane). Likewise, the size of a profile in a section may reveal little about the size of the structure from which it arose. Also bear in mind that very thin sections (say 75 nm thick) may not provide a representative sample of all the organelles present in a cell (which may be 1 000–100 000 nm in width).

Always relate what you see to what you know; at a minimum, you should take account of your basic knowledge of animal, microbial or plant cell structure. Note that you cannot be sure in the absence of a defining characteristic about the nature of the specimen because it may simply be chance that the section does not include the feature.

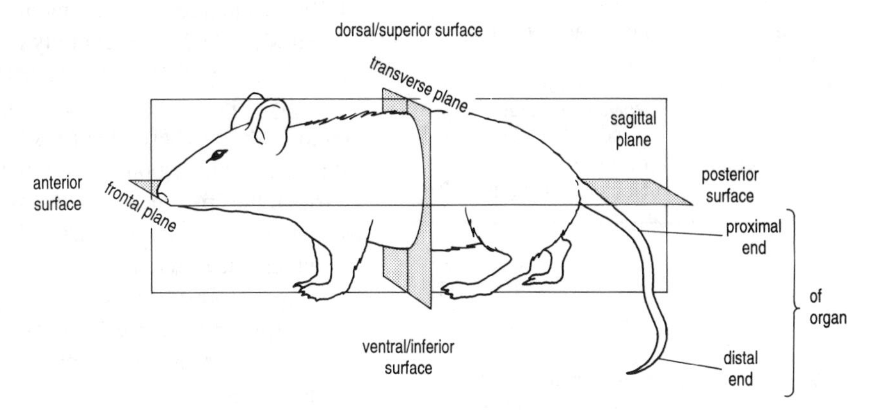

Fig. 29.1 The terms used for various parts and sections through the animal body.

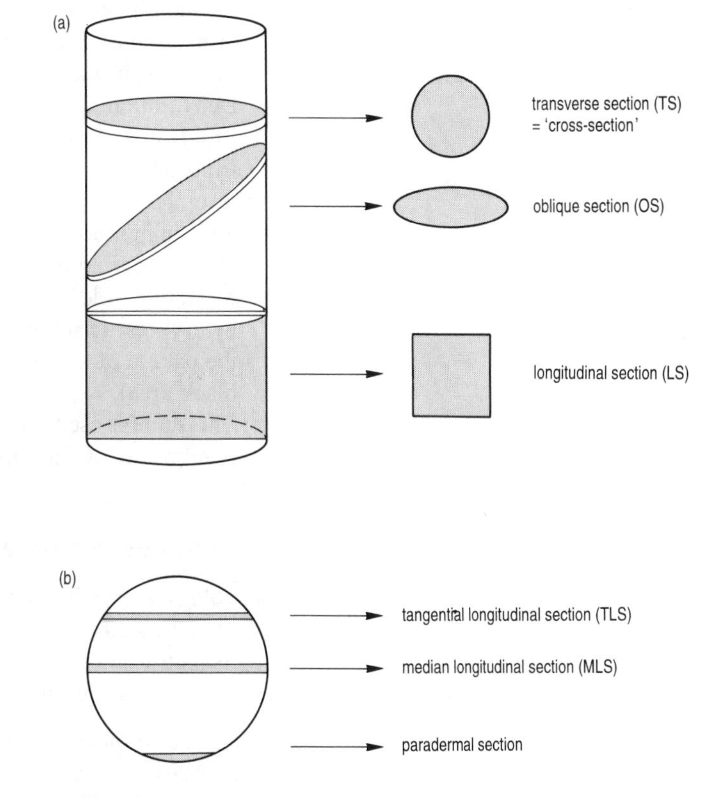

Fig. 29.2 The main section planes: (a) sections through a cylindrical object; (b) various types of longitudinal sections as seen in transverse orientation.

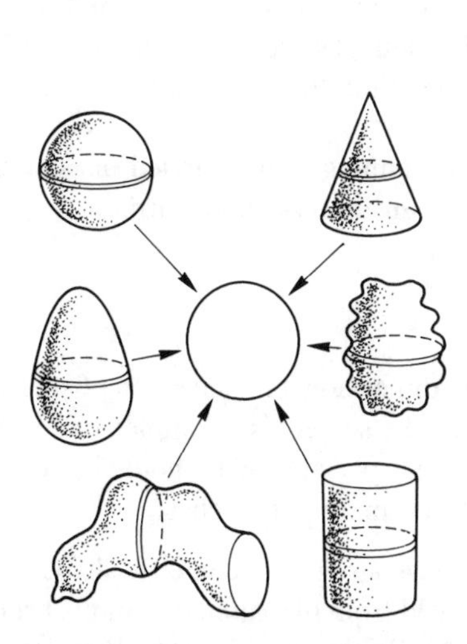

Fig. 29.3 A two-dimensional section may be representative of many different three-dimensional structures.

Table 29.1 The main stains used in electron microscopy

Stain	What it stains
Lead citrate	Lipoproteins Ribonucleoproteins Glycogen
Osmium tetroxide	Lipids and proteins
Phosphotungstic acid	Proteins (at pH >3) Polysaccharides, glycoproteins (at pH <3)
Uranyl acetate	Proteins Nucleic acids (at pH 3.5)

Comparing methods — it is informative to compare the same tissue prepared and viewed by different methods.

Accounting for methodology

As discussed on pp. 113–19, a great deal of preparative work is required before specimens can be viewed under LM or TEM. Under normal circumstances, it is hoped that the stain has the greatest effect on the image. The colours produced in various cell types and structures by the main stains used in light microscopy are tabulated on pp. 117–18. You should bear in mind that colours are subjective and that dyes may exhibit different colours in different chemical environments (e.g. at different pH values).

Stains for TEM are generally compounds or ions containing elements of high atomic number, such as the heavy metals, because these scatter electrons strongly. On a positive micrograph from a positively stained specimen, dark regions show where electrons have been scattered, i.e. where the stain has attached. Light regions show where electrons have passed relatively unaffected through the specimen, i.e. where the stain has not been taken up. Table 29.1 shows the main stains used in TEM and indicates the specific cell components to which they attach.

Images from SEM are relatively easy to interpret because the brain is used to interpreting surface features. However, it is possible for optical illusions to occur — for instance, it may be difficult to decide whether a shadowed feature projects out of a surface or represents a depression in it.

Artefacts

No preparative procedure gives a 'perfect' image. Artefacts can be introduced at the following stages:

- Fixation: swelling or shrinkage may occur, often due to the use of hypotonic or hypertonic fixative solutions. Glutaraldehyde-fixed tissue is particularly likely to shrink if this fixative is used at high concentration.
- Dehydration: shrinkage may occur as water is withdrawn from the tissue. Lipids may dissolve in the solvent used and be lost from the tissue.
- Embedding: difficulties in sectioning may be caused by incorrect hardness of embedding resins.
- Sectioning: too thick a section can result in a blurred image. Compression effects, tearing and knife chattering may occur.
- Mounting: folds may be introduced into the section. These can be detected by the presence of discontinuities in features which cross the fold. In TEM, the pattern of the metal grid may sometimes obscure the image (as a solid black area).
- The stain(s) used on the section: if staining is not carried out correctly, precipitates of stain may appear, often in crystalline form.

Stereological studies

Stereology allows the study of the three-dimensional organization of specimens based on two-dimensional information. The subject is a complex one, relying on geometric and statistical principles, and specialist texts should be consulted for detailed background. There are three main approaches:

- Cells or organelles are assumed to have a regular geometric shape, whose surface area or volume can be estimated from measurements of the principal axes (see Table 44.3). Cell and organelle shapes are generally so irregular as to make such estimates very approximate.
- A physical or computer-generated model of structures in three dimensions

is built up from a set of serial sections. Considerable expertise is required to obtain the basic information required.

- Representative sections are sampled and information derived from the profiles seen. This method can be based on simple count data and is surprisingly informative. Three parameters of interest are:

(a) Numerical density, N_V (the numbers of a component found per unit volume). For spherical components, this can be estimated from:

$$N_V = N_A/\bar{D} \qquad [29.3]$$

where N_A is the number of components per unit area and $\bar{D}$ is their mean diameter. N_A is simply obtained by counting component profiles in defined areas. The mean diameter (of spheres) will be approximately equal to the largest diameters seen in the sample, or may be calculated from $\bar{D} = 4/\pi\bar{d}$, where $\bar{d}$ is the mean component diameter of the sample. The situation for non-spherical components is more complex.

(b) Surface density, S_V (the surface area of the component per unit containing volume). For any shape of object, this can be estimated from:

$$S_V = 4/\pi B_A \qquad [29.4]$$

where B_A is the length of profile boundaries per unit area of section. Profile boundary lengths can be measured by fitting a thread or by using a map-measuring wheel or a planimeter. An even simpler method is to count the number of intersections of the bounding membrane I_i with test lines of total length L_T on an overlay (Fig. 29.4). Then S_V is obtained from:

$$S_V = 2I_i/L_T \qquad [29.5]$$

Note that this method depends on the components being randomly orientated with respect to the lines on the overlay; this can be tested by comparing results with the lines placed in different orientations.

(c) Volume density, V_V (the volume of the component as a proportion of the total volume). This is numerically equivalent to A_A, the component density on the test area. A_A can be most simply estimated:

(i) Destructively, by the cut-and-weigh method. Here, the total area to be considered is cut out, weighed, then the component profiles are cut out and weighed: the ratio of weights equals A_A.

(ii) Non-destructively, by point counting, where a patterned overlay or lattice (e.g. Fig. 29.4) is placed over the micrograph and counts made of the number of points overlaying the feature of interest and the total area. A_A is estimated by the ratio of these point counts.

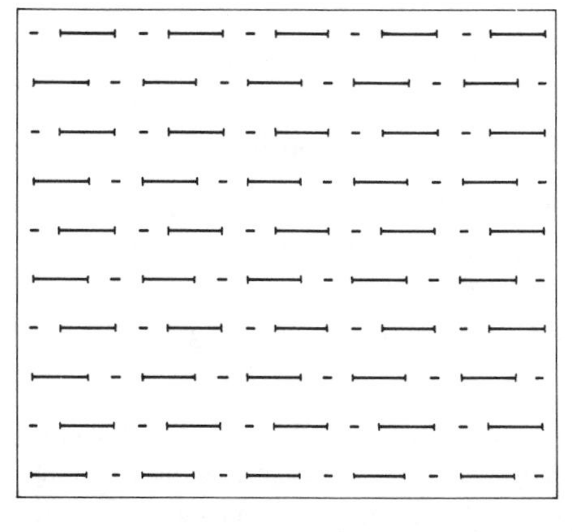

Fig. 29.4 Weibel's multipurpose test overlay for stereological studies. The number of intersections of a membrane with the solid lines can be used when estimating surface densities, while point counts for estimates of profile density on test areas can be made using the ends of the lines.

Both of these methods of estimating A_A require that a reasonably large representative area is sampled. Other methods for estimating A_A include planimetry and linear integration.

30 Photography and photographs

Photography is a valuable technique used primarily for:

- documentation and validation of observations for projects or publications;
- preparation of material for illustrating talks;
- preparation of poster displays;
- producing 'hard copy' of microscopical observations which need to be examined in greater detail and/or interpreted; for example, for making area measurements (pp. 20 and 138).

As a student, you may be able to request professional technical assistance for some or all of this type of work but there is an important role for you in the selection of the precise requirements: this will be improved by a basic understanding of the key factors which affect the final product.

Your photography will usually be carried out in one of three general situations:

- laboratory or studio environments where conditions are under your control and problems relatively easily overcome;
- field situations where many factors may be difficult to control;
- using specialized equipment such as photo-microscopes or electron microscopes, where the scope for your control of the photographic process is very limited and largely managed by the instrument's operating system.

Successful photography depends upon an appropriate choice of equipment and materials for your particular requirements.

Types of photographic film

There are four important decisions to make when choosing a film:

- what size of film you require in terms of both negative size and number of negatives on the film — this depends upon the camera system used;
- whether to use colour or black and white film;
- whether to use transparency (slide or reversal) film or negative film;
- what speed of film to use.

Definitions
Saturation — the term used to describe the intensity of a colour — a saturated colour is an intense colour.
Contrast — the degree of gradation between colours or the number of grey shades in black and white film: the higher the contrast, the sharper is the gradation.

Film is classified by its sensitivity to light (its 'speed') and this is measured in either ASA or DIN units: do not confuse them. Black and white films and colour negative films use an emulsion which contains crystals and, therefore, have 'grain'. Table 30.1 summarizes the relationship between speed, grain and definition for such films. Colour transparency films, although classified using the same speed criteria, are based upon colour dyes and are effectively grainless. Slow film is used when fine detail and/or saturated colour is required, such as in microphotography, or when considerable enlargement may be

Table 30.1 The speed and grain relationship in film

Speed of film	ASA number	DIN number	Grain	Definition
Slow	25–64	15–19	Very fine	Very sharp
Medium	64–200	19–24	Fine	Sharp
Fast	200–400	24–27	Medium	Medium
Extra fast	4000–1600	27–32	Coarse	Poor to medium

required. Use fast film when light levels are low, but remember that this results in reduction of contrast and detail. Films faster than 200 ASA (24 DIN) are not recommended for use except in exceptional circumstances.

The choice of film is often a compromise between speed and detail and a good general choice is film of about 100–160 ASA (21–23 DIN). There are many subtle differences between different makes of films, particularly with respect to colour balance, saturation and contrast, and often the choice of film is a matter of personal preference. Slow films have inherently more contrast than fast films.

Black and white films

All the usual types of modern films are panchromatic, i.e. sensitive to ultraviolet light and all the colours of the visible spectrum. Special films are available which respond only to selected wavelengths (such as infra-red or X-ray) or are orthochromatic, responding only to blue, green and yellow but not to red light; such films are often used in copying and graphics work.

Using colour film — colour film is highly sensitive to environmental factors such as heat and humidity which cause changes in film speed and colour rendition. Make sure that your film has a sufficiently long expiry date. Store film for extended periods in a refrigerator or freezer: if stored in a freezer, allow a 24 h thawing period before use.

Colour films

The main classification is into positive (reversal) film and negative film: the former is used to produce slides (transparencies) and the latter to produce colour prints. Prints can be produced from transparencies and slides from negatives but the former is the better process. Most negative films are colour masked giving them an overall orange tint when developed — this makes the colours purer when printed. Colour films must be balanced for the colour temperature of the light source used. Colour temperature is measured in kelvin (K; see Fig. 30.1). There are two main types of film: (a) daylight film is balanced for daylight conditions (5 400 K) and for electronic flash; and (b) artificial light types A (3 400 K) or B (3 200 K), designed for studio lighting conditions. Filters must be used if colour temperature corrections are necessary.

Key point
To use a camera properly, you should understand the relationship between aperture (f-number), shutter speed and depth of field. Use bracketing of exposures to ensure good results.

Type of lighting

The quantity and quality of the light is a critical factor in all photography except in electron microscopy. The quantity of light is measured by a photographic light meter which may be external or built-in. The more light there is, the smaller the lens aperture you can use (larger 'f' number), and the greater will be the depth of field (= depth of focus). Therefore, the more light available, the better! By using the camera on a tripod, you can use slow shutter speeds and allow larger 'f' numbers to be used to maximize depth of focus. The use of an electronic flash system to provide some or all of the lighting makes this even easier since the effective shutter speed with electronic flash is extremely short (1 ms). Modern electronic flash systems are computerized making them easy to use.

Adjusting the lens aperture — *shallow* depth of field is sometimes required to restrict the area of sharp focus; use small f numbers to achieve this.

Tips for better photography —

Use your camera on a tripod whenever possible, and use high shutter speed to minimize the effect of movements.

Use electronic flash wherever possible as it provides uniformity of colour balance. Unless you want shadowless lighting, do not place the flash on the camera (hot-shoe) connection.

Shadowless lighting of smaller objects is usefully obtained through a ring-flash system.

The quality of light refers to the colour balance of the film being used. Even black and white film has different sensitivity to different parts of the visible spectrum: this might be important under artificial light conditions when the spectrum can be different from that of sunlight. Always be sure to know the quality of light required for your film.

Remember that your choice of lighting arrangement will affect the quality of the picture: shadowless lighting is appropriate in some situations, but often shadows help to give three-dimensional form to the objects. Give some thought to this before taking your picture. In general, the use of more than one light source is advisable to prevent hard shadows.

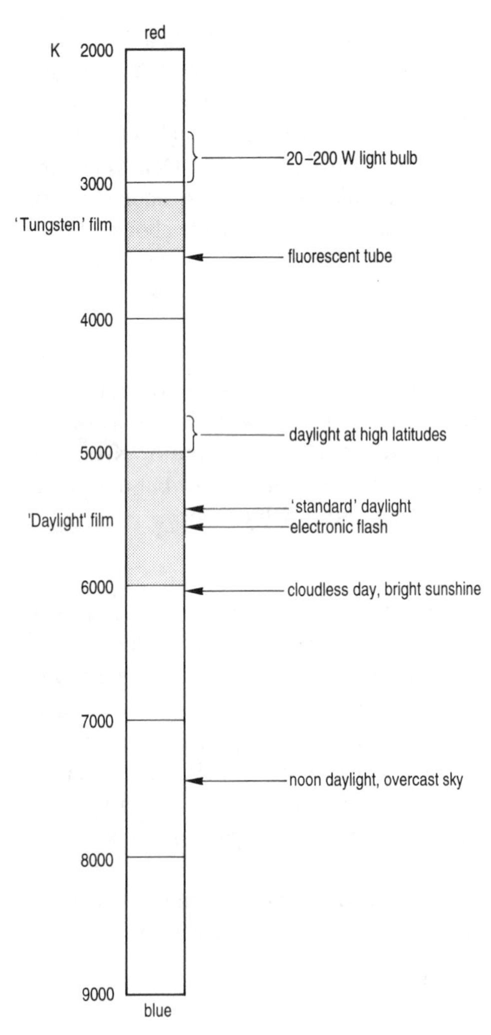

Fig. 30.1 Colour temperature and the Kelvin scale.

Optimizing print quality — remember that the quality of your negative is vital: you cannot get a good print from a poor quality negative!

Film development (black and white)

Remember that the image quality is influenced not only by the conditions under which the film was exposed, but also by development conditions: the main factors include the type of developer used, the temperature and the duration of development. It is possible, therefore, to modify development when it is known that a film has been exposed under less than ideal light conditions and thus maximize the quality of the negative produced. If this is necessary, tell the photographer exactly what the conditions of exposure were so that adjustment of the development process can be considered. Do this only with the help of a professional photographer, however, as it is just as easy to ruin a film as it is to improve it!

Photomicrography

This requires the use of special equipment mounted on a microscope. Consult the manual(s) for the particular system you are using since most of the operations will be semi-automatic. The important operations for successful photomicrography are outlined below:

1. Carefully prepare the object: ensure thorough cleanliness of any slides or coverslips used.
2. Choose the correct film type and any filters required for alteration of colour balance, depending upon the type of light source available.
3. Decide on the magnification to be used: make sure that you know how this relates to the magnification of the negative/transparency.
4. Carefully focus the object onto the film plane: there will be a system-specific method for doing this but sometimes it requires practice to get it right.
5. Make extra exposures above and below the one indicated (called 'bracketing'), even when using an automatic exposure system: exposures of at least +1 and −1 stop are recommended, especially for colour photography.
6. Include a photograph of a stage micrometer so that the final magnification can be calculated and shown on the photograph.

Type of print

The type of print used depends upon its purpose. Glossy finish prints generally appear sharper than matt or other finishes and are usually required for publication: the addition of lettering and scales is often only possible on the smooth surface of a glossy print. However, if preparing prints for display on a poster, the matt/velvet finishes are often preferable. The contrast of the image is determined by the choice of paper, which comes in a variety of grades of contrast. If your photo has too much contrast, reprint using a 'softer' grade of paper. You may be able to learn to develop and print your own black and white films but colour printing is particularly difficult and best left to professionals.

Adding scales and labels to photographic prints

Having acquired a suitable print, it is often necessary to add information to it.

- Use transfer letters (p. 6) to add lettering, scales and labels.
- Choose a simple font type that will not distract and a size which is legible

Fig. 30.2 Use of white areas for labelling dark prints, illustrated by an SEM micrograph of a developing aesthete on the shell of a chiton (mollusc).

but not overpowering: prints for publication should be prepared bearing in mind any reductions which may take place during subsequent operations.
- An 18 point label usually works well since published text is usually 10 point.
- Do not mix font types in related sets of prints.
- Choose an appropriate part of the photograph for the lettering, i.e. a black area for a white letter. If this is too variable, add a background label on which to place the letters (see Fig. 30.2).

Storing and mounting photographs

Both slides and negatives should be stored and maintained in good condition and be well organized. Avoid dampness, which is very destructive to all photographic materials: use silica gel desiccant in damp climates/environments. Record-keeping should be done carefully and include all relevant details, both of the subject and of the relevant processing procedures.

- Transparencies (slides) should be mounted in plastic mounts but not between glass, as this often causes more problems than it solves. Beware of cheap plastic filing materials as some contain residual chemicals which can damage transparencies over long time periods. Labelling of transparencies is best done on the mount, using small, self-adhesive labels.
- Negative strips should be stored in transparent or translucent paper filing sheets. Obtain a set of contact prints for each film and store this with the negatives for easy reference.
- Prints should be stored flat in boxes or in albums. When used for display, mount on stiff board either with modern photographic adhesives or dry-mounting tissue.

Using biological systems

31 Measuring the growth and responses of organisms

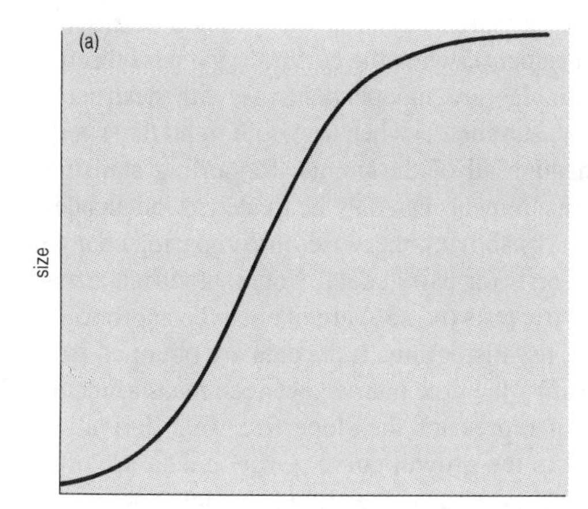

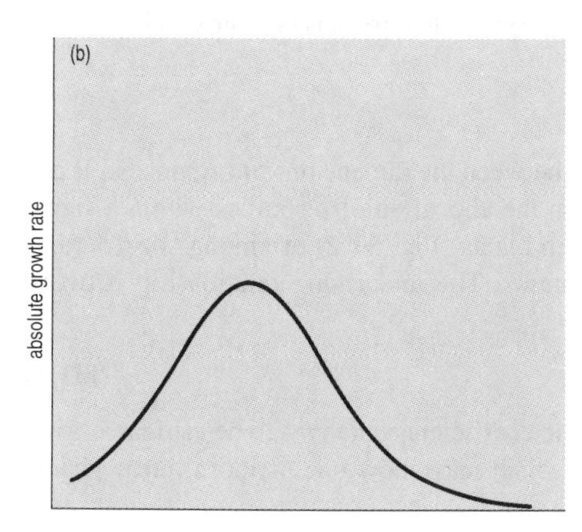

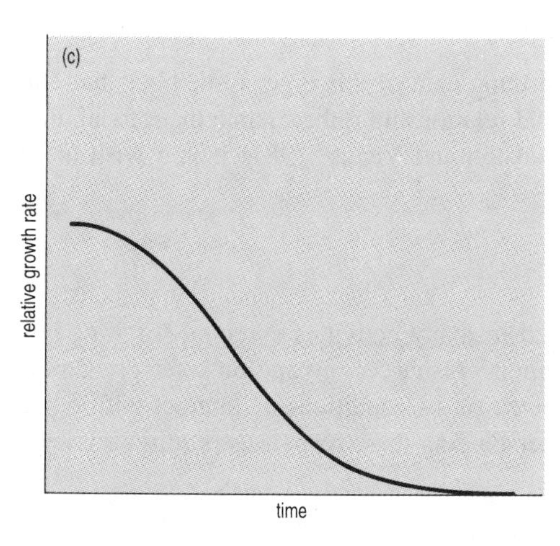

Fig. 31.1 Different ways of analysing growth: (a) simple plot of size against time; (b) plot of absolute growth rate for the same data; (c) plot of relative growth rate for the same data.

The ability of organisms to grow, develop and respond to stimuli is fundamental to life. The aims of studies concerned with these phenomena include:

- investigating developmental processes;
- comparing developmental patterns;
- understanding behavioural responses;
- characterizing physiological processes;
- measuring the levels of a particular compound (bioassays);
- assessing sensitivity to a stimulus.

Growth and development

The growth and development of a multicellular organism is a complex phenomenon which cannot be easily measured by a single variable. An overall picture emerges from changes in size, weight, number, complexity and function evaluated at cell, tissue and organ levels.

Length and area are relatively easy to measure (see p. 20 for techniques) though curved organisms may present difficulties. Volume measurements can be made from the displacement of water employing Archimedes' principle, or derived from fresh weight if the density of the organism can be estimated. Growth may not occur smoothly through time (e.g. plants increase in height mainly at night, when cell turgor is highest). For rapidly growing organisms it is desirable to have a continuous photographic or video record, though this may only be feasible with sedentary organisms. Care is required to ensure that focus is maintained on the points that will be measured, or your data will be inaccurate. If you are interested in growth of different parts of the organism, use reference points, e.g. recognizable morphological features such as hairs or marks made with ink or paint.

Fresh weight might at first sight appear a useful measure of growth: however, water content may vary according to availability and consumption. Dry weight (p. 21) does not have this disadvantage, but it may decrease when respiration dominates metabolism (e.g. during seed germination), despite the fact that growth is obviously occurring. Protein or nitrogen content may be better correlated with 'mass of protoplasm', but they are both complex to measure.

Whichever of the above variables is measured, growth generally shows a sigmoidal increase in time (Fig. 31.1a) which is reminiscent of the population growth curve of cells in batch culture (p. 126). The slope of this curve at any given time represents the absolute growth rate, with units of increment in size per unit time. Plotting this against time generally shows a bell-shaped curve (Fig. 31.1b). The relative growth rate or rate of increase in size per unit of size measures the 'efficiency' of growth (Fig. 31.1c).

Cell number is a common measure of the growth of unicellular suspensions (p. 128). For multicellular organisms, this can be estimated from measurements of average cell size and total volume. Numbers of organs (e.g. leaves), specialized cells (e.g. hairs) or organelles (e.g. mitochondria) can indicate the progress of differentiation and development, but in general these are difficult processes to characterize quantitatively. Changes in numbers can also be expressed as absolute and relative rates, as above.

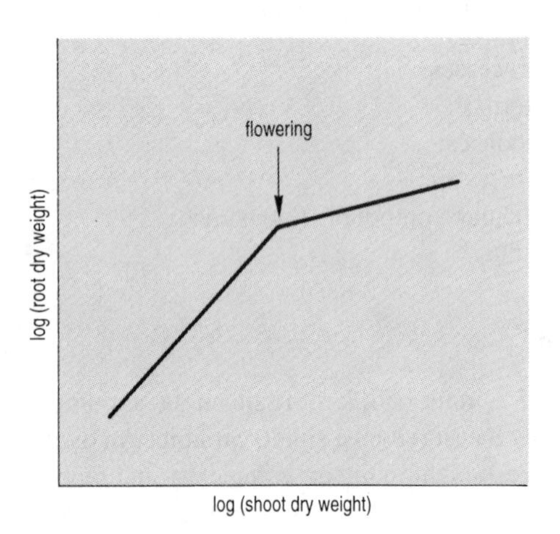

Fig. 31.2 Example showing changes in an allometric relationship based on a published analysis of the growth of a grass. Both root and shoot dry weight increase as the plant ages. The onset of flowering causes a clear change in their relationship.

Analysing allometric data — when fitting eqn [31.2] to allometric data, you should use type II regression as the variables are interdependent (see p. 241). Note that for monophasic log–log relationships, a can be determined from the intercept (log a) and b from the slope (log b).

Display and analysis of growth data

Care should be taken when plotting a growth curve. When plotting change in size against time for a single organism, a smooth curve passing through all measurement points is appropriate (i.e. measurement error is assumed to be negligible). If dealing with a sample of organisms, the approach depends on whether (a) the same sample is measured each time (as with non-destructive harvesting, e.g. for linear measurement), when the curve should pass through each mean point, or (b) the samples are independent (as with destructive harvesting, e.g. for dry weight measurement), when a smooth trend line should be drawn that need not pass through all of the points. Regarding statistical analysis, note that time-series measurements can only be treated as independent if the experimental design is as in (b) above; otherwise, in comparing changes in size with time, you should use tests for paired data. Note that variance may increase with size, so non-parametric tests (p. 236) are likely to be appropriate.

When presenting growth rates, use histograms if the data are obtained from differences in mean values divided by the time interval between measurements. A smooth curve can be used if it represents the slope (i.e. first derivative) of a function that has been fitted to the growth curve. Growth data are often log-normally distributed, so curves should be fitted to logarithmically transformed data. Fitting functions to growth curves is complex and specialist advice should be sought.

Allometry

This is the study of relationships between the dimensions of organisms. It can be used for describing changes in the allocation of resources within a single species (e.g. root-to-shoot ratio in plants, Fig. 31.2) or among species (e.g. brain size in relation to body mass). The underlying relationship between dimensions x and y is generally:

$$y = ax^b \qquad [31.1]$$

where a is a positive constant. The coefficients a and b can be estimated from a log–log plot (e.g. Fig 31.2), which may show one or more linear phases of the form:

$$\log y = \log a + b \log x \qquad [31.2]$$

Care should be taken in interpreting data of this type: remember that eqn [31.1] is an experimentally derived relationship rather than a theoretical one. Consult a specialist text (e.g. Causton and Venus, 1981) if you wish to fit a line or lines to such data.

Responses to stimuli

Responses to stimuli are involved in many activities essential for life. For example, they are required to capture resources, escape or deter predators, move towards or away from a given set of conditions or interact with other individuals in a community. Understanding these responses requires answers to the following questions:

- What is the precise response and what causes it?
- How does the response develop?
- What is the function of the response?
- How might the response have evolved?
- How much of the response is inherited and how much learned?

Definition
Anthropomorphism — attributing human characteristics to animals.

The first stage when investigating a response is to describe it fully (preferably under conditions as near to those in nature as possible) and determine the precise stimuli that cause it. For instance, a study of plant phototropism might start with research into which cells perceive the light stimulus and which respond; effective wavelengths and doses of light might then be investigated. When working with animals, avoid anthropomorphism and describe objectively what happens. All circumstances of potential relevance (e.g. weather, time of day) should be recorded until a clear pattern emerges. A full description of most responses obviously requires physiological and biochemical techniques whose scope is too wide to be covered here.

You may wish to follow an individual animal in space or time and this will require either recognizing its features or marking it in some way that does not interfere with the response (e.g. leg-ringing birds or painting symbols on insects). When recording the elements of a behavioural pattern, a coding system may be helpful (e.g. 1 = flaps wings, 2 = drops neck, etc.). As with growth, photography and video techniques are good aids for recording movements and analysing them later. Tape recorders can be used to record a description of responses so long as your voice does not interfere with the response. They can also record the sound component of a response and might be used later to elicit a response under controlled conditions (e.g. for a suspected alarm cry).

The second question listed above requires a study of the response throughout the developmental stages of the organism. The third and fourth questions require a certain amount of interpretive skills but, as with the fifth question, are amenable to experimentation. You may be presented with a natural experiment if conditions change in the right way in the course of your observations. Experiments in a laboratory give you control over conditions but this is inevitably at the expense of realism (see p. 49).

Distinguishing between innate and learned behaviour — innate behaviour is genetically determined; it is always executed perfectly and in its entirety the first and every subsequent time the stimulus occurs. Learned behaviour is modified by experience.

In designing experiments to investigate animal behaviour, you must take account of the fact that an animal's behaviour may change according to its experience. If not actually investigating learning *per se*, you need to ensure that the opportunity for a subject to learn is the same for each treatment, or bias will occur. Ways round this problem are either to use naïve animals for every replicate or to adopt a Latin square design (p. 54) so that experiences are shared evenly among treatments (fully randomized or randomized block designs may not achieve this).

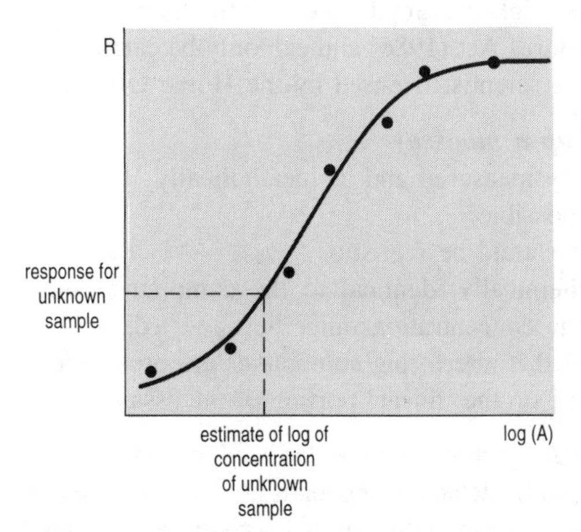

Fig. 31.3 Typical bioassay response curve, showing estimation of an unknown sample. Symbols as noted in text. Closed circles represent responses with standard samples.

Bioassays

A bioassay is a method of quantifying a chemical substance (analyte) by measuring its effect on a biological system under controlled conditions. The hypothetical underlying phenomena are summarized by the relationship:

$$A + Rec \rightleftharpoons ARec \longrightarrow R \qquad [31.3]$$

where A is the analyte, Rec the receptor, ARec the analyte–receptor complex and R the response. This relation is analogous to the formation of product from an enzyme–substrate complex and, using similar mathematical arguments to those of enzyme kinetics, it can be shown that the expected relationship between [A] and rate of response is hyperbolic (sigmoidal in a log–linear plot like Fig. 31.3). This pattern of response is usually observed in practice if a wide enough range of [A] is tested.

To carry out the assay, the response elicited by the unknown sample is compared to the response obtained for differing concentrations of the substance, as shown in Fig. 31.3. When fitting a curve to standard points and estimating unknowns, the available methods, in order of increasing accuracy, are:

1. fitting by eye;
2. using linear regression on a restricted 'linear' portion of the assay curve;
3. linearization followed by regression (e.g. by probit transformation);
4. non-linear regression (e.g. to the Morgan–Mercer–Flodin equation).

In general, bioassay techniques have more potential faults than physico-chemical methods of quantification. These may include the following:

- A greater level of variability: error in the estimate of the unknown compound will result because no two organisms will respond in exactly the same way. Assay curves vary through time, and because they are non-linear, a full standard curve is required each time the assay is carried out.
- Lack of chemical information: bioassays provide information about *biological* activity; they say little about the chemical structure of an unknown compound. The presence of a specific compound should ideally be confirmed with physico-chemical evidence (e.g. mass spectroscopy).
- Possibility of interference: while many bioassays are very specific, it is possible that different chemicals in the extract may influence results.

Despite these problems, bioassays are still much used. They are 'low-tech' and generally cheap to set up. They often allow detection at very low concentrations. Bioassays also provide the means to assess the biological activity of chemicals and to study changes in sensitivity to a chemical, which physico-chemical techniques cannot do. Changes in sensitivity may be evident in the shape of the dose–response curve and its position on the concentration axis.

Bioassays can involve responses of whole organisms or parts of organisms. 'Isolated' responding systems (e.g. excised tissues or cells) decrease the possibility of interference from other parts of the organism. Disadvantages include disruption of nutrition and wound damage during excision. Isolation can continue down to the molecular level, as in immunoassays (p. 153).

Bioassays are the basis for characterizing the efficacy of drugs and the toxicity of chemicals. Here, response is treated as a quantal (all-or-nothing) event. The E_{50} is defined as that concentration of a compound causing 50% of the organisms to respond. Where death is the observed response, the LD_{50} describes the concentration of a chemical that would cause 50% of the test organisms to die within a specified period under a specified set of conditions. Where such experiments involve 'higher' animals, they are tightly controlled by the Animals Scientific Procedures Act (1986) and can only be carried out under the direct supervision of a scientist licensed by the Home Office.

Considerations when setting up a bioassay

- The response should be easily measured and as metabolically 'close' to the initial binding event as possible.
- The experimental conditions should be realistic.
- The standards should be chemically identical to the compound being measured and spread over the concentration range being tested.
- The samples should be purified if interfering compounds are present and diluted so the response will be on the 'linear' portion of the assay curve.

To check for interference, the assay should be standardized against another method (preferably physico-chemical). Related compounds known to be present in the analyte solution should be shown to have minimal activity in the assay. If an interfering compound is present, this may show up if a known amount of standard is added to sample vials — the result will not be the sum of independently determined results for the standard and sample.

32 Immunological assays

> **Definitions**
> **Antibody** — a protein produced in response to an antigen (an antibody-generating foreign macromolecule).
> **Epitope** — a site on the antigen that determines its interaction with a particular antibody.
> **Hapten** — a substance which contains at least one epitope, but is too small to induce antibody formation unless it is linked to a macromolecule.
> **Ligand** — a molecule or chemical group that binds to a particular site on another molecule.

> **Key point**
> **The presence of two antigen-binding sites on a single flexible antibody molecule is relevant to many immunological assays, especially the agglutination and precipitation reactions.**

Antibodies are an important component of the immune system, which protects animals against certain diseases (see Roitt, 1991). They are produced in response to foreign macromolecules (antigens). A particular antibody will bind to a site on a specific antigen, forming an antigen–antibody complex (immune complex). Immunological assays use the specificity of this interaction for:

- identifying macromolecules, cellular components or whole cells;
- quantifying a particular substance.

Antibody structure

An antibody is a complex globular protein, or immunoglobulin (Ig). While there are several types, IgG (Fig. 32.1) is the major soluble antibody in vertebrates and is used in most immunological assays. Its main features are:

- Shape: IgG is a Y-shaped molecule, with two antigen-binding sites.
- Specificity: variation in amino acid composition at the antigen-binding sites explains the specificity of the antigen–antibody interaction.
- Flexibility: each IgG molecule can interact with epitopes which are different distances apart, including those on different antigen molecules.
- Labelling: regions other than the antigen-binding sites can be labelled, e.g. using a radioisotope or enzyme (p. 157).

Antibody production

> Producing polyclonal antibodies — this is controlled by Government regulations, since it involves vertebrate animals: personnel must be licensed by the Home Office and must operate in accordance with the Animals Scientific Procedures Act (1986).

Polyclonal antibodies

These are commonly used at undergraduate level. They are produced by repeated injection of antigen into a laboratory animal. After a suitable period (3–4 weeks) blood is removed and allowed to clot, leaving a liquid phase (polyclonal antiserum) containing many different IgG antibodies, resulting in:

- cross-reaction with other antigens or haptens;
- batch variation, as individual animals produce slightly different antibodies in response to the same antigen;
- non-specificity, as the antiserum will contain many other antibodies.

Standardization of polyclonal antisera therefore is difficult. You may need to assess the amount of cross-reaction, inter-batch variation or non-specific binding using appropriate controls, assayed at the same time as the test samples.

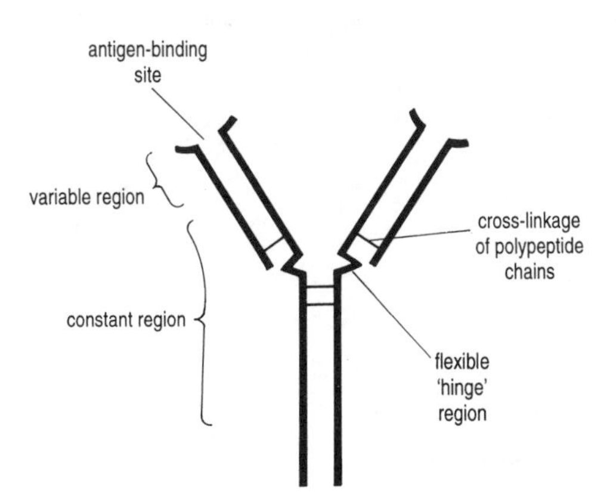

Fig. 32.1 Diagrammatic representation of IgG (antibody).

Monoclonal antibodies

These are specific to a single epitope and are produced from individual clones of cells, grown using cell culture techniques (p. 125). Such cultures provide a stable source of antibodies of known, uniform specificity. While monoclonal antibodies are likely to be used increasingly in future years, polyclonal antisera are currently employed for most routine immunological assays.

Agglutination tests

When antibodies interact with a suspension of a particulate antigen, e.g. cells or latex particles, the formation of immune complexes (Fig. 32.2) causes visible

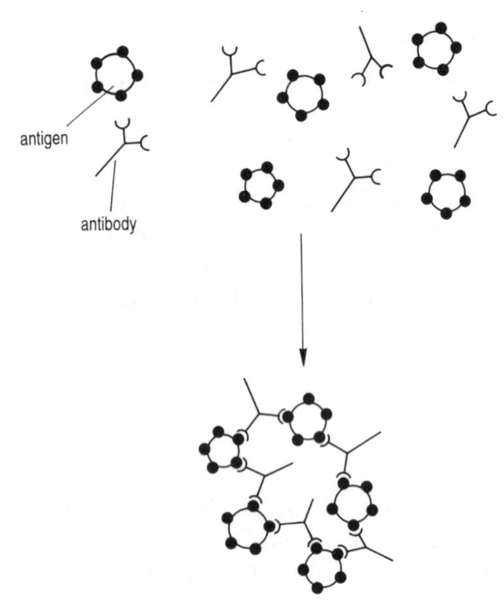

Fig. 32.2 Formation of an antigen–antibody complex.

clumping, termed agglutination. Agglutination tests are used in several ways:

- Microbial identification: at the species or subspecies level (serotyping), e.g. mixing an unknown bacterium with the appropriate antiserum will cause the cells to agglutinate.
- Latex agglutination (bound antigens): by coating soluble (non-particulate) antigens onto microscopic latex spheres, their reaction with a particular antibody can be visualized.
- Latex agglutination (bound antibodies): antibodies can be bound to latex microspheres, leaving their antigen-binding sites free to react with soluble antigen.
- Haemagglutination: red blood cells can be used as agglutinating particles. However, in some instances, such reactions do not involve antibody interactions (e.g. some animal viruses may haemagglutinate unmodified red blood cells).

Precipitin tests

Immune complexes of antibodies and soluble antigens (or haptens) usually settle out of solution as a visible precipitate: this is termed a precipitin test, or precipitation test. The formation of visible immune complexes in agglutination and precipitation reactions only occurs if antibody and antigen are present in an optimal ratio (Fig. 32.3). It is important to appreciate the shape of this curve: cross-linkage is maximal in the zone of equivalence, decreasing if either component is present in excess. The quantitative precipitin test can be used to measure the antibody content of a solution (Clausen (1988) gives details). Visual assessment of precipitation reactions forms the basis of several other techniques, described below.

Immunodiffusion assays

These techniques are easier to perform and interpret than the quantitative precipitin test. Precipitation of antibody and antigen occurs within an agarose

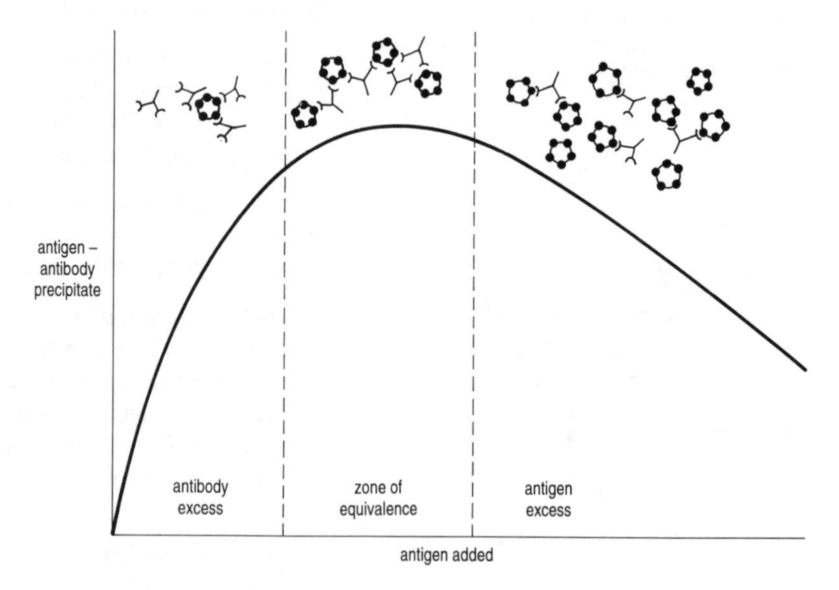

Fig. 32.3 Precipitation curve for an antigen titrated against a fixed amount of antibody

gel, giving a visible line corresponding to the zone of equivalence (Fig. 32.3). The most widespread techniques are detailed below.

Preparing wells for RID — cut your wells carefully. They should have straight sides and the agarose must not be torn or lifted from the glass plate. All wells should be filled to the top, with a flat meniscus, to ensure identical diffusion characteristics. Non-circular precipitin rings, resulting from poor technique, should not be included in your analysis.

Single radial immunodiffusion (RID) (Mancini technique)

This is used to quantify the amount of antigen in a test solution, as follows:

1. Prepare an agarose gel (1.5% w/v), containing a fixed amount of antibody: allow to set on a glass slide or plate, on a level surface.
2. Cut several circular wells in the gel. These should be of a fixed size between 2 and 4 mm in diameter (see Fig. 32.4a).
3. Add a known amount of the appropriate antigen or test solution to each well.
4. Incubate on a level surface at room temperature in a moist chamber: diffusion of antigen into the gel produces a precipitin ring. This is usually measured after 2–7 days, depending on the molecular mass of the antigen.
5. Examine the plates against a black background (with side illumination), or stain using a protein dye (e.g. Coomassie blue).
6. Measure the diameter of the precipitin ring, e.g. using Vernier calipers (p. 20).
7. Prepare a calibration curve from the samples containing known amounts of antigen (Fig. 32.4b): the squared diameter of the precipitin ring is directly proportional to the amount of antigen in the well.
8. Use the calibration curve to quantify the amount of antigen in your test solutions, assayed at the same time.

Drawing RID calibration curves — note the non-zero intercept of the calibration curve (Fig. 32.3b), corresponding to the square of the well diameter: do not force such calibration lines through the origin.

Double diffusion immunoassay (Ouchterlony technique)

This technique is widely used to detect particular antigens in a test solution, or to look for cross-reaction between different antigens.

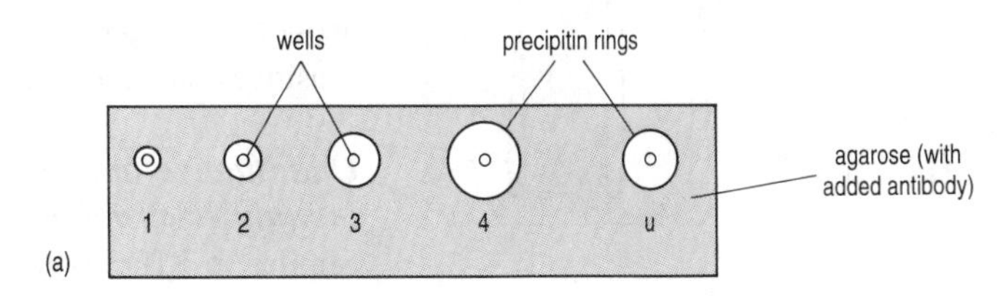

Fig. 32.4 Single radial immunodiffusion (RID). (a) Assay: four standards are shown (wells 1 to 4, each one double the strength of the previous standard), and an unknown (u), run at the same time. (b) Calibration curve. The unknown contains 6.25 μg of antigen.

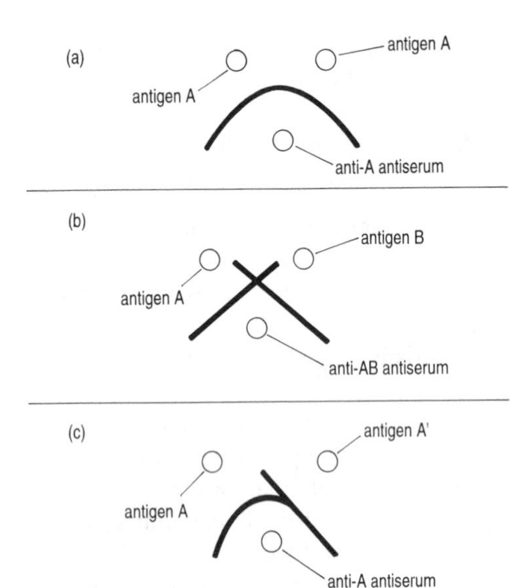

Fig. 32.5 Precipitin reactions in double diffusion immunoassay: (a) identity; (b) non-identity; (c) partial identity.

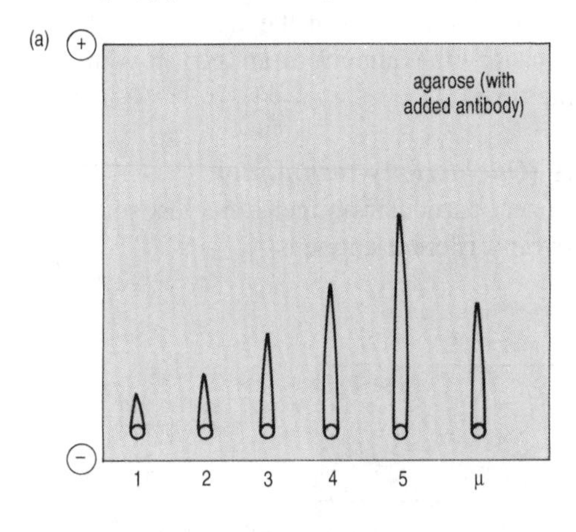

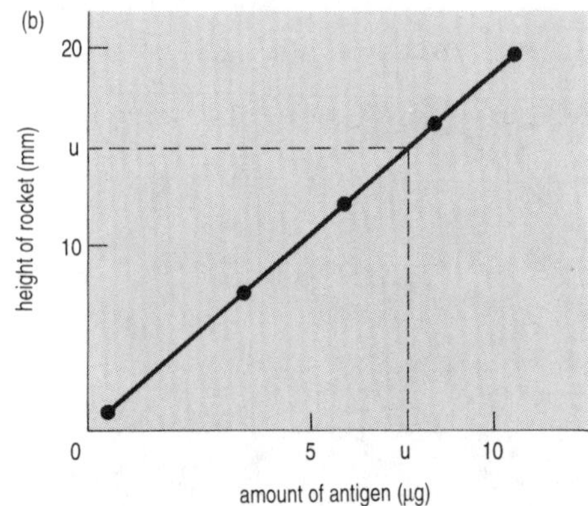

Fig. 32.6 Laurell rocket immunoelectrophoresis. (a) Assay: precipitin rockets are formed by electrophoresis of five standards of increasing concentration (wells 1 to 5) and an unknown (u). (b) Calibration curve: the unknown sample contains 7.4 μg of antigen.

1. Prepare an agarose gel (1.5% w/v) on a level glass slide or plate: allow to set.
2. Cut several circular wells in the gel.
3. Add test solutions of antigen or polyclonal antiserum to adjacent wells. Both solutions diffuse outwards, forming visible precipitin lines where antigen and corresponding antibody are present in optimal ratio (Fig. 32.5).

The various reactions between antigen and antiserum are:

- Identity: two wells containing the same antigen, or antigens with identical epitopes, will give a fused precipitin line (identical interaction between the antiserum and the test antigens, Fig. 32.5a).
- Non-identity: where the antiserum contains antibodies to two different antigens, each with its own distinct epitopes, giving two precipitin lines which intersect without any interaction (no cross-reaction, Fig. 32.5b).
- Partial identity: where two antigens have at least one epitope in common, but where other epitopes are present, giving a fused precipitin line with a spur (cross-reaction, Fig. 32.5c).

Immunoelectrophoretic assays

These methods combine the precipitin reaction with electrophoretic migration, providing sensitive, rapid assays with increased separation and resolution. The principal techniques are:

Cross-over electrophoresis (counter-current electrophoresis)

Similar to the Ouchterlony technique, since antigen and antibody are in separate wells. However, the movement of antigen and antibody towards each other is driven by a voltage gradient (p. 166): most antigens migrate towards the anode, while IgG migrates towards the cathode. This method is faster and more sensitive than double immunodiffusion, taking 15–20 min to reach completion.

Quantitative immunoelectrophoresis (Laurell rocket immunoelectrophoresis)

Similar to RID, as the antibody is incorporated into an agarose plate while the antigen is placed in a well. However, a voltage gradient moves the antigen into the gel, usually towards the anode, while the antibody moves towards the cathode, giving a sharply peaked, rocket-shaped precipitin line, once equivalence is reached (within 2–10 h). The height of each rocket shape at equivalence is directly proportional to the amount of antigen added to each well. A calibration curve for samples containing known amount of antigen can be used to quantify the amount of antigen present in test samples (Fig. 32.6).

Radioimmunological methods

These methods use radioisotopes to detect and quantify the antigen–antibody interaction, giving improved sensitivity over agglutination and precipitation methods. The principal techniques are described here.

Radioimmunoassay (RIA)

This is based on competition between a radioactively labelled antigen (or hapten) and an unlabelled antigen for the binding sites on a limited amount of antibody. The quantity of antigen in a test solution can be determined using a known amount of radiolabelled antigen and a fixed amount of antibody (Fig. 32.7).

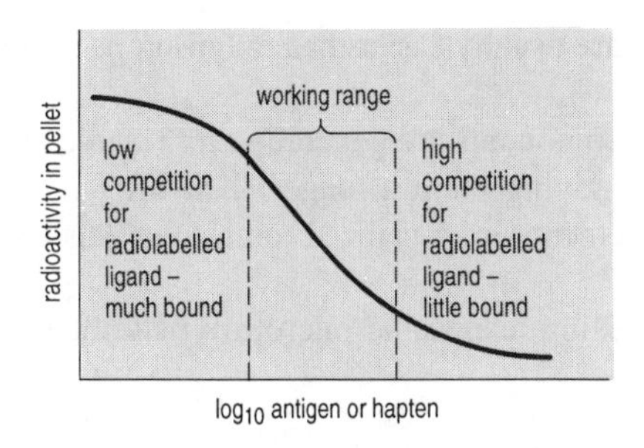

Fig. 32.7 Radioimmunoassay calibration curve. Note that the assay is insensitive at very low and very high antigen levels.

As with other immunoassay methods, it is important to perform appropriate controls, to screen for potentially interfering compounds. The basic procedure for RIA is:

1. Add appropriate volumes of a sample to a series of small test tubes. Prepare a further set of tubes containing known quantities of the substance to be assayed to provide a standard curve.
2. Add a known amount of radiolabelled antigen (or hapten) to each tube (sample and standard).
3. Add a fixed amount of antibody to each tube (the antibody must be present in limited quantity).
4. Leave at constant temperature for a fixed time (usually 24 h), to allow antigen–antibody complexes to form.
5. Precipitate the antibody and bound antigen using saturated ammonium sulphate, followed by centrifugation.
6. Determine the radioactivity of the supernatant or the precipitate (p. 179).
7. Prepare a calibration curve of radioactivity against $\log_{10}$ antigen (Fig. 32.7). The curve is most accurate in the central region, so adjust the amount of antigen in your test sample to fall within this range.

Note the following:

- You must be registered to work with radioactivity: check with the Departmental Radiation Protection Supervisor, if necessary (p. 183).
- Measure all volumes as accurately as possible as the end result depends on the volumetric quantities of unlabelled (sample) antigen, radiolabelled antigen and antibody: an error in any of these reagents will invalidate the assay.
- Incorporate replicates, so that errors can be quantified.
- Seek your supervisor's advice about fitting a curve to your data: this can be a complex process (see p. 150).

Immunoradiometric assay (IRMA)

This technique uses radiolabelled antibody, rather than antigen, for direct measurement of the amount of antigen (or hapten) in a sample. Most immunoradiometric assays are similar to the double antibody sandwich method described below (Fig. 32.8), except that the second antibody is labelled using a radioisotope. The important advantages over RIA are:

- linear relationship between amount of radioactivity and test antigen;
- wider working range for test substance;
- improved stability/longer shelf-life.

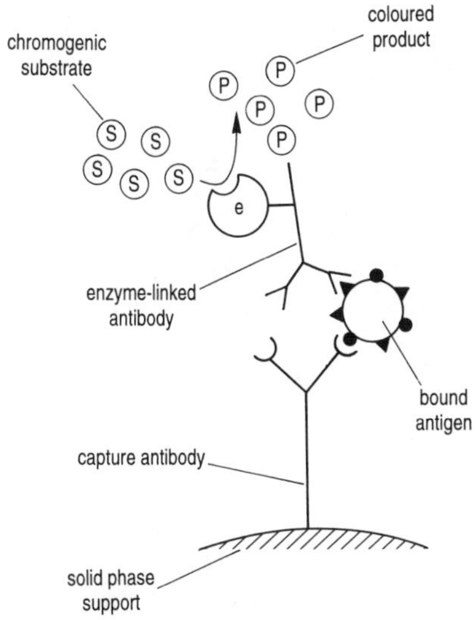

Fig. 32.8 Double antibody sandwich ELISA.

Enzyme immunoassays (EIA)

These techniques are also known as enzyme-linked immunosorbent assays (ELISA). They combine the specificity of the antibody–antigen interaction with the sensitivity of enzyme assays using either an antibody or an antigen conjugated (linked) to an enzyme at a site which does not affect the activity of either component. The enzyme is measured by adding an appropriate substrate which yields a coloured product. Enzymes offer the following advantages over radioisotopic labels:

- Increased sensitivity: a single enzyme molecule can produce many product molecules, amplifying the signal.

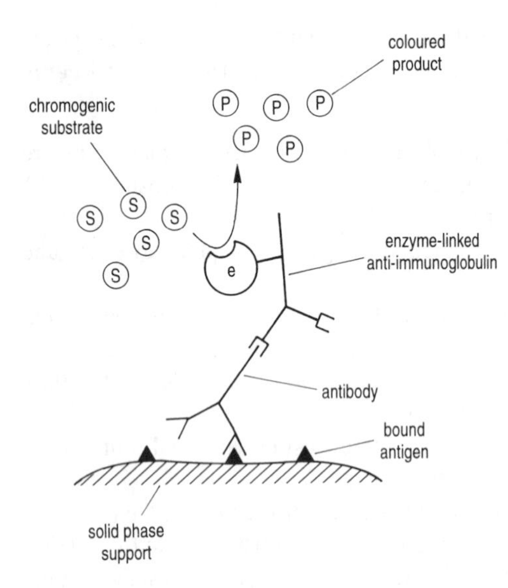

Fig. 32.9 Indirect ELISA.

- Simplified assay: enzyme assays are usually easier than radioisotope assays (p. 161).
- Improved stability of reagents: components are generally more stable than their radiolabelled counterparts, giving them a longer shelf-life.
- No radiological hazard: no requirement for specialized containment/disposal facilities.
- Automation is straightforward: using disposable microtitre plates and an optical scanner.

The principal techniques are:

Double antibody sandwich ELISA

This is used to detect specific antigens, involving a three-component complex between a capture antibody linked to a solid support, the antigen, and a second, enzyme-linked antibody (Fig. 32.8). This can be used to detect a particular antigen, e.g. a virus in a clinical sample, or to quantify the amount of antigen.

Indirect ELISA

This is used for antibody detection, with a specific antigen attached to a solid support. When the appropriate antibody is added, it binds to the antigen and will not be washed away during rinsing. Bound antibody is then detected using an enzyme-linked anti-immunoglobulin, e.g. a rabbit IgG antibody raised against human IgG (Fig. 32.9). One advantage of the indirect assay is that a single enzyme-linked anti-immunoglobulin can be used to detect several different antibodies, since the specificity is provided by the bound antigen.

33 Enzyme studies

Enzymes are globular proteins which increase the rate of specific biochemical reactions. Each operates on a limited number of closely related substrates to generate products under well-defined conditions of concentration, pH, temperature, etc. Metabolic pathways are composed of several enzymes, acting sequentially, e.g. glucose breakdown to pyruvate in the glycolytic pathway.

Enzymes are categorized according to the chemical reactions involved, leading to a four-figure Enzyme Commission code number and systematic name for each enzyme. Most enzymes also have a recommended trivial name, most often signified by the suffix 'ase'.

Example
Enzyme EC 1.1.1.1 is usually known by its trivial name, alcohol dehydrogenase.

Measuring enzyme reactions

Activity

This is measured in terms of the rate of enzyme reaction. Activity may be expressed directly as amount of substrate utilized per unit time (e.g. nmol min^{-1}, etc.), or in terms of the non-SI international unit (U, or sometimes IU), defined as the amount of enzyme which will convert 1 μmol of substrate to product(s) in 1 min under specified conditions. However, the recommended (SI) unit of enzyme activity is the katal (kat), which is the amount of enzyme which will convert 1 mol of substrate to product(s) in 1 s under optimal conditions, determined from the following equation:

$$\text{enzyme activity (kat)} = \frac{\text{amount of substrate converted (mol)}}{\text{time (s)}} \quad [33.1]$$

This unit is relatively large (1 kat = 6×10^7 U) so SI prefixes are often used, e.g. nkat or pkat (p. 41). Note that the units are amount of substrate (mol), not concentration (mol l^{-1} or mol m^{-3}).

Example
The hydrolysis of 1 molecule of maltose to give 2 glucose molecules by α-glucosidase means that enzyme activity specified in terms of substrate consumption (nmol maltose) would be half the value expressed with respect to product formation (nmol glucose).

For enzymes with macromolecular substrates of unknown molecular weight (e.g. deoxyribonuclease, amylase), activity can be expressed as the mass of substrate consumed (e.g. ng DNA min^{-1}), or amount of product formed (e.g. nmol glucose min^{-1}). You must ensure that your units clearly specify the substrate or product used, especially when the enzyme transformations involve different numbers of substrate or product molecules.

The turnover number of an enzyme is the *amount of substrate* (mol) converted to product in 1 s by 1 mol of enzyme operating at maximum rate under optimum conditions. In practice, this requires information on the molecular weight of the enzyme, the amount of enzyme present and its maximum activity (p. 163).

Purity

The specific activity of an enzyme preparation expresses the enzyme activity in terms of the quantity of protein present:

$$\text{specific activity} = \frac{\text{enzyme activity (kat)}}{\text{mass of protein (kg)}} \quad [33.2]$$

Alternative units may be used, e.g. U mg^{-1}, ng min^{-1} mg^{-1}. The specific activity is a useful way of comparing the purity of different enzyme preparations (purified enzyme preparations have high specific activities), comparing the stages in purification of an enzyme (specific activity increases as other proteins are eliminated), and assessing enzyme stability (an unstable enzyme will show a decrease in specific activity with time). The protein content of a solution can be determined by various means (see Box 33.1).

Box 33.1 Methods of determining the amount of protein in an aqueous solution

Most assays for protein content do not give an absolute value, but require standard solutions, containing appropriate concentrations of a particular protein, to be analysed at the same time, so that a standard curve can be constructed. Bovine serum albumin (BSA) is commonly used as a protein standard. However, you may need to use an alternative standard if the protein you are assaying has an amino acid composition which is markedly different from that of BSA, depending on your chosen method. The most common methods are:

Direct measurement of UV absorbance (Warburg–Christian method).

Proteins absorb electromagnetic radiation maximally at 280 nm (due to the presence of aromatic amino acids, especially tyrosine and threonine) and this forms the basis of the method. The principal advantages of this approach are its simplicity and the fact that the assay is non-destructive. The most common interfering substances are nucleic acids, which can be corrected for by measuring the absorbance at 260 nm (where nucleic acids have a stronger absorption and proteins have a weaker absorption): a pure solution of protein will have a ratio of absorption (A_{280}/A_{260}) of approximately 1.8, decreasing with increasing nucleic acid contamination. Note also that any free aromatic amino acids in your solution will absorb at 280 nm, leading to an overestimation of protein content. The simplest procedure, which includes a correction for small amounts of nucleic acid, is as follows (use quartz cuvettes throughout):

1. Measure the absorbance of your solution at 280 nm (A_{280}): if A_{280} is greater than 1, dilute by an appropriate amount and re-measure (see p. 196).
2. Repeat at 260 nm (A_{260}).
3. Estimate the approximate protein content using the following relationship:

$$[\text{protein}]\ \text{mg ml}^{-1} = 1.45\,A_{280} - 0.74\,A_{260} \qquad (33.3)$$

This equation is based on the work of Warburg and Christian (1942), for the enzyme enolase. For other proteins, it should not be used for quantitative work, since it gives only a rough approximation of the amount of protein present, due to variations in aromatic amino acid composition. A better approach is to establish the extinction coefficient of your particular protein at 280 nm (e.g. using the pure substance and a spectrophotometer) and then use the Beer–Lambert relationship (p. 194) to determine the amount of protein in your samples — this is only feasible if you have a purified sample, free from significant nucleic acid and amino acid contamination.

For all of the following methods, the amounts are appropriate for semi-micro cuvettes (1.5 ml volume, path length 1 cm). The Lowry and Bradford methods can also be used in unmodified form for standard 4.5 ml cuvettes. Appropriate controls (blanks) must be analysed, to assess for possible interference, (e.g. due to buffers, etc.) in your protein extract.

Biuret method

This is based on the specific reaction between cupric ions (Cu^{++}) in alkaline solution and two adjacent peptide bonds, as found in proteins. As such, it is not significantly affected by differences in amino acid composition between proteins, in contrast to the Warburg– Christian (UV absorption) method.

1. Prepare protein standards over an appropriate range (typically, between 1–20 mg ml^{-1}).
2. Add 1 ml of each standard solution to a separate test tube. Prepare a reagent blank, using 1 ml of distilled water, or an appropriate solution.
3. Add 1 ml of each unknown solution to a separate test tube.
4. Add 1 ml of Biuret reagent (1.5 g $CuSO_4.5H_2O$, 6.0 g sodium potassium tartrate in 300 ml of 10 w/v NaOH) to all standard and unknown tubes and to the reagent blank.
5. Incubate at 37°C for 15 min.
6. Read the absorbance of each solution at 520 nm against the reagent blank. The colour is stable for several hours.

The main limitation of the Biuret method is its lack of sensitivity — it is unsuitable for solutions with a protein content of less than 1 mg ml^{-1}.

Lowry (Folin–Ciocalteau) method

This is a colorimetric assay, based on a combination of the Biuret method, described above, and the reduction of tyrosine and tryptophan residues with Folin and Ciocalteau's reagent to give a blue–purple colour. The method is extremely sensitive (down to a protein content of 20 μg ml^{-1}), but is subject to interference from a wide range of non-protein substances, including many organic buffers (e.g. TRIS, HEPES, etc.), EDTA, urea and certain sugars. Choice of an appropriate standard is important, as the intensity of colour produced for a particular protein is dependent on the amount of aromatic amino acids present.

Box 33.1 continued

1. Prepare protein standards within an appropriate range for your samples (the method can be used from 0.02–1.00 mg ml^{-1}).
2. Add 1 ml of each standard solution to a separate test tube. Prepare a reagent blank, using 1 ml of distilled water, or an appropriate solution.
3. Add 1 ml of each of your unknown solutions to a separate test tube.
4. Then, add 5 ml of alkaline sodium carbonate solution (2 g Na_2CO_3 in 100 ml of 0.1 mol l^{-1} NaOH). Mix thoroughly and allow to stand for at least 10 min.
5. Add 0.5 ml of Folin–Ciocalteau reagent (commercial reagent, diluted 1:1 with distilled water on the day of use). Mix rapidly and thoroughly and then allow to stand for 30 min.
6. Read the absorbance of each sample at 600 nm.

Dye-binding (Bradford) method

Coomassie brilliant blue combines with proteins to give a dye–protein complex with an absorption maximum of 595 nm. This provides a simple and sensitive means of measuring protein content, with few interferences. However, the formation of dye–protein complex is affected by the number of basic amino acids within a protein, so the choice of an appropriate standard is important. The method is sensitive down to a protein content of approximately 5 μg ml^{-1} but the relationship between absorbance and concentration is often non-linear, particularly at high protein content.

1. Prepare protein standards over an appropriate range (between 5 and 100 μg ml^{-1}).
2. Add 100 μl of each standard solution to a separate test tube. Prepare a reagent blank, using 100 μl of distilled water, or an appropriate solution (note that these small volumes must be accurately dispensed, e.g. using a calibrated pipettor, p. 14).
3. Add 100 μl of your unknown solutions to a separate test tube.
4. Add 5.0 ml of Coomassie brilliant blue G250 solution (0.1 g l^{-1}).
5. Mix and incubate for at least 5 min: read the absorbance of each solution at 595 nm.

Other methods are less widely used. They include determination of the total amount of nitrogen in solution (e.g. using the Kjeldahl technique) and calculating the protein content, assuming a nitrogen content of 16%. An alternative approach is to precipitate the protein (e.g. using trichloracetic acid, tannic acid, or salicylic acid) and then measure the turbidity of the resulting precipitate (using a spectrophotometer, or a nephelometer, p. 196).

Types of assay

The rate of substrate utilization or product formation must be measured under controlled conditions, using some characteristic which changes in direct proportion to the concentration of the test substance.

Example
α-Glucosidase activity can be measured using the artificial substrate p-nitrophenol-α-D-glucose, which liberates p-nitrophenol (yellow) with an absorption maximum at 404 nm.

Spectrophotometric assays

Many substrates and products absorb visible or UV light and the change in absorbance at a particular wavelength provides a convenient assay method (p. 194). Artificial substrates are used where no suitable spectrophotometric assay is available for the natural substrates. In most cases, these are chromogenic analogues of the natural substrate, producing coloured products. In other cases, a product may be measured by a colorimetric chemical reaction.

Several assays are based on interconversion of the nicotinamide adenine dinucleotide coenzymes NAD^+ or $NADP^+$ which are reversibly reduced by many enzymes. The reduced form (either NADH or NADPH) can be detected at 340 nm, where the oxidized form has negligible absorbance (p. 197). An alternative approach is to use a coupled enzyme assay, where the product of the test enzyme is used as a substrate for a second enzyme reaction which involves oxidation/reduction of nucleotide coenzymes. Such assays are particularly useful for continuous monitoring of enzyme activity (p. 162) and for reactions where the product from the test substance is too low to detect by other methods, since coupled assays are more sensitive.

Fluorimetric assays

Certain substrates liberate fluorescent products as a result of enzyme activity, providing a highly sensitive assay method. Fluorogenic substrates include fluorescein and methylumbelliferone derivatives. Care is required, since impurities in the enzyme preparation may produce background fluorescence, or quenching (reduction) of the signal.

Radioisotopic assays

These are useful where the substrate and product can be easily separated, e.g. in decarboxylase assays using a ^{14}C-labelled substrate, where gaseous $^{14}CO_2$ is produced.

Potentiometric assays

Enzyme reactions involving acids and bases can be monitored using a pH electrode (p. 30), though the change in pH will affect the activity of the test enzyme. An alternative approach is to measure the amount of acid or alkali required to maintain a constant pre-selected pH in a pH-stat.

An oxygen electrode can be used if O_2 is a substrate or a product (p. 185). Other ion-specific electrodes can monitor ammonia, nitrate, etc.

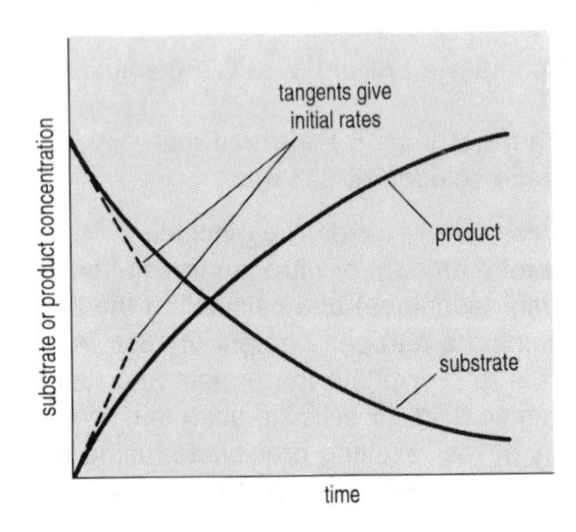

Fig. 33.1 Enzyme reaction progress curve: substrate utilization/product formation as a function of time.

Methods of monitoring substrate utilization/product formation

Continuous assays (kinetic assays)

The change in substrate or product is monitored as a function of time, to provide a progress curve for the reaction (Fig. 33.1). This curve starts off in a near-linear manner, decreasing in slope as the reaction proceeds and substrate is used up. The initial velocity of the reaction (v_0) is obtained by drawing a tangent to the curve at zero time and measuring its slope. Continuous monitoring can be used when the test substance can be assayed rapidly (and non-destructively), e.g. using a chromogenic substrate.

Discontinuous assays (fixed time assays)

It is sometimes necessary to measure the amount of substrate consumed or product formed after a fixed time period, e.g. where the test substance is assayed by a (destructive) colorimetric chemical method. It is vital that the time period is kept as short as possible, with the substrate change limited to around 10%, so that the assay is within the linear part of the progress curve (Fig. 33.1). This may need to be established in a preliminary experiment, e.g. using a continuous assay.

Enzyme purification

The purification of an enzyme usually involves several stages:

1. Tissue/cell disruption, often using mechanical/ultrasonic homogenization.
2. Differential centrifugation, removing larger particulate components (p. 190).
3. Ammonium sulphate fractionation: selective precipitation by the stepwise addition of ammonium sulphate at low temperature ($<10°C$). The precipitated protein is then redissolved in a fresh buffer solution.
4. Chromatographic and/or electrophoretic separation, including ion exchange and gel filtration chromatography and polyacrylamide gel electrophoresis or isoelectric focusing (p. 166). This stage may involve several individual steps and can lead to a fully purified enzyme (purification to homogeneity), with maximum specific activity.

Example
If an original preparation of 50 ml had an activity of 100 pkat ml^{-1} (total 5 nkat) and a subsequent fraction of 5 ml volume had an activity of 200 pkat ml^{-1} (total 1 nkat), this would represent a yield of 1/5 = 0.2 (or 20%).

Example
An increase in specific activity from 10 nkat (mg protein)$^{-1}$ to 120 nkat (mg protein)$^{-1}$ represents a twelve-fold purification.

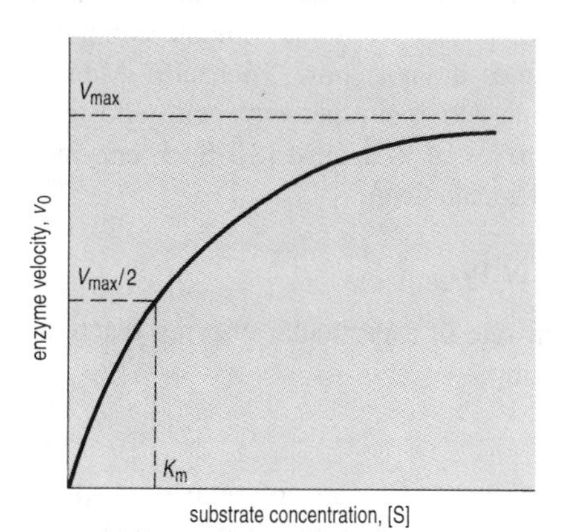

Fig. 33.2 Effect of substrate concentration on enzyme activity.

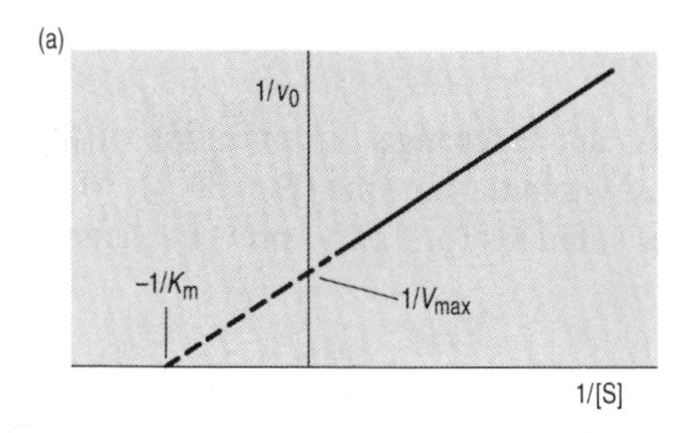

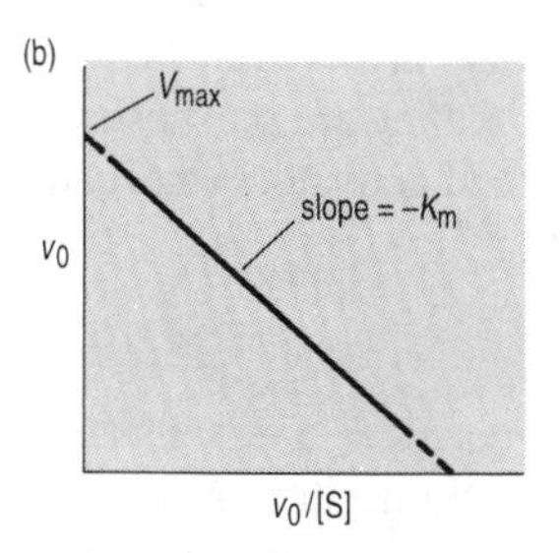

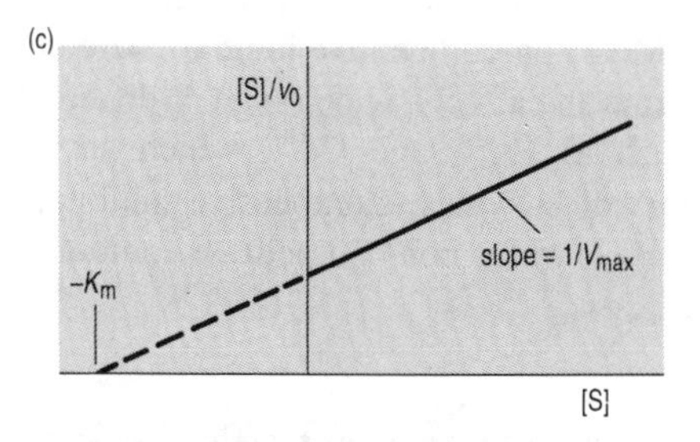

Fig. 33.3 Graphical transformations for determining the kinetic constants of an enzyme. (a) Lineweaver–Burk plot. (b) Eadie–Hofstee plot. (c) Hanes–Woolf plot.

At each step in the procedure the yield of enzyme in an individual fraction can be determined:

$$\text{yield} = \frac{\text{amount of enzyme in fraction (kat or U)}}{\text{amount of enzyme in original preparation (kat or U)}} \qquad [33.4]$$

Note that the yield equation does not use a concentration-based measure of enzyme activity, as this would be affected by the volumes of the solutions involved. The relative purification of an enzyme is usually expressed as 'n-fold purification', where:

$$n = \frac{\text{specific activity of fraction}}{\text{specific activity of original preparation}} \qquad [33.5]$$

Enzyme kinetics

For most enzymes, when the initial rate (v_0) of a fixed amount of an enzyme is plotted as a function of the concentration of a single substrate [S] with all other substrates present in excess, a rectangular hyperbola is obtained (Fig. 33.2). At low substrate concentrations v_0 is directly proportional to [S], with a decreasing response as substrate concentration is increased until saturation is achieved. The shape of this plot can be described by a mathematical relationship, known as the Michaelis–Menten equation:

$$v_0 = \frac{V_{max}\,[S]}{K_m + [S]} \qquad [33.6]$$

This equation makes use of two kinetic constants:

- V_{max}, the maximum velocity of the reaction (at infinite substrate concentration).
- K_m, the Michaelis constant, equivalent to the substrate concentration where $v_0 = \frac{1}{2} V_{max}$.

V_{max} is a function of the amount of enzyme and is the appropriate rate to use when determining the specific activity of a purified enzyme (p. 159). The Michaelis constant provides a measure of the substrate affinity of an enzyme and is an important characteristic of a particular enzyme. Thus, an enzyme with a large K_m has a low affinity for its substrate while an enzyme with a small K_m has a high affinity.

Your first step in determining the kinetic constants for a particular enzyme is to measure the rate of reaction at several substrate concentrations, as in Fig. 33.2. There are various ways to obtain K_m and V_{max} from such data, mostly involving drawing a graph representing a linear transformation of eqn [33.6]:

- The Lineweaver–Burk plot: a graph of the reciprocal of the reaction rate ($1/v_0$) against the reciprocal of the substrate concentration (1/[S]) gives $-1/K_m$ as the intercept of the x axis and $1/V_{max}$ as the intercept of the y axis (Fig. 33.3a). The slope of the plot is most affected by the least accurate values, i.e. those measured at low substrate concentration.
- The Eadie–Hofstee plot: v_0 against v_0/[S], where the intercept on the y axis gives V_{max} and the slope equals $-K_m$ (Fig. 33.3b).
- The Hanes–Woolf plot: [S]/v_0 against [S], giving $-K_m$ as the intercept of the x axis and $1/V_{max}$ from the slope (Fig. 33.3c).

There are several computer packages that will plot the above relationships and calculate the kinetic constants from a given set of data using linear regression analysis (p. 240). While the Eadie–Hofstee and Hanes–Woolf plots distribute the data points more evenly than the Lineweaver–Burk plot, the best approach to such data is to use non-linear regression on untransformed data, although this is usually outside the scope of the simpler computer packages. Note also that some enzymes do not show Michaelis–Menten kinetics, e.g. those with more than one active site per molecule (allosteric enzymes) which often give sigmoid curves of v_0 against [S]. Such enzymes are usually involved in the control of metabolism.

Factors affecting enzyme activity

If you want to measure the maximum rate of a particular enzyme reaction, you will need to optimize the following:

When removing a bottle of freeze dried enzyme from a freezer or fridge, do not open it until it has been warmed to room temperature or water may condense on the contents — this will make weighing inaccurate and may lead to loss of enzyme activity.

Temperature

Enzyme activity increases with temperature, until an optimum is reached. Above this point, activity decreases as a result of protein denaturation (Fig. 33.4). Note that the optimum temperature for enzyme *activity* may not be the same as that for maximum *stability* (enzymes are usually stored at temperatures near to or below 0°C, to maximize stability).

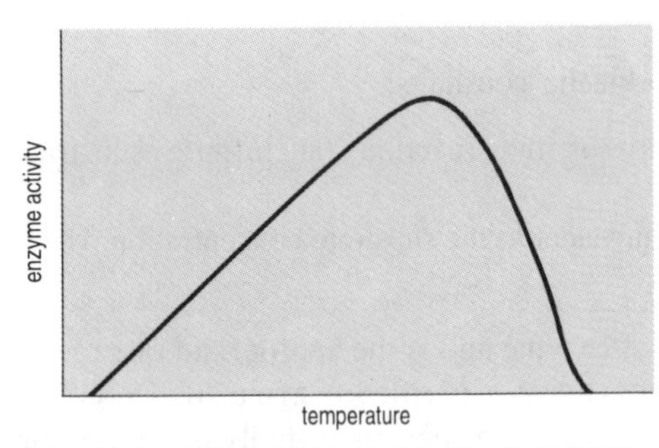

Fig. 33.4 Effect of temperature on enzyme activity.

pH

Enzymes work best at a particular pH, due to changes in ionization of the substrates or of the amino acid residues within the enzyme (Fig. 33.5). Most enzyme assays are performed in buffer solutions (p. 32), to prevent changes in pH during the assay.

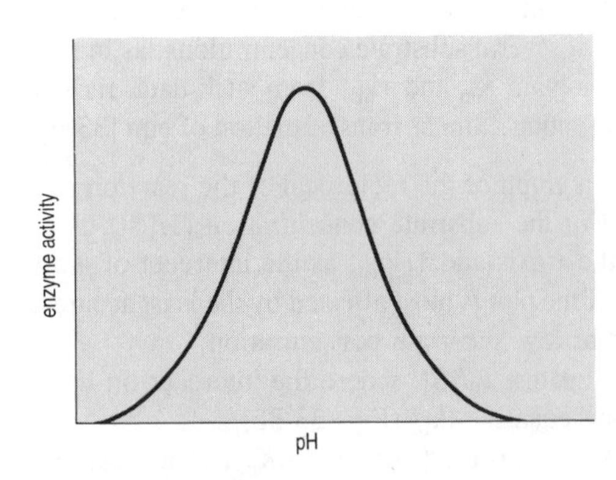

Fig. 33.5 Effect of pH on enzyme activity.

Cofactors

Many enzymes require appropriate concentrations of specific cofactors for maximum activity. These are sub-divided into coenzymes (soluble, low molecular weight organic compounds which are actively involved in catalysis by accepting or donating specific chemical groups, i.e. they are co-substrates of the enzyme; examples include NAD^+ and ADP); and activators (inorganic metal ions, required for maximal activity), e.g. Mg^{2+}, K^+.

Inhibitors

Many compounds can reduce the rate of an enzyme reaction, e.g. substances which compete for the active site of the enzyme due to similarities in chemical structure with the natural substrate (competitive inhibitors), or substances which bind to the enzyme at other sites, inhibiting the normal functioning of the enzyme (non-competitive inhibitors). Most non-competitive inhibitors are not chemically related to the natural substrate and affect a wide range of different enzymes, e.g. heavy metal ions, such as Hg^{2+} and Cd^+, which act as irreversible, non-competitive inhibitors of enzymes. Competitive inhibitors increase K_m but have no effect on V_{max} while non-competitive inhibitors reduce V_{max} but have no effect on K_m.

Substrate concentration

Substrates must be present in excess, to ensure maximum reaction velocity. Equation [33.6] shows that a substrate concentration equivalent to $10 \times K_m$ will give 91% of V_{max}, while a concentration of $100 \times K_m$ will give 99% of the maximum rate.

34 Genetic engineering techniques

Deoxyribonucleic acid (DNA) is the genetic material of all cellular organisms. The sequential arrangement of nucleotide sub-units represents a genetic code for the synthesis of specific cellular proteins. A portion of DNA which encodes the information for a single polypeptide (or protein) is usually referred to as a gene while the entire genetic information of an organism is known as its genome.

Recent advances in the manipulation of nucleic acids in the laboratory have increased our understanding of the structure and function of genes at the molecular level. Additionally, these techniques can be used to alter the genome of an organism (genetic engineering), e.g. to create a bacterium capable of synthesizing a foreign protein such as a potentially valuable hormone or vaccine, or an unusual variant of a normal protein. Such procedures are carried out under strict containment, in accordance with the Genetic Manipulation Regulations (1989). However, the general techniques involved in the isolation and manipulation of DNA are often used at undergraduate level, to illustrate the techniques involved.

Basic principles

Genetic engineering involves several steps:

Definintion
Plasmids — circular molecules of DNA which are capable of autonomous replication. They can be isolated, manipulated and then reintroduced into bacterial cells.

1. Isolation of the DNA sequence (gene) of interest from the genome of an organism. This usually involves DNA purification followed by enzymic digestion, to liberate the target DNA sequence.
2. Creation of an artificial recombinant DNA molecule (sometimes referred to as rDNA), by inserting the (foreign) gene into a DNA molecule capable of replicating in a host cell, i.e. a 'cloning vector'. Suitable cloning vectors for bacterial cells include plasmids and bacteriophages (bacterial viruses).
3. Introduction of the recombinant DNA molecule into a suitable host, e.g. *E. coli*. The process is termed transformation when a plasmid is used or transfection for a recombinant virus vector.
4. Selection and growth of the transformed (or transfected) cell, using the techniques of microbial culture (p. 86). Since a single transformed host cell can be grown to give a clone of genetically identical cells, each carrying the gene of interest, the technique is often referred to as gene cloning, or molecular cloning.

Extraction and purification of DNA

Specific details of the steps involved in the isolation of DNA vary, depending upon the source material. However, the following sequence shows the principal stages in the purification of plasmid DNA from bacterial cells.

Preparing glassware — all glassware for DNA purification must be siliconized, to prevent adsorption of DNA. All glass and plastic items must be sterilized before use.

1. Cell wall digestion: incubation of bacteria in a lysozyme solution will remove the peptidoglycan cell wall. This is often carried out under isotonic conditions, to stop the cells from bursting open and releasing chromosomal DNA. Note that Gram-negative bacteria are relatively insensitive to lysozyme, requiring additional treatment to allow the enzyme to reach the cell wall layer, e.g. osmotic shock, or incubation with a chelating agent, e.g. ethylenediaminetetra-acetate (EDTA). The latter treatment will also

inactivate any bacterial deoxyribonucleases in the solution, preventing enzymic degradation of plasmid DNA during extraction.

2. Lysis using strong alkali (NaOH) and a detergent, e.g. sodium dodecyl sulphate, to solubilize the cellular membranes and partially denature the proteins. Neutralization of this solution (e.g. using potassium acetate) causes the chromosomal DNA to aggregate as an insoluble mass, leaving the plasmid DNA in solution.
3. Removal of other macromolecules, particularly RNA and proteins by, for example, enzymic digestion using ribonuclease and proteinase. Additional chemical purification steps give further increases in purity, e.g. proteins can be removed by mixing the extract with water-saturated phenol (50% v/v), or a phenol/chloroform mixture. On centrifugation, the DNA remains in the upper aqueous layer, while the proteins partition into the lower organic layer. Repeated cycles of phenol/chloroform extraction can be used to minimize the carry-over of these macromolecules. Additional purification can be obtained using isopycnic density gradient centrifugation in CsCl (p. 190).
4. Precipitation of DNA using around 70% v/v ethanol: water (produced by adding two volumes of 95% v/v ethanol to one volume of aqueous extract), followed by centrifugation, to recover the DNA pellet. Further rinsing with 70% v/v ethanol: water will remove any salt contamination from the previous stages. The extracted DNA can then be either frozen, for future use, or redissolved in buffer solution.

Safe working practices — note that phenol is toxic and corrosive while chloroform is highly flammable. Take appropriate safety precautions (e.g. wear gloves, extinguish all naked flames, use a fume hood, where available).

Separation of DNA using agarose gel electrophoresis

Electrophoresis is the term used to describe the movement of ions in an applied electrical field, as shown in Fig. 34.1. DNA molecules are negatively charged anions, migrating through an agarose gel towards the anode at a rate which is dependent upon molecular size — smaller, compact DNA molecules can pass through the sieve-like agarose matrix more easily than large, extended fragments. Electrophoresis of plasmid DNA is usually carried out using a submerged agarose gel. The amount of agarose is adjusted, depending on the size of the DNA molecules to be separated, e.g. 0.3% w/v agarose is used

Maximizing recovery of DNA — these large molecules are easily damaged by mechanical forces, e.g. vigorous shaking or stirring during extraction. In addition, all glassware must be scrupulously cleaned and gloves must be worn, to prevent deoxyribonuclease contamination of solutions.

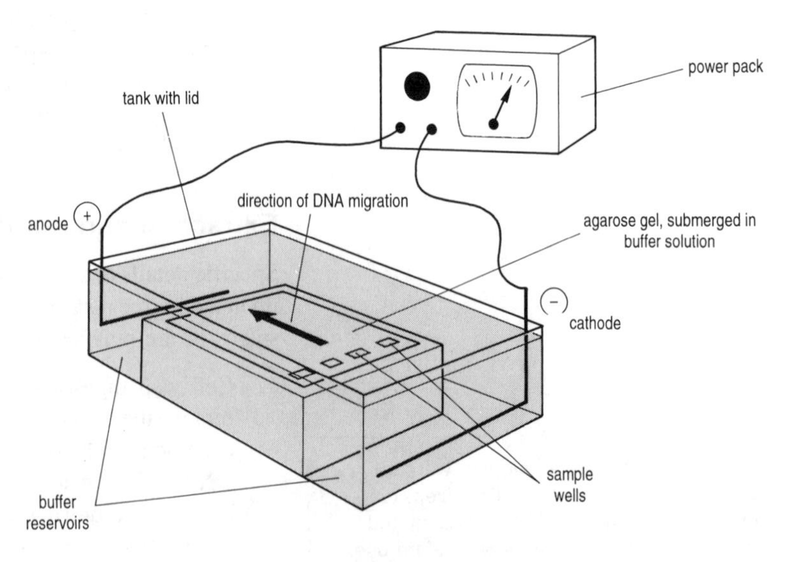

Fig. 34.1 Agarose gel electrophoresis of DNA.

for large fragments (>20 000 bases) while 0.8% is used for smaller fragments. Note the following:

- Individual samples are added to pre-formed wells using a pipettor. The volume of sample added to each well is usually less than 25 μl so a steady hand and careful dispensing are needed to pipette each sample.
- The density of the samples is usually increased by adding a small amount of sucrose, so that each sample is retained within the appropriate well.

Using molecular weight standards — for accurate molecular weight determination your standards must have the same conformation as the DNA in your sample, i.e. linear DNA standards for linear (restriction) fragments and closed circular standards for plasmid DNA.

- A water-soluble anionic tracking dye (e.g. bromophenol blue) is also added to each sample, so that migration can be followed visually.
- Molecular weight standards are added to one or more wells. After electrophoresis, the relative position of bands of known molecular weight can be used to prepare a calibration curve (usually, by plotting the $\log_{10}$ of the molecular weight of each band against the distance travelled).
- The gel should be run until the tracking dye has migrated across 80% of the gel (see manufacturer's instructions for appropriate voltages/times). Note that the tank cover must be in position during electrophoresis, to prevent evaporation and to reduce the possibility of electric shock.

Safe handling of ethidium bromide — ethidium bromide is carcinogenic so always use gloves when handling stained gels and make sure you do not spill any staining solution.

- After electrophoresis, the bands of DNA can be visualized by soaking the gel for around 5 min in 1 mg l^{-1} ethidium bromide, which binds to DNA by intercalation between the paired nucleotides of the double helix.
- Under UV light, bands of DNA are visible due to the intense orange–red fluorescence of the ethidium bromide. The limit of detection using this method is around 10 ng DNA per band. Suitable plastic safety glasses or goggles will protect your eyes from the UV light. The migration of each band can be measured from the well using a ruler. Alternatively, a Polaroid® photograph can be taken, using a special camera and adaptor.
- If a particular band is required for further study (e.g. a plasmid), the piece of gel containing that band is cut from the gel using a scalpel and the DNA extracted.

Identification of specific DNA molecules using Southern blotting

After electrophoretic separation in agarose gel, the fragments of DNA can be immobilized on a filter using a technique named after E.M. Southern:

1. The fragments are first denatured using concentrated alkali, giving single-stranded DNA.
2. A nitrocellulose filter is then placed directly onto the gel, followed by several layers of absorbent paper. The DNA is 'blotted' onto the filter as the buffer solution is pulled through.
3. Specific sequences can be identified by incubation with radiolabelled complementary probes of single-stranded DNA, which will hybridize with a particular sequence, followed by autoradiography.

DNA amplification using the polymerase chain reaction

A small amount of DNA can be detected by Southern blotting after *in vitro* amplification, using the following procedure:

1. DNA is heated (e.g. 94°C for 1 min), to separate the individual strands.
2. Oligonucleotide primers are added: these are complementary to part of the target sequence, and they will hybridize (anneal) at the appropriate position, when the temperature is reduced, e.g. 55°C for 2 min.
3. The primers are extended by a heat-resistant DNA polymerase (*Taq* polymerase, from the thermophilic bacterium *Thermus aquaticus*). This will create a complementary strand to each of the original strands, doubling the amount of target DNA, e.g. at 72°C for 2 min.
4. Repeated cycles of strand separation, annealing and primer extension lead to exponential amplification of the desired sequence.

The polymerase chain reaction is particularly useful for amplifying small amounts of DNA, e.g. viruses in clinical samples, or DNA from a small tissue sample in forensic science.

DNA assay

The simplest approach to quantifying the amount of nucleic acid in an aqueous solution is to measure the absorbance of the solution at 260 nm using a spectrophotometer, as detailed on p. 197. Note that the A_{260} value applies to purified DNA, whereas a plasmid DNA extract prepared using the protocol on p. 165 will contain a substantial amount of contaminating RNA, with similar absorption characteristics to DNA. Any contaminating protein would also invalidate the calculation; the protein can be detected by measuring the absorbance of the solution at 280 nm. Purified nucleic acids have a value for A_{260}/A_{280} of around 1.9 and contaminating protein will give a lower ratio. If your solution gives a ratio substantially lower than this, you should repeat the phenol/chloroform extraction steps (p. 166).

Enzymic cleavage and ligation of DNA

Type II restriction endonucleases (commonly called restriction enzymes) recognize and cleave a particular sequence of double-stranded DNA (usually, four or six nucleotide pairs), known as the restriction site. Each enzyme is given a code name derived from the name of the bacterium from which it is isolated, e.g. *Hin* dIII was the third restriction enzyme isolated from *Haemophilus influenzae* strain Rd (Fig. 34.2). Most enzymes cleave each strand at a different position, producing short, single-stranded regions known as cohesive ends, or 'sticky ends'. A few enzymes cleave DNA to give blunt-ended fragments. Restriction enzymes which cleave DNA to give 'sticky ends' are widely used in genetic engineering, since two DNA molecules cut with

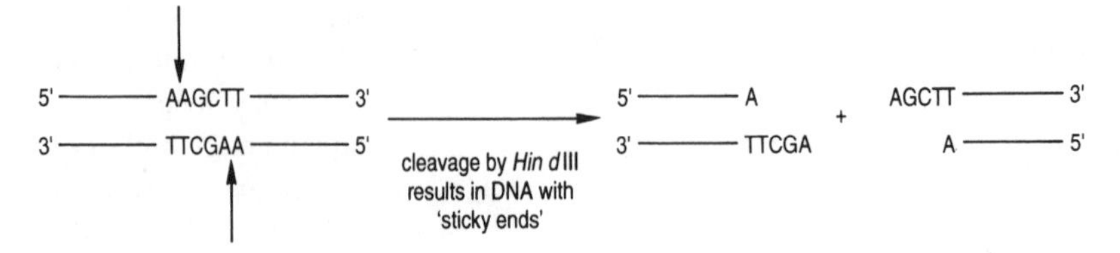

Fig. 34.2 Recognition site for the restriction enzyme *Hin* dIII. This is the conventional representation of double-stranded DNA, showing the individual bases, where A is adenine, C cytosine, G guanine and T thymine. The cleavage site on each strand is shown by an arrow.

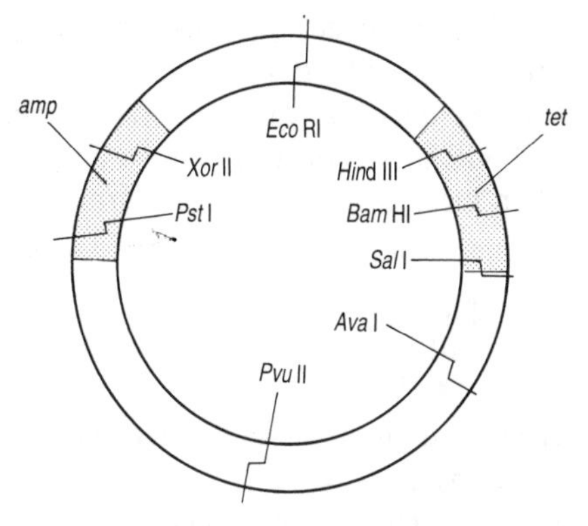

Fig. 34.3(a) Restriction map of the plasmid pBR 322. The position of individual restriction sites is shown together with the genes for ampicillin resistance (*amp*) and tetracycline resistance (*tet*).

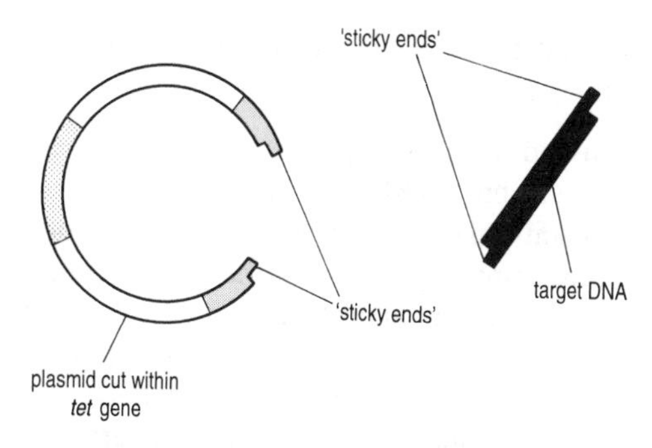

Fig. 34.3(b) Restriction of plasmid and foreign DNA with *Hin*dIII.

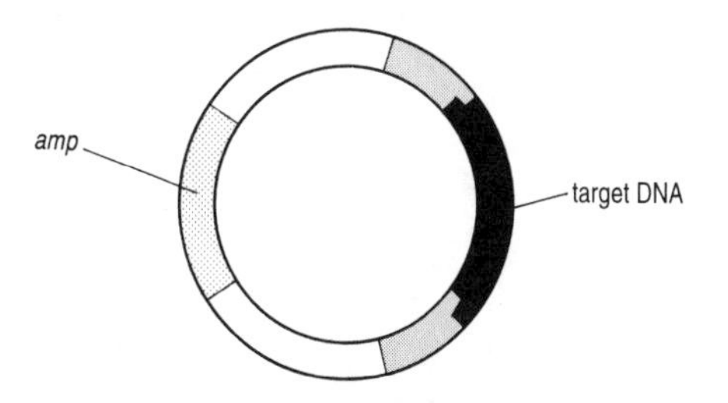

Fig. 34.3(c) Annealing and ligation of plasmid and target DNA to give a recombinant plasmid which confers resistance to ampicillin only. As target DNA has been inserted within the tc^r gene, the gene is now discontinuous and inactive.

the same restriction enzyme will have complementary single-stranded regions, allowing them to anneal (base pair), due to the formation of hydrogen bonds between individual bases within this region.

Restriction enzymes have two important applications in molecular genetics:

- Mapping. A DNA molecule can be cleaved into several restriction fragments whose number and size can be determined using agarose gel electrophoresis. The position of individual restriction sites can be used to create a diagnostic restriction enzyme map for a particular molecule, e.g. a plasmid (Fig. 34.3a).
- Genetic engineering. Two restriction fragments cut using the same enzyme (Fig. 34.3b) and annealed by complementary base pairing can be permanently joined together using another bacterial enzyme (DNA ligase), which forms covalent bonds between the annealed strands, creating the recombinant molecule (Fig. 34.3c). When the two molecules involved are the cloning vector and the target DNA, the size of the recombinant plasmid can be predicted (e.g. a plasmid of 4 500 base pairs, plus a target DNA fragment of 2 500 base pairs will give a recombinant molecule of 7 000 base pairs), allowing separation by electrophoresis. Note that most of the plasmids used in genetic engineering code for two or more easily detectable marker genes (e.g. antibiotic resistance), with single restriction sites on each plasmid (Fig. 34.3a).

Transformation of a suitable host cell

Once a recombinant plasmid has been produced *in vitro*, it must be introduced into a suitable host cell (e.g. *E. coli*). Several procedures can be used:

- Pre-treatment with $CaCl_2$ at low temperature: exponential phase cells (p. 126) are incubated with hypotonic $CaCl_2$ at 4°C for around 30 min, followed by a brief heat shock (e.g. 42°C for 2 min). The low-temperature incubation allows DNA to adhere to the cells, while the heat shock promotes DNA uptake.
- Electroporation: cells or protoplasts are subjected to electric shock treatment (typically, up to 800 kV m^{-1}) for short periods (<1 s).

These treatments cause a temporary increase in membrane permeability, leading to the uptake of plasmid DNA from the external medium. Such systems are often inefficient, with fewer than 0.1% of all cells showing stable transformation. However, this is not usually a significant problem, since a single viable transformant can be grown to give a large number of identical cells, using standard microbiological plating techniques (p. 123).

To maximize the transformation efficiency of *E. coli* in the heat shock/ $CaCl_2$ procedure, use the minimum volume of solution in a thin-walled glass tube, so that the cells experience a rapid change in temperature.

Selection of transformants

Many of the plasmids used in genetic engineering code for antibiotic resistance, e.g. pBR322 carries separate genes for ampicillin resistance (*amp*) and tetracycline resistance (*tet*) (Fig. 34.3a). These genes act as markers. One gene (e.g. *amp*) can be used to select for transformants, which form colonies on agar-based media in the presence of the antibiotic while non-transformed cells will be killed. The other gene (e.g. *tet*) is used as a marker for the recombinant plasmid, since the ligation of the target DNA sequence into this gene causes insertional inactiviation (Fig. 34.3c). Thus cells transformed with the recombinant plasmid will be resistant to ampicillin only, while cells transformed with the recircularized (native) plasmid will be resistant to both antibiotics.

One method of distinguishing between these two types of transformant is to use replica plating:

1. Transformants are first grown on Petri plates containing agar medium with added ampicillin, where both types will produce single colonies after overnight incubation.
2. A sterile velvet pad is then lightly pressed on the surface of the plate, so that some cells are transferred to the pad.
3. This is then used to inoculate a second agar medium containing tetracycline (i.e. a replica plate): any colonies which grow only on the first plate must be derived from cells containing the recombinant plasmid. These colonies can then be screened for the particular target DNA, e.g. using a radioactive probe and Southern blotting (p. 167), or its protein product, e.g. using immunological methods (Chapter 32).

Plasmids of the pUC series are now more widely used than pBR322, having the following practical advantages:

- High copy number: several thousand identical copies of the plasmid are present in each bacterial cell, giving improved yield of plasmid DNA.
- Single-step selection of recombinants: the plasmid carries the gene for ampicillin resistance and a multiple cloning site within the gene for the enzyme β-galactosidase. Insertional inactivation of β-galactosidase can be detected by including a suitable enzyme inducer (e.g. isopropylthio-galactoside) and the chromogenic substrate 5-bromo-4-chloro-3-indolyl-β-*D*-galactoside (Xgal) within the agar medium. A colony of cells derived from the native plasmid will be blue, while a clone containing the recombinant molecule will be white.

Recognizing transformants — after plating bacteria onto medium containing ampicillin, you may notice a few small 'feeder' colonies surrounding a single larger (transformant) colony. The feeder colonies are derived from non-transformed cells which survive due to the breakdown of antibiotic in the medium around the transformant colony, and should not be selected for sub-culture.

Advanced laboratory techniques

35 Light measurement

Light is most strictly defined as that part of the spectrum of electromagnetic radiation detected by the human eye. However, the term is also applied to radiation just outside that visible range (e.g. UV and infra-red 'light'). Light measurement is directly relevant to several aspects of biology, including:

- photosynthesis, photomorphogenesis and photoperiodism in plants;
- perception and thermoregulation in animals;
- aquatic biology.

The nature of light

Electromagnetic radiation is emitted by the sun and by other sources (e.g. an incandescent lamp) and the electromagnetic spectrum is a broad band of radiation, ranging from cosmic rays to radio waves (Fig. 35.1). Most biological experiments involve measurements within the UV, visible and infra-red regions (generally, within the wavelength range 200–1 000 nm). Light has the characteristics of a particle and of a vibrating wave. Radiation travels in discrete particulate units, or 'packets', termed photons: a quantum is the amount of energy contained within a single photon (it is important not to confuse these

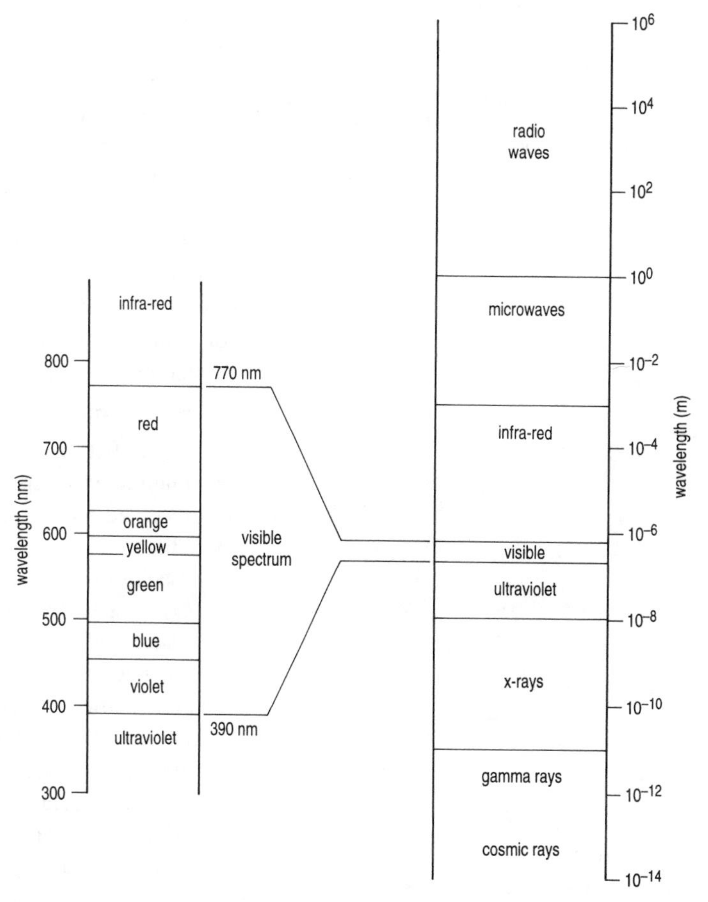

Fig. 35.1 The electromagnetic spectrum.

two terms, although they are sometimes used interchangeably in the literature). In some circumstances, it is appropriate to measure light in terms of the number of photons, usually expressed directly in moles (6.02×10^{23} photons = 1 mol) older textbooks may use the redundant term Einstein as the unit of measurement, where 1 Einstein = 1 mol photons). Alternatively, the energy content of the photon exposure may be measured (e.g. in W m^{-2}). Radiation also behaves as a vibrating electrical and magnetic field moving in a particular direction, with the magnetic and electrical components vibrating perpendicular to one another and perpendicular to the direction of travel. The wave nature of radiation gives rise to the concepts of wavelength (λ, usually measured in nm), frequency (ν, measured in s^{-1}), speed (c, the speed of electromagnetic radiation, which is 3×10^8 m s^{-1} in a vacuum), and direction. In other words, radiation is a vector quantity, where

$$c = \lambda\nu \quad [35.1]$$

Photometric and radiometric measurements

Photometric measurements

These are based on the energy perceived by a 'standard' human eye, with maximum sensitivity in the yellow–green region, around 555 nm (Fig. 35.2). The unit of measurement is the candela, which is a base unit in the SI system, defined in terms of the visual appearance of the radiation emitted by a specific quantity of platinum at its freezing point. Derived units are used for the luminous flow (lumen) and luminous flow per unit area (lux). The latter units were once used in photobiology and you may come across them in older literature. However, it is now recognized that such measurements are of little direct relevance to biologists, including even those who may wish to study the visual responses of the human eye, because they are not based on fundamental physical principles.

Judging light quality by eye — the human eye rapidly compensates for changes in light climate by varying the size of the pupil and is a poor guide to light quantity.

Radiometric measurements

The radiometric system is based on the absolute physical properties of the electromagnetic radiation itself, expressed either as the number of photons, or their energy content. The following terms are used (units of measurement are shown in parentheses):

- Photon flux (mol photons s^{-1}) is the number of photons arriving at an object within the specified time interval.
- Photon exposure (mol photons m^{-2}) is the total number of photons received by an object, usually expressed per unit surface area.
- Photon flux density (mol photons m^{-2} s^{-1}) or PFD is the most commonly used term to describe the number of photons arriving at a particular surface, expressed per unit surface area and per unit time interval.
- Photosynthetically active radiation or PAR is radiation within the waveband 400–700 nm, since the photosynthetic pigments (chlorophylls, carotenoids, etc.) show maximum absorption within this band.
- Photosynthetic photon flux density (mol photons m^{-2} s^{-1}) or PPFD is the number of photons within the waveband 400–700 nm arriving at a particular surface, expressed per unit surface area and per unit time interval. Often, this term is used interchangeably with PFD.
- Irradiance (J m^{-2} s^{-1} = W m^{-2}) is the amount of energy arriving at a surface, expressed per unit surface area and per unit time interval.

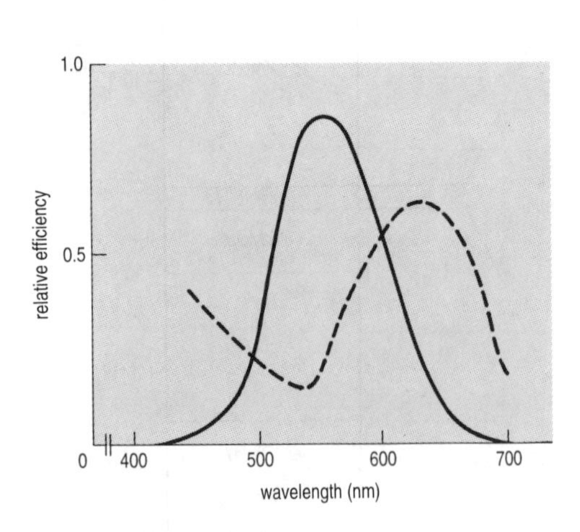

Fig. 35.2 Relative efficiency of vision (solid line) and photosynthesis (dotted line) as a function of wavelength.

- Photosynthetic irradiance (W m^{-2}) or PI is the energy of radiation within the waveband 400–700 nm arriving at a surface, expressed per unit surface area and per unit time interval.

Choice of measurement scale

Photon flux density

This is often the most appropriate unit of measurement for biological systems where individual photons are involved in the underlying process, e.g. in photosynthetic studies, where PPFD is measured, since each photochemical reaction involves the absorption of a single photon by a pigment molecule. Most modern light-measuring instruments (radiometers) can measure this quantity, giving a reading in μmol photons m^{-2} s^{-1}.

Table 35.1 Approximate conversion factors for a photosynthetic irradiance of 1 W m^{-2} to photosynthetic photon flux density (PPFD)

Source	PPFD (μmol photons m^{-2} s^{-1})
Sunlight	4.6
'Cool white' fluorescent tube	4.6
Osram 'daylight' fluorescent tube	4.6
Quartz-iodine lamp	5.0
Tungsten bulb	5.0

(*Source*: Lüning 1981)

Irradiance

This is appropriate if you are interested in the energy content of the light, e.g. if you are studying energy balance, or thermal effects. Many radiometers measure photosynthetic irradiance within the waveband 400–700 nm, giving a reading in W m^{-2}.

It is possible to make an approximate conversion between photon flux density and irradiance measurements, providing the spectral properties of the light source are known (see Table 35.1).

Spectral distribution

This can be determined using a spectroradiometer, e.g. in a study of light as a function of depth in an underwater habitat, or to compare different artificial light sources. A spectroradiometer measures irradiance or photon flux density in specific wavebands. The instrument consists of a monochromator (p. 196) to allow separate narrow wavebands (5–25 nm bandwidth) to be measured by a detector; some instruments provide a plot of the spectral characteristics of the source.

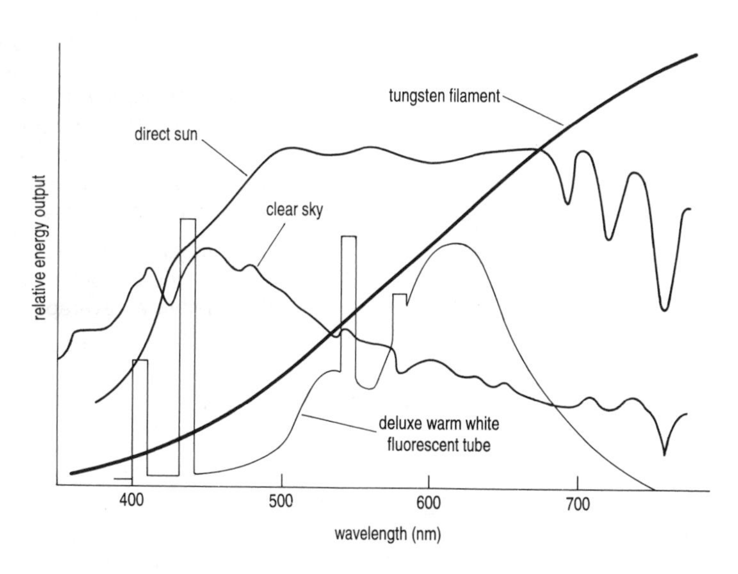

Fig. 35.3 Spectral distribution of energy output from various sources. (Adapted from Golterman *et al.* (1978).)

Using a radiometer

The main components of a radiometer are:

- Receiver: either flat-plate, hemispherical or spherical, depending upon your requirements. Most receivers incorporate a diffuser, to reduce reflection and protect the detector beneath.
- Detector: either thermoelectric or photoelectric. Some photoelectric detectors suffer from fatigue, with a decreasing response on prolonged exposure to light: check the manufacturer's handbook for recommended exposure times.
- Processor and readout device to convert the output from the detector into a visible reading, usually in analogue form (galvanometer).

Box 35.1 gives practical details of the steps involved in measuring photon flux density or irradiance using a radiometer.

Box 35.1 Measuring photon flux density or irradiance using a battery-powered radiometer

1. **Check the battery**. Most instruments have a setting that gives a direct readout of battery voltage. Recharge if necessary before use.
2. **Select the appropriate type of measurement** (e.g. photon flux density or irradiance over the PAR waveband, or an alternative range): the simpler instruments have a selection dial for this purpose.
3. **Place the sensor in the correct location and position** for the measurement: it may be appropriate to make several measurements at different positions, and take an average.
4. **Choose the most appropriate scale** for the readout device: for needle-type meters, the choice of maximum reading is usually selected by a dial, within the range 0.3 to 30 000. During use, start at a high range and work your way down until the reading is on the scale. Your final scale should be chosen to provide the most accurate reading, e.g. a reading of 15 μmol photons $m^{-2} s^{-1}$ should be made using the 0–30 scale, rather than a higher range.
5. **Read the value from the meter**. For needle-type instruments there will be two scales, the upper one marked from 0 to 10 and the lower one from 0 to 3: make sure you use the correct one, e.g. a half-scale deflection on the 0–30 scale is 15.
6. **Check that the answer is realistic**, e.g. full sunlight has a PPFD of up to 2 000 μmol photons $m^{-2} s^{-1}$, though the value will be far lower on a dull or cloudy day, while the PPFD at a distance of 1 m from a mercury lamp is around 150 μmol photons $m^{-2} s^{-1}$, and 50 μmol photons $m^{-2} s^{-1}$ at the same distance from a fluorescent lamp.

In your write-up give full details of how the measurement was made, e.g. the type of light source, instrument used, where the sensor was placed, whether an average was calculated, etc.

36 Radioactive isotopes and their use

Examples
^{12}C, ^{13}C and ^{14}C are three of the isotopes of carbon. About 98.9% of naturally occurring carbon is in the stable ^{12}C form. ^{13}C is also a stable isotope but it only occurs at 1.1% natural abundance. Trace amounts of radioactive ^{14}C are found naturally; this is a negatron-emitting radioisotope (see Table 36.2).

The isotopes of a particular element have the same number of protons in the nucleus but a different number of neutrons, giving them the same atomic number (i.e. number of protons) but a different mass number (i.e. number of protons + number of neutrons). Isotopes may be stable or radioactive. Radioactive isotopes (radioisotopes) disintegrate spontaneously at random to yield radiation and a decay product.

Radioactive decay

There are three forms of radioactivity (Table 36.1) arising from three main types of nuclear decay:

Table 36.1 Types of radioactivity and their properties

Radiation	Range of maximum energies (MeV*)	Penetration range in air (m)	Suitable shielding material
Alpha (α)	4–8	0.025–0.080	Unnecessary
Beta (β)	0.01–3	0.150–16	Plastic (e.g. Perspex®)
Gamma (γ)	0.03–3	1.3–13†	Lead

*Note that 1 MeV = 1.6×10^{-13} J.
†Distance at which radiation intensity is reduced to half.

- Alpha decay involves the loss of a particle equivalent to a helium nucleus. Alpha (α) particles, being large and positively charged, do not penetrate far in living tissue, but they do cause ionization damage and this makes them generally unsuitable for tracer studies.
- Beta decay involves the loss or gain of an electron or its positive counterpart, the positron. There are three sub-types:
  (a) Negatron (β^-) emission: loss of an electron from the nucleus when a neutron transforms into a proton. This is the most important form of decay for radioactive tracers used in biology. Negatron-emitting isotopes of biological importance include ^3H, ^{14}C, ^{32}P and ^{35}S.
  (b) Positron (β^+) emission: loss of a positron when a proton transforms into a neutron. This only occurs when sufficient energy is available from the transition and may involve the production of gamma rays when the positron is later annihilated by collision with an electron, as occurs with ^{22}Na.
  (c) Electron capture (EC): when a proton 'captures' an electron and transforms into a neutron. This may involve the production of X-rays as electrons 'shuffle' about in the atom (as with ^{125}I) and it frequently involves electron emission.
- Internal transition involves the emission of electromagnetic radiation in the form of gamma (γ) rays from a nucleus in a metastable state and always follows initial alpha or beta decay. Emission of gamma radiation leads to no change in atomic number or mass.

Note from the above that more than one type of radiation may be emitted when a radioisotope decays. The main radioisotopes used in biology and their properties are listed in Table 36.2.

Table 36.2 Properties of some isotopes used commonly in biology. Physical data obtained from Lide (1990)

Isotope	Emission(s)	Maximum energy (MeV)	Half-life	Main uses	Advantages	Disadvantages
^{3}H	β^-	0.018 61	12.3 years	Suitable for labelling organic molecules in wide range of positions at high specific activity	Relatively safe	Low efficiency of detection, high isotope effect, high rate of exchange with environment
^{14}C	β^-	0.156 48	5 715 years	Suitable for labelling organic molecules in a wide range of positions	Relatively safe, low rate of exchange with environment	Low specific activity
^{22}Na	β^+ (90%), EC	2.842 (β^+)	2.60 years	Transport studies	High specific activity	Short half-life, hazardous
^{32}P	β^-	1.710	14.3 days	Labelling proteins and nucleotides (e.g. DNA)	High specific activity, ease of detection	Short half-life, hazardous
^{35}S	β^-	0.167	87.2 days	Labelling proteins and nucleotides	Low isotope effect	Low specific activity
^{36}Cl	β^-, β^+, EC	0.709 (β^-) 1.142 (β^+, EC)	300 000 years	Transport studies	Low isotope effect	Low specific activity, hazardous
^{125}I	EC + γ	0.178 (EC)	59.9 days	Labelling proteins and nucleotides	High specific activity	Hazardous
^{131}I	β^- + γ	0.971 (β^-)	8.04 days	Labelling proteins and nucleotides	High specific activity	Hazardous

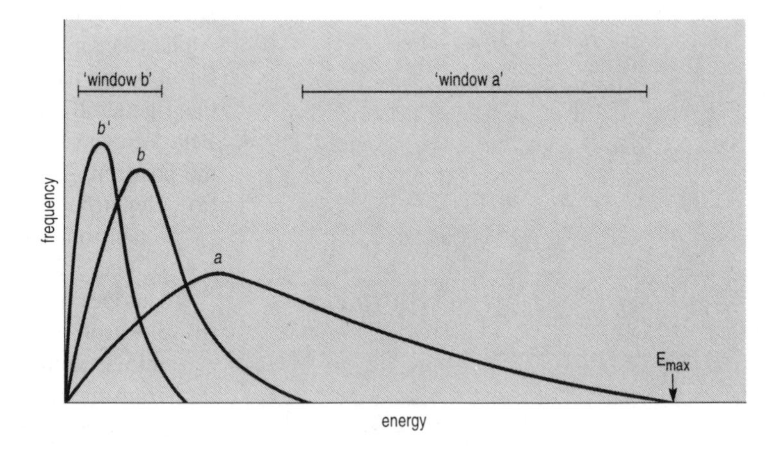

Fig. 36.1 Energy spectra for three radioactive samples as detected using a scintillation counter. Sample *a* contains a high-energy β-emitter while sample *b* contains a low-energy β-emitter, giving a lower spectral range. Sample *b'* contains the same amount of the low-energy β-emitter, but with quenching. Note that the spectral distribution is shifted to a lower energy band. The counter can be set up to record disintegration with energies that fall within a selected range (a 'window'). In this example, 'window a' could be used to count isotope *a*, while 'window b' could give a value for isotope *b* (by subtracting the counts due to isotope *a* based on the data from 'window a' and the spectral distribution of the isotope). Such dual counting allows experiments to be carried out using two isotopes simultaneously (double labelling).

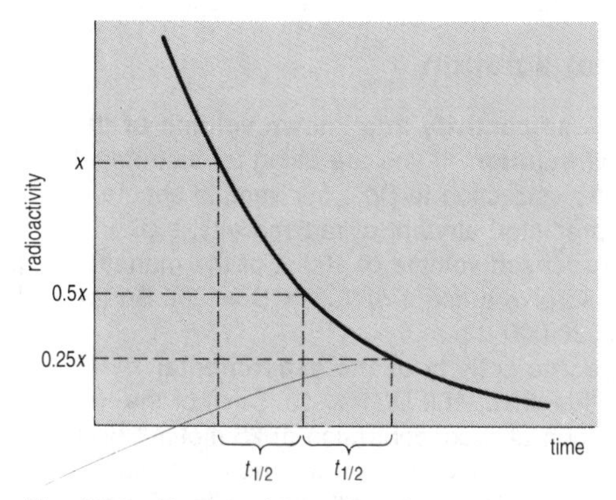

Fig. 36.2 Decay of a radioactive isotope with time. The time taken for the radioactivity to decline from x to $0.5x$ is the same as the time taken for the radioactivity to decline from $0.5x$ to $0.25x$, and so on. This time is the half-life ($t_{1/2}$) of the isotope.

Each radioactive particle or ray carries energy, usually measured in electron volts (eV). The particles or rays emitted by a particular radioisotope exhibit a range of energies, termed an energy spectrum, characterized by the maximum energy of the radiation produced, E_{max} (Fig. 36.1). The energy spectrum of a particular radioisotope is relevant to the following:

- Safety: isotopes with the highest maximum energies will have the greatest penetrating power, requiring appropriate shielding (Tables 36.1 and 36.2).
- Detection: different instruments vary in their ability to detect isotopes with different energies (Fig. 36.1).
- Discrimination: some instruments can distinguish between isotopes, based on the energy spectrum of the radiation produced.

The decay of an individual atom (a 'disintegration') occurs at random, but that of a population of atoms occurs in a predictable manner. The radioactivity decays exponentially, having a characteristic half-life ($t_{1/2}$). This is the time taken for the radioactivity to fall from a given value to half that value (Fig. 36.2). The $t_{1/2}$ values of different radioisotopes range from fractions of a second to more than 10^{19} years (see also Table 36.2). If the $t_{1/2}$ is very short, as with ^{15}O ($t_{1/2} \approx 2$ min), then it is generally impractical to use the isotope in experiments because you would need to account for the decay during the experiment and counting period.

Key point
The fraction f of the original radioactivity left at time t is given by:

$$f = e^x, \text{ where } x = \frac{-0.693t}{t_{1/2}} \quad [36.1]$$

The same units must be used for t and $t_{1/2}$.

Measuring radioactivity

The SI unit of radioactivity is the becquerel (Bq), equivalent to one disintegration per second (d.p.s.), but disintegrations per minute (d.p.m.) are also used. The curie (Ci) is a non-SI unit equivalent to the number of disintegrations produced by 1 g of radium (37 GBq). Table 36.3 shows the relationships between these units. In practice, most instruments are not able to detect all of the disintegrations from a particular sample, i.e. their efficiency is less than 100% and the rate of decay may be presented as counts min^{-1} (c.p.m.) or counts s^{-1} (c.p.s.). Most modern instruments correct for background radiation and inefficiencies in counting, converting count data to d.p.m. Alternatively, the results may be presented as the measured count rate, although this is only valid where the efficiency of counting does not vary greatly among samples.

The specific activity is a measure of the radioactivity present in a known amount of the substance:

$$\text{specific activity} = \frac{\text{quantity of radioactivity (Bq, Ci, d.p.m., etc.)}}{\text{amount of substance (mol, g, etc.)}} \quad [36.2]$$

This is an important concept in practical work involving radioisotopes, since it allows inter-conversion of disintegrations (activity) and amount of substance (see Box 36.1).

Two SI units refer to doses of radioactivity and these are used when calculating exposure levels for a particular source. The sievert (Sv) is the amount of radioactivity giving a dose in man equivalent to 1 gray (Gy) of X-rays: 1 Gy = an energy absorption of 1 J kg^{-1}. The dose received in most biological experiments is a negligible fraction of the maximum permitted exposure limit. Conversion factors from older units are given in Table 36.3.

Table 36.3
Relationships between units of radioactivity

1 Bq = 1 d.p.s.
1 Bq = 60 d.p.m.
1 Bq = 27 pCi
1 d.p.s. = 1 Bq
1 d.p.m. = 0.016 7 Bq
1 Ci = 37 GBq
1 mCi = 37 MBq
1 μCi = 37 kBq
1 Sv = 100 rem
1 Gy = 100 rad
1 Gy ≈ 100 roentgen
1 rem = 0.01 Sv
1 rad = 0.01 Gy
1 roentgen ≈ 0.01 Gy

Box 36.1 Determining the specific activity of an experimental solution

Suppose you need to make up a certain volume of an experimental solution, to contain a particular amount of radioactivity. For example, 50 ml of a mannitol solution at a concentration of 25 mol m^{-3}, to contain 5 Bq μl^{-1} — using a manufacturer's stock solution of ^{14}C-labelled mannitol (specific activity = 0.4 Ci/mmol^{-1}).

1. **Calculate the total amount of radioactivity in the experimental solution,** in this example 5 × 1 000 (to convert μl to ml) × 50 (50 ml required) = 2.5 × 10^5 Bq (i.e. 250 kBq).
2. **Establish the volume of stock radioisotope solution required:** for example, a manufacturer's stock solution of ^{14}C-labelled mannitol contains 50 μCi of radioisotope in 1 ml of 90% v/v ethanol: water. Using Table 36.3, this is equivalent to an activity of 50 × 37 = 1 850 kBq. So, the volume of solution required is 250/1850 of the stock volume, i.e. 0.135 1 ml (135 μl).
3. **Calculate the amount of non-radioactive substance required** as for any calculation involving concentration (see pp. 18, 24), e.g. 50 ml (50 × 10^{-6} m^3) of a 25 mol m^{-3} mannitol (formula mass 182.17) will contain 50 × 10^{-6} × 25 × 182.17 = 0.227 7 g.
4. **Check the amount of radioactive isotope to be added.** In most cases, this represents a negligible amount of substance, e.g. in this instance, 250 kBq of stock solution at a specific activity of 14.8 × 10^6 kBq/mmol^{-1} (converted from 0.4 Ci mmol^{-1} using Table 36.3) is equal to 250/14 800 000 = 16.89 nmol, equivalent to approximately 3 μg mannitol. This can be ignored in calculating the mannitol concentration of the experimental solution.
5. **Make up the experimental solution** by adding the appropriate amount of non-radioactive substance and the correct volume of stock solution.
6. **Measure the radioactivity in a known volume of the experimental solution.** If you are using an instrument with automatic correction to Bq, your sample should contain the predicted amount of radioactivity, e.g. an accurately dispensed volume of 100 μl of the mannitol solution should give a count of 100 × 5 = 500 Bq (or 500 × 60 = 30 000 d.p.m.).
7. **Note the specific activity of the experimental solution**: in this case, 100 μl (1 × 10^{-7} m^3) of the mannitol solution at a concentration of 25 mol m^{-3} will contain 25 × 10^{-7} mol (2.5 μmol) mannitol. Dividing the radioactivity in this volume (30 000 d.p.m.) by the amount of substance (eqn [36.2]) gives a specific activity of 30 000/2.5 = 12 000 d.p.m. μmol^{-1}, or 12 d.p.m. nmol^{-1}. This value can be used:
   - (a) To assess the accuracy of your protocol for preparing the experimental solution: if the measured activity is substantially different from the predicted value, you may have made an error in making up the solution.
   - (b) To determine the counting efficiency of an instrument; by comparing the measured count rate with the value predicted by your calculations.
   - (c) To interconvert activity and amount of substance: the most important practical application of specific activity is the conversion of experimental data from counts (activity) into amounts of substance. This is only possible where the substance has not been metabolized or otherwise converted into another form; e.g., a tissue sample incubated in the experimental solution described above with a measured activity of 245 d.p.m. can be converted to nmol mannitol by dividing by the specific activity, expressed in the correct form. Thus 245/12 = 20.417 nmol mannitol.

The most important methods of measuring radioactivity for biological purposes are described below.

The Geiger–Müller (G–M) tube

This operates by detecting radiation when it ionizes gas between a pair of electrodes across which a voltage has been applied. You should use a hand-held Geiger–Müller tube for routine checking for contamination (although it will not pick up ^3H activity).

The scintillation counter

This operates by detecting the scintillations (fluorescence 'flashes') produced when radiation interacts with certain chemicals called fluors. In solid (or external) scintillation counters (often referred to as 'gamma counters') the radioactivity causes scintillations in a crystal of fluorescent material held close to the sample. This method is only suitable for radioisotopes producing penetrating radiation.

Liquid scintillation counters are mainly used for detecting beta decay and they are especially useful in biology. The sample is dissolved in a suitable solvent containing the fluor(s) — the 'scintillation cocktail'. The radiation first interacts with the solvent, and the energy from this interaction is passed to the fluors which produce detectable light. The scintillations are measured by photomultiplier tubes which turn the light pulses into electronic pulses, the magnitude of which is directly related to the energy of the original radioactive event. The spectrum of electronic pulses is thus related to the energy spectrum of the radioisotope.

Modern liquid scintillation counters use a series of electronic 'windows' to split the pulse spectrum into two or three components. This may allow more than one isotope to be detected in a single sample, provided their energy spectra are sufficiently different (Fig. 36.1). A complication of this approach is that the energy spectrum can be altered by pigments and chemicals in the sample, which absorb scintillations or interfere with the transfer of energy to the fluor; this is known as quenching (Fig. 36.1). Most instruments have computer-operated quench correction facilities (based on measurements of standards of known activity and energy spectrum) which correct for such changes in counting efficiency.

Correcting for quenching — find out how your instrument corrects for quenching and check the quench indication parameter (QIP) on the printout, which measures the extent of quenching of each sample. Large differences in the QIP would indicate that quenching is variable among samples and might give you cause for concern.

Box 36.2 Tips for preparing samples for liquid scintillation counting

Modern scintillation counters are very simple to operate; problems are more likely to be due to inadequate sample preparation than to incorrect operation of the machine. Common pitfalls are the following:

- Incomplete dispersal of the radioactive compound in the scintillation cocktail. This may lead to underestimation of the true amount of radioactivity present:
  (a) Water-based samples may not mix with the scintillation cocktail — change to an emulsifier-based cocktail. Take care to observe the recommended limits, upper and lower, for amounts of water to be added or the cocktail may not emulsify properly.
  (b) Solid specimens may absorb disintegrations or scintillations: extract radiochemicals using an intermediate solvent like ethanol (ideally within the scintillation vial) and then add the cocktail. Tissue solubilizing compounds such as Soluene® are effective, particularly for animal material, but extremely toxic, so the manufacturer's instructions should be followed closely. Radioactive compounds on slices of agarose or polyacrylamide gels may be extracted using a product such as Protosol® . Agarose gels can be dissolved in a small volume of boiling water.
  (c) Particulate samples may sediment to the bottom of the scintillation vial — suspend them by forming a gel. This can be done with certain emulsifier-based cocktails by adding a specific amount of water.
- Chemiluminescence. This is where a chemical reacts with the fluors in the scintillation cocktail causing spurious scintillations, a particular risk with solutions containing strong bases or oxidizing agents. Symptoms include very high initial counts which decrease through time. Possible remedies are:
  (a) Leave the vials at room temperature for a time before counting. Check with a suitable blank that counts have dropped to an acceptable level.
  (b) Neutralize basic samples with acid (e.g. acetic acid or HCl).
  (c) Use a scintillation cocktail that resists chemiluminescence such as Hiconicfluor® .
  (d) Raise the energy of the lower counts detected to about 8 keV — most chemiluminescence pulses are weak (0–7 keV). This method is not suitable for ^3H.

Many liquid scintillation counters treat the first sample as a 'background', subtracting whatever value is obtained from the subsequent measurements as part of the procedure for converting to d.p.m. Make sure that you use an appropriate background sample, identical in all respects to your radioactive sample but with no added radioisotope, in the correct position within the machine. Check that the background reading is reasonable (15–30 c.p.m.). Tips for preparing samples for liquid scintillation counting are given in Box 36.2.

Autoradiography

This is a method where photographic film is exposed to the isotope. It is used mainly to locate radioactive tracers in thin sections of an organism or on chromatography papers or gels, but quantitative work is possible. The radiation interacts with the film in a similar way to light, silver grains being formed in the developed film where the particles or rays have passed through. The radiation must have enough energy to penetrate into the film, but if it has too much energy the grain formation may be too distant from the point where the isotope was located to identify precisely the point of origin (e.g. high-energy β emitters). Autoradiography is a relatively specialized method and specific lab protocols should be followed.

Biological applications for radioactive isotopes

The main advantages from using radioactive isotopes in biological experiments are:

- Radioactivity is readily detected. Methods of detection are sufficiently sensitive to measure extremely small amounts of radioactive substances.
- Studies can be carried out on intact, living organisms. If care is taken, minimal disruption of normal conditions will occur when radio-labelled compounds are introduced.
- Protocols are simple compared to equivalent methods for chemical analysis.

The main disadvantages are:

- The 'isotope effect'. Molecules containing different isotopes of the same atom may react at slightly different rates and behave in slightly different ways to the natural isotope. The isotope effect is more extreme the smaller the atom and is most important for ^{3}H-labelled compounds of low molecular mass.
- The possibility of mistaken identity. The presence of radioactivity does not tell you anything about the compound in which the radioactivity is present: it could be different from the one in which it was applied, due to metabolism or spontaneous breakdown of a ^{14}C-containing organic compound.

The main types of experiments are:

- Investigations of metabolic pathways: a radioactively labelled substrate is added (often to an *in vitro* experimental 'system' rather than a whole organism) and samples taken at different time intervals. By identifying the labelled compounds and plotting their appearance through time, an indication of the pathway of metabolism can be obtained.
- Translocation studies: radioisotopes are used to follow the fate of molecules within an organism. Uptake and translocation rates can be determined with relative ease.

- Radio-dating: the age of plant or mineral samples can be determined by measuring the amount of a radioisotope in the sample. The age of the specimen can be found using the $t_{1/2}$ by assuming how much was originally incorporated.
- Ecological studies: radioisotopic tracers provide a convenient method for determining food web interrelationships and for investigating behaviour patterns, while environmental monitoring may involve following the 'spectral signature' of isotopes deliberately or accidentally released.
- Mutagenesis and sterilization: radioactive sources can be used to induce mutations, particularly in microorganisms. Gamma emitters of high energy will kill microbes and are used to sterilize equipment such as disposable Petri dishes.
- Assays: radioisotopes are used in several quantitative detection methods of value to biologists. Radioimmunoassay is described on p. 156. Isotope dilution analysis works on the assumption that introduced radio-labelled molecules will equilibrate with unlabelled molecules present in the specimen. The amount of substance initially present can be worked out from the change in specific activity of the radioisotope when it is diluted by the 'cold' material. A method is required whereby the substance can be purified from the sample and sufficient substance must be present for its mass to be measured accurately.

Billington *et al.* (1992) give further details and practical advice on using radioisotopes in biological experiments.

Working practices when using radioactive isotopes

By law, undergraduate work with radioactive isotopes must be very closely supervised. In practical classes, the protocols will be clearly outlined, but in project work you may have the opportunity to plan and carry out your own experiments, albeit under supervision. Some of the factors that you should take into account, based on the assumption that your department and laboratory are registered under the Radioactive Substances Act (1960), are discussed below:

1. Must you use radioactivity? If not, it is a legal requirement that you use the alternative method.
2. Have you registered for radioactive work? Normal practice is for all users to register with a local Radiation Protection Supervisor (Ionizing Radiations Act, 1985). Details of the project may have to be approved by the appropriate administrator(s). You may have to have a short medical examination before you can start work.
3. What labelled compound will you use? Radioactive isotopes must be ordered well in advance through your department's Radiation Protection Supervisor. Aspects that need to be considered include:
   (a) The radionuclide. With many organic compounds this will be confined to ^{3}H and ^{14}C (but see Table 36.2). The risk of a significant 'isotope effect' may influence this decision (see above).
   (b) The labelling position. This may be a crucial part of a metabolic study. Specifically labelled compounds are normally more expensive than those that are uniformly ('generally') labelled.
   (c) The specific activity. The upper limit for this is defined by the isotope's half-life, but below this the higher the specific activity, the more expensive the compound.

4. Are suitable facilities available? You'll need a suitable work area, preferably out of the way of general lab traffic and within a fume cupboard for those cases where volatile radioactive substances are used or may be produced.

Each new experiment should be planned carefully and precise experimental protocols laid down in advance so you work as safely as possible and do not waste expensive radioactively labelled compounds. In conjunction with your supervisor, decide whether your method of application will introduce enough radioactivity into the system, how you will account for any loss of radioactivity during recovery of the isotope and whether there will be enough activity to count at the end. You should be able to predict approximately the amount of radioactivity in your samples, based on the specific activity of the isotope used, the expected rate of uptake/exchange and the amount of sample to be counted. Use the isotope's specific activity to estimate whether the non-radioactive ('cold') compound introduced with the radio-labelled ('hot') compound may lead to excessive concentrations being administered. Advice for handing data is given in Box 36.1.

Carrying out a 'dry run' — consider doing this before working with radioactive compounds, perhaps using a dye to show the movement or dilution of introduced liquids, as this will lessen the risks of accident and improve your technique.

Safety and procedural aspects

Make sure the bench surface is one that can be easily decontaminated by washing (e.g. Formica®) and always use a disposable surfacing material such as Benchkote®. It is good practice to carry out as many operations as possible within a Benchkote®-lined plastic tray so that any spillages are contained. You will need a lab coat to be used exclusively for work with radioactivity, safety spectacles and a supply of thin latex or vinyl disposable gloves. Suitable vessels for liquid waste disposal will be required and special plastic bags for solids — make sure you know beforehand the disposal procedures for liquid and solid wastes. Wash your hands after handling a vessel containing a radioactive solution and again before removing your gloves. Gloves should be placed in the appropriate disposal bag as soon as your experimental procedures are complete.

Using Benchkote® — the correct way to use Benchkote® and similar products is with the waxed surface down (to protect the bench or tray surface) and the absorbent surface up (to absorb any spillage). Write the date in the corner when you put down a new piece. Monitor using a G–M tube and replace regularly under normal circumstances. If you are aware of spillage, replace immediately and dispose of correctly.

It is important to comply with the following guidelines:

- Read and obey the local rules for safe usage of radiochemicals.
- Maximize the distance between you and the source as much as possible.
- Minimize the duration of exposure.
- Wear protective clothing (properly fastened lab coat, safety glasses, gloves) at all times.
- Use appropriate shielding at all times (Table 36.1).
- Monitor your working area for contamination frequently.
- Mark all glassware, trays, bench work areas, etc., with tape incorporating the international symbol for radioactivity (Fig. 36.3).
- Keep adequate records of what you have done with a radioisotope — the stock remaining and that disposed of in waste form must agree.
- Store radio-labelled compounds appropriately and return them to storage areas immediately after use.
- Dispose of waste promptly and with due regard for local rules.
- Make the necessary reports about waste disposal, etc., to your Departmental Radiation Protection Supervisor.
- Clear up after you have finished each experiment.
- Wash thoroughly after using radioactivity.
- Monitor the work area and your body when finished.

Fig. 36.3 Tape showing the international symbol for radioactivity

37 Polarography and manometry

Oxygen electrodes

These instruments measure oxygen in solution using the polarographic principle, i.e. by monitoring the current flowing between two electrodes when a voltage is applied. The most widespread electrode is the Clark type (Fig. 37.1), manufactured by Rank Bros, Cambridge, which is suitable for measuring O_2 concentrations in cell, organelle and enzyme suspensions. Platinum and silver electrodes are in contact with a solution of electrolyte (normally saturated KCl). The electrodes are separated from the medium by a Teflon® membrane, permeable to O_2. When an electrical potential is applied to the electrodes, this generates a current proportional to the O_2 concentration. The reactions can be summed up as:

$4Ag \rightarrow 4Ag^+ + 4e^-$ (at silver anode)

$O_2 + 2e^- + 2H^+ \rightleftharpoons H_2O_2$ (in electrolyte solution; O_2 replenished by diffusion from test solution)

$H_2O_2 + 2e^- + 2H^+ \rightleftharpoons 2H_2O$ (at platinum cathode)

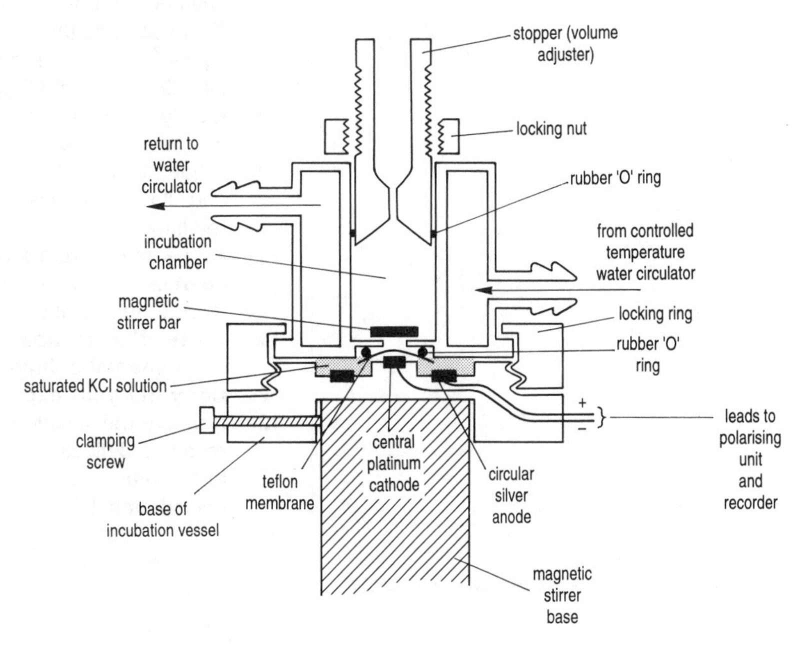

Fig. 37.1 Transverse section through a Clark (Rank) oxygen electrode.

Setting up and using a Clark (Rank) oxygen electrode

Box 37.1 describes the steps involved: if you are setting up from scratch, perform steps 1–13; if a satisfactory membrane is already in place, start at step 7.

The temperature of the incubation vessel can be controlled by passing water (e.g. from a water bath) through the outer chamber. Cells or organelles may be present in the solution added to the incubation chamber or can be added via the hole in the stopper using a syringe, as can any chemicals you wish

Box 37.1 How to set up a Clark (Rank) oxygen electrode

1. **Detach the base of the incubation vessel** (see Fig. 37.1) by unscrewing the locking ring.
2. **Add enough saturated KCl to cover the electrodes.**
3. **Cut a 1 mm square hole in the centre of a 10 × 10 mm square of lens tissue** and place this so that the hole is over the central platinum cathode.
4. **Cut a 10 × 10 mm square of Teflon® membrane and place over the lens tissue**; seal by gently lowering the incubation vessel and tightening the locking ring, making sure that the rubber O-ring is correctly positioned over the membrane.
   (a) Do not overtighten the locking ring.
   (b) Take care not to trap air bubbles beneath the membrane.
   (c) Make sure that the membrane does not become twisted.
5. **Clamp the electrode over the magnetic stirrer base using the clamping screw.**
6. **Connect the electrode leads to the polarizing unit/recording device** (silver anode to positive, platinum cathode to negative). In most cases, the output will be passed to a chart recorder, to give a readout of oxygen status as a function of time.
7. **Add air-saturated experimental solution and a small Teflon® -coated magnetic stirrer bar to the chamber.** The volume of the incubation chamber can be adjusted by moving the locking nut on the stopper. To adjust, add the appropriate amount of liquid to the chamber using a pipette, insert the stopper and screw the locking nut until the solution just fills the incubation chamber.
8. **Gently push the stopper (volume adjuster) into position**, making sure that no air bubbles are trapped in the chamber, and switch on the stirrer.
9. **Adjust the sensitivity control of the recorder** to give a suitable reading and allow to stabilize — this may take 5–10 min. This is the air-saturated reading.
10. **Add a few crystals of sodium dithionite to absorb all the O_2 in solution**; this gives a zero reading (an alternative is to bubble N_2 through the solution for several minutes).
11. **Rinse the incubation chamber thoroughly and add fresh experimental solution.**
12. **Carry out your experiment.**
13. **Remove the solution and check the calibration** (steps 7 to 10). If the recorder deflections are different, the electrode's sensitivity or the temperature may have changed and you may need to repeat the measurement.

to introduce, such as metabolic substrates or inhibitors. Take care not to introduce air bubbles and remove any that appear by gently raising and lowering the stopper. The electrode can be used repeatedly, providing the membrane is satisfactory: remove solutions carefully (e.g. using a pipette, or vacuum line and trap). Keep water in the chamber when not in use. Replace the membrane if:

- the reading becomes noisy;
- the electrode will not zero after adding sodium dithionite;
- the response becomes too slow (check by switching off the stirrer — the oxygen concentration should drop rapidly as the available oxygen is consumed).

Changing the Teflon® membrane — make sure that the electrodes are clean. Use a mild abrasive cleaning paste.

Examples
At 37°C, air-saturated distilled water contains O_2 at 0.212 mol m^{-3} (Table 37.1), so in 5×10^{-6} m^3 of medium, there will be 1.06 μmol O_2. If the deflection between zero and fully saturated solution is 67 divisions, each division will correspond to 1.06/67 = 0.0158 μmol O_2 (15.8 nmol per division).

For the above calibration, for a trace with a gradient of −14 units over 2 cm of chart paper at a speed of 0.2 cm min^{-1} (10 min), the O_2 consumption rate would be 14/10 × 15.8 = 22.1 nmol min^{-1}.

If there were 1.2×10^9 cells in the electrode, the O_2 consumption rate on a cellular basis would be $22.1 \times 10^{-9}/1.2 \times 10^9$ = 18.4 amol O_2 min^{-1} $cell^{-1}$.

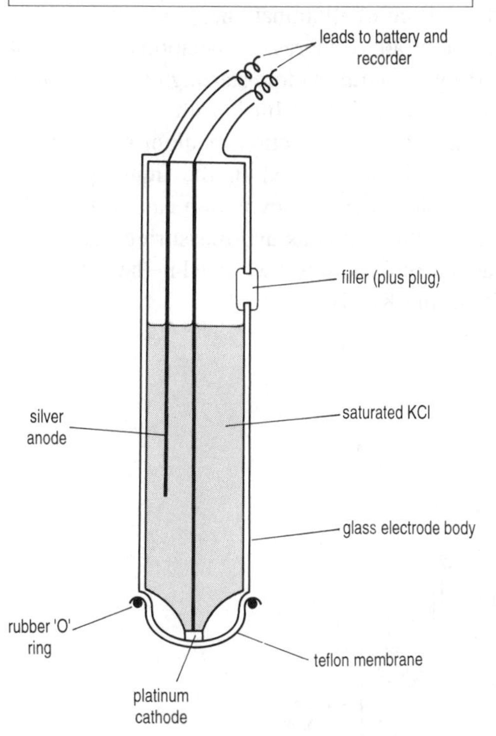

Fig. 37.2 A Clark-type oxygen probe.

Definition
Respiratory quotient (RQ) — the ratio of CO_2 evolved to O_2 consumed; this ratio is useful because it can give an indication of the type of substrate being used in metabolism. The complete oxidation of carbohydrates should result in RQ values of about 1.0, whereas fat metabolism will give a value of about 0.7 and protein metabolism a value of about 0.8.

Converting chart traces to rates of O_2 consumption or evolution

1. Calibrate the chart recorder using the zero and fully saturated values as reference points. The concentration of O_2 in saturated aqueous solution changes as a function of temperature and salt concentration (see Table 37.1) so multiply the appropriate value by the volume of the solution in the incubation chamber (in m^3) to obtain the number of moles of O_2 present. Divide by the number of chart divisions to find the number of moles per chart division.
2. Convert the gradient of the trace into a rate of oxygen consumption or evolution. Only use portions of the trace showing a stable rate for at least 5 min. Work out the gradient as divisions per unit time, taking account of the speed of the chart recorder. Use the calibration to convert this to a rate of O_2 consumption or evolution.
3. Express the rate on a cellular, chlorophyll or protein basis. Find the number of cells in the suspension (e.g. by using a haemocytometer, p. 129) and divide the figure obtained in (2) by this. Alternatively, find the chlorophyll or protein concentration (pp. 197 and 160 respectively) and divide by this value.

Oxygen probes

Clark-type oxygen electrodes are also available in probe form for immersion in the test solution (Fig. 37.2). Such probes are particularly useful for field studies since they allow direct measurement of oxygen status *in situ*, in contrast to chemical assays (e.g. the Winkler method). The main point to note is that the solution must be stirred during measurement, to replenish the oxygen consumed by the electrode. Golterman *et al.* (1978) discuss field sampling protocols for measuring dissolved gases in fresh waters.

Manometric techniques

These involve measurements of pressure or volume change as gases are produced or consumed. They are principally used to study the exchange of O_2 and CO_2 between an organism and the environment and to measure respiratory quotients. The reaction of interest takes place in a flask attached to a manometer (Fig. 37.3); the flask is kept in a water bath to control temperature (precise control is required, i.e ± 0.05°C) with constant shaking, to ensure equilibration of gases between the gas and liquid phases. The calculations involved in determining gas production or consumption are complex and specific to individual instruments, so follow the manufacturer's guidelines.

- The Warburg manometer operates by keeping the volume of the flask–manometer system constant and measuring *pressure* changes via the manometer.
- The Gilson differential manometer operates by maintaining constant pressure within the flask–manometer system and measuring *volume* changes via the manometer.

Manometry is relatively insensitive and subject to relatively large measurement errors, but it is useful for measurements on intact organisms (e.g. microbial suspensions, seeds).

Basic procedure

1. Place the experimental material in a clean flask; any reactants (e.g. a substrate or metabolic inhibitor) are added to the side arm.

Table 37.1 O_2 saturation values for distilled and sea water at standard atmospheric pressure and a range of temperatures. (*Source*: Green and Carritt, 1967)

Temperature (°C)	O_2 saturation concentration ($mol\ m^{-3}$)	
	Distilled water	Sea water (salinity 35‰)
0	0.460	0.359
2	0.435	0.342
4	0.413	0.326
6	0.392	0.311
8	0.373	0.298
10	0.355	0.285
12	0.339	0.273
14	0.324	0.261
16	0.310	0.251
18	0.297	0.241
20	0.285	0.232
22	0.274	0.224
24	0.263	0.215
26	0.253	0.208
28	0.244	0.200
30	0.235	0.193
37	0.212	0.174

Note: Tabulated values assume atmospheric pressure = 101.3 kPa; for more accurate work, a correction for any deviation from this value can be made by multiplying the appropriate figures by the ratio of the real pressure to the assumed value.

2. Set up control flasks in which CO_2 is absorbed by an alkali-saturated filter paper wick placed in the central well (a reference flask is used as a control, to compensate for any temperature and pressure changes that occur during the experiment).
3. Attach the flask to the apparatus by the (well-greased) ground glass joint, hold in position with an elastic band and switch on the shaking mechanism.
4. Equilibrate the flask for at least 10 min and check the ground glass joints for leakage before resetting the manometer to zero.
5. Mix the reactants after a further short equilibration period, by disconnecting the flask and tipping the contents of the side-arm into the experimental solution. Take an initial reading and switch on the shaking mechanism.
6. Stop at appropriate intervals and take further readings.

Take care over the following points:

- Check the accuracy of your settings and readings on the manometer scale; this is vital, as it is the most important source of error.
- If studying photosynthesis, you will need to use a manometer with a transparent water bath and a source of illumination.
- If studying the effects of different atmospheres on metabolism, the flask will have to be flushed with the gas mixture before starting the experiment, e.g. for studying the effects of anaerobiosis, flush with N_2.
- If studying CO_2 exchange, the pH of the reaction medium must be less than 5, otherwise some CO_2 will be retained in the liquid phase as bicarbonate. If this is not possible, total CO_2 evolution can be found by adding strong acid to the alkali-saturated wick and measuring the change in the manometer reading (a control flask must be treated in the same way, to take account of any CO_2 in the KOH).

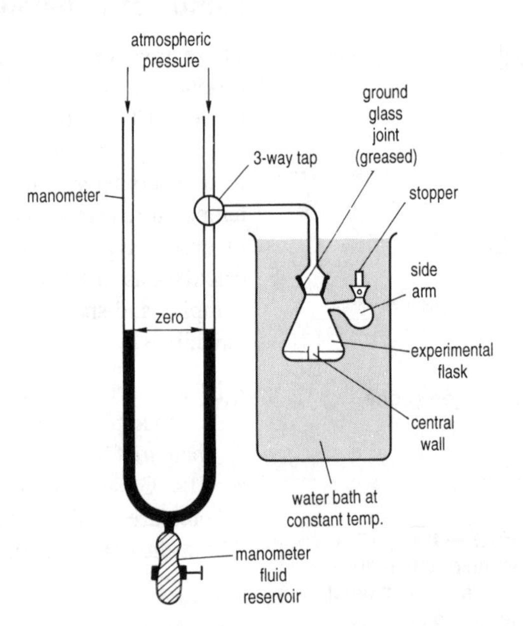

Fig. 37.3 A manometer and flask from a Warburg manometer.

38 Centrifugation

Particles suspended in a liquid will move at a rate which depends on:

- the applied force — particles in a stationary test tube will move in response to the earth's gravity;
- the density difference between the particles and the liquid — particles less dense than the liquid will float upwards while particles denser than the liquid will sink;
- the size and shape of the particles;
- the viscosity of the medium.

For most biological particles (cells, organelles or molecules) the rate of flotation or sedimentation in response to the earth's gravity is too slow to be of practical use in separation. A centrifuge is an instrument designed to produce a centrifugal force far greater than the earth's gravity, by spinning the sample about a central axis (Fig. 38.1). Particles of different size, shape or density will thereby sediment at different rates, depending on the speed of rotation and their distance from the central axis.

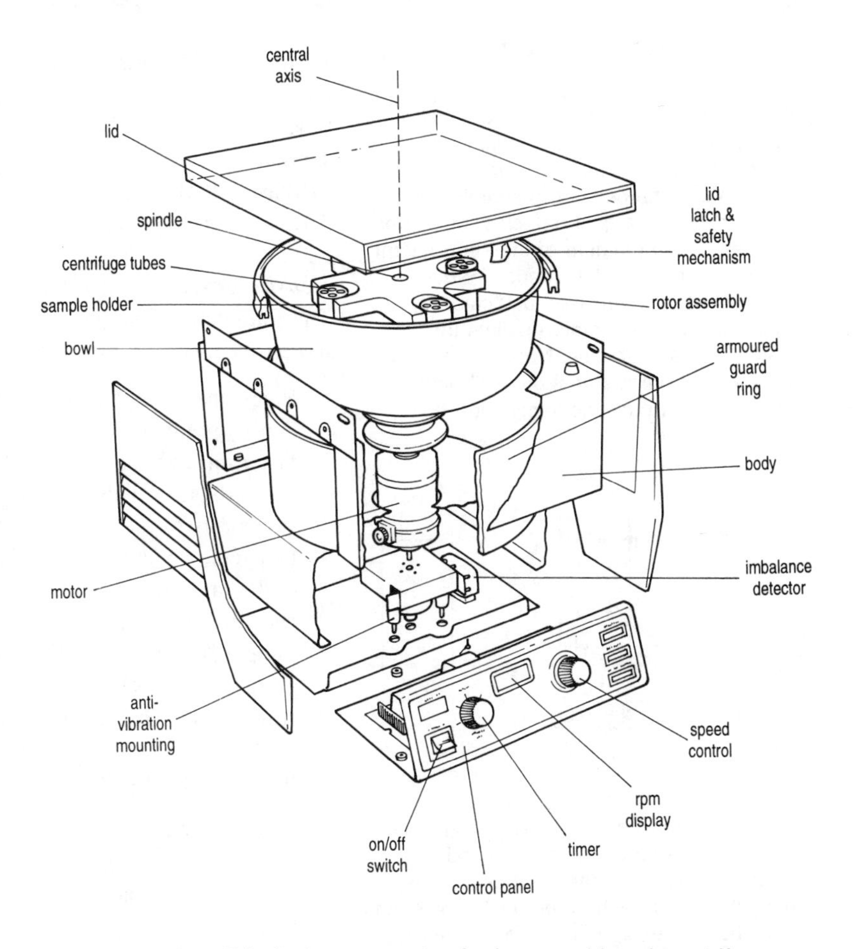

Fig. 38.1 Principal components of a low-speed bench centrifuge.

To convert RCF to SI acceleration, multiply by 9.80 m s^{-2}.

Examples
Suppose you wanted to calculate the RCF of a bench centrifuge with a rotor of r_{av} = 95 mm running at a speed of 3 000 r.p.m. Using eqn [38.1] the RCF would be: 1.118 × 95 × $(3)^2$ = 956 g.

You might wish to calculate the speed (r.p.m.) required to produce a relative centrifugal field of 2000 g using a rotor of r_{av} = 85 mm. Using eqn [38.2] the speed would be: 945.7 √(2000/85) = 4 587 r.p.m.

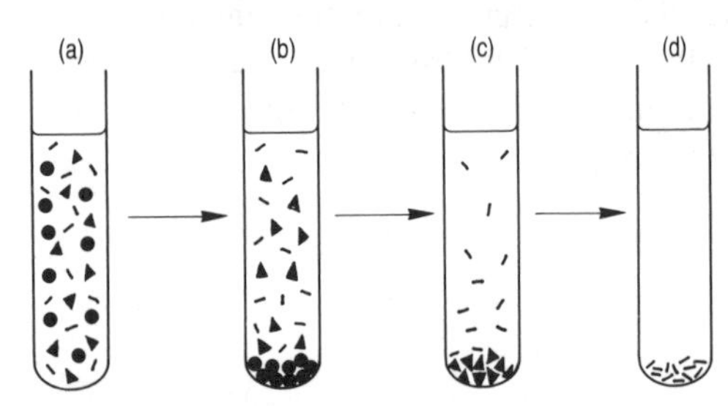

Fig. 38.2 Differential sedimentation. (a) Before centrifugation, the tube contains a mixed suspension of large, medium and small particles. (b) After low-speed centrifugation, the pellet is predominantly composed of the largest particles. (c) Further high-speed centrifugation of the supernatant will give a second pellet, predominantly composed of medium-sized particles. (d) A final ultracentrifugation step pellets the remaining small particles. Note that all of the pellets apart from the final one will have some degree of cross-contamination of smaller particles.

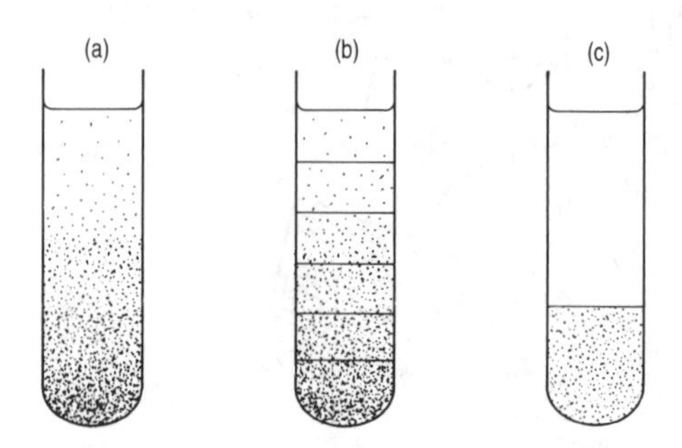

Fig. 38.3 Density gradients. (a) A continuous (linear) density gradient. (b) A discontinuous (stepwise) density gradient, formed by layering solutions of decreasing density on top of each other. (c) A single-step density barrier, designed to allow selective sedimentation of one type of particle.

How to calculate centrifugal acceleration

The acceleration of a centrifuge is usually expressed as a multiple of the acceleration due to gravity (g = 9.80 m s^{-2}), termed the relative centrifugal field (RCF, or 'g value'). The RCF depends on the speed of the rotor (n, in revolutions per minute, r.p.m.) and the radius of rotation (r, in mm) where:

$$\text{RCF} = 1.118\, r \left(\frac{n}{1\,000}\right)^2 \qquad [38.1]$$

This relationship can be rearranged, to calculate the speed (r.p.m.) for specific values of r and RCF:

$$n = 945.7 \sqrt{\left(\frac{\text{RCF}}{r}\right)} \qquad [38.2]$$

However, you should note that RCF is not uniform within a centrifuge tube: it is highest near the outside of the rotor (r_{max}) and lowest near the central axis (r_{min}). In practice, it is customary to report the RCF calculated from the average radius of rotation (r_{av}), as shown in Fig. 38.5. It is also worth noting that RCF varies as a *squared* function of the speed: thus the RCF will be doubled by an increase in speed of approximately 41%.

Centrifugal separation methods

Differential sedimentation (pelleting)

By centrifuging a mixed suspension of particles at a specific RCF for a particular time, the mixture will be separated into a pellet and a supernatant (Fig. 38.2). The successive pelleting of a suspension by spinning for a fixed time at increasing RCF is widely used to separate organelles from cell homogenates. The same principle applies when cells are harvested from a liquid medium.

Density gradient centrifugation

The following techniques use a density gradient, a solution which increases in density from the top to the bottom of a centrifuge tube (Fig. 38.3).

- Rate-zonal centrifugation. By layering a sample onto a shallow pre-formed density gradient, followed by centrifugation, the larger particles will move faster through the gradient than the smaller ones, forming several distinct zones (bands). This method is time dependent, and centrifugation must be stopped before any band reaches the bottom of the tube (Fig. 38.4).
- Isopycnic centrifugation. This technique separates particles on the basis of their buoyant density. Several substances form density gradients during centrifugation (e.g. sucrose, CsCl, Ficoll®, Percoll®, Nycodenz®). The sample is mixed with the appropriate substance and then centrifuged — particles form bands where their density corresponds to that of the medium (Fig. 38.4). This method requires a steep gradient and sufficient time to allow gradient formation and particle redistribution, but is unaffected by further centrifugation.

Bands within a density gradient can be sampled using a fine Pasteur pipette, or a syringe with a long, fine needle. Alternatively, the tube may be punctured and the contents (fractions) collected dropwise in several tubes. For accurate work, an upward displacement technique can be used (see Ford and Graham, 1991).

Working with silicone oil — the density of silicone oil is temperature-sensitive so work in a location with a known, stable temperature or the technique may fail.

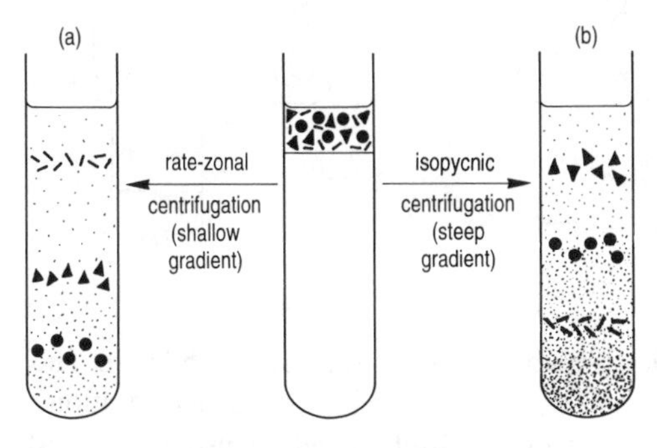

Fig. 38.4 Density gradient centrifugation. The central tube shows the position of the sample prior to centrifugation, as a layer on top of the density gradient medium. Note that particles sediment on the basis of size during rate-zonal centrifugation (a), but form bands in order of their densities during isopycnic centrifugation (b). •, large particles, intermediate density; ▲, medium-sized particles, low density; —, small particles, high density.

Density barrier centrifugation

A single step density barrier (Fig. 38.3c) can be used to separate cells from their surrounding fluid, e.g. using a layer of silicone oil adjusted to the correct density using dinonyl phthalate. Blood cell types can be separated using a density barrier of e.g. Ficoll®

Types of centrifuge and their uses

Low-speed centrifuges

These are bench-top instruments for routine use, with a maximum speed of 3 000–6 000 rp.m. and RCF up to 6 000 *g* (Fig. 38.1). They are used to harvest cells, larger organelles (e.g. nuclei, chloroplasts) and coarse precipitates (e.g. antibody–antigen complexes, p. 154). Most modern machines also have a sensor that detects any imbalance when the rotor is spinning and cuts off the power supply (Fig. 38.1). However, some of the older models do not, and must be switched off as soon as any vibration is noticed, to prevent damage to the rotor or harm to the operator.

High-speed centrifuges

These are usually larger, free-standing instruments with a maximum speed of up to 25 000 r.p.m. and RCF up to 60 000*g*. They are used for microbial cells, many organelles (e.g. mitochondria, lysosomes) and protein precipitates. They often have a refrigeration system to keep the rotor cool at high speed. You would normally use such instruments only under direct supervision.

Ultracentrifuges

These are the most powerful machines, having maximum speeds in excess of 30 000 r.p.m. and RCF up to 600 000*g*, with sophisticated refrigeration and vacuum systems. They are used for smaller organelles (e.g. ribosomes, membrane vesicles) and biological macromolecules. You would not normally use an ultracentrifuge, though your samples may be run by a member of staff.

Most departments have a log book for recording usage (samples/speeds/times) of high-speed centrifuges/ultracentrifuges: make sure you record these details, as the information is important for servicing and replacement of rotors.

Microcentrifuges (microfuges)

These are bench-top machines, capable of rapid acceleration up to 12 000 r.p.m. and 10 000*g*. They are used to sediment small sample volumes (up to 1.5 ml) of larger particles (e.g. cells, precipitates) over short time-scales (typically, 0.5–15 min). They are particularly useful for the rapid separation of cells from a liquid medium, e.g. silicone oil microcentrifugation.

Continuous flow centrifuges

Useful for harvesting large volumes of cells from their growth medium. During centrifugation, the particles are sedimented as the liquid flows through the rotor.

Rotors

Many centrifuges can be used with tubes of different size and capacity, either by changing the rotor, or by using a single rotor with different buckets/adaptors.

- Swing-out rotors: sample tubes are placed in buckets which pivot as the rotor accelerates (Fig. 38.5a). Swing-out rotors are used on many low-

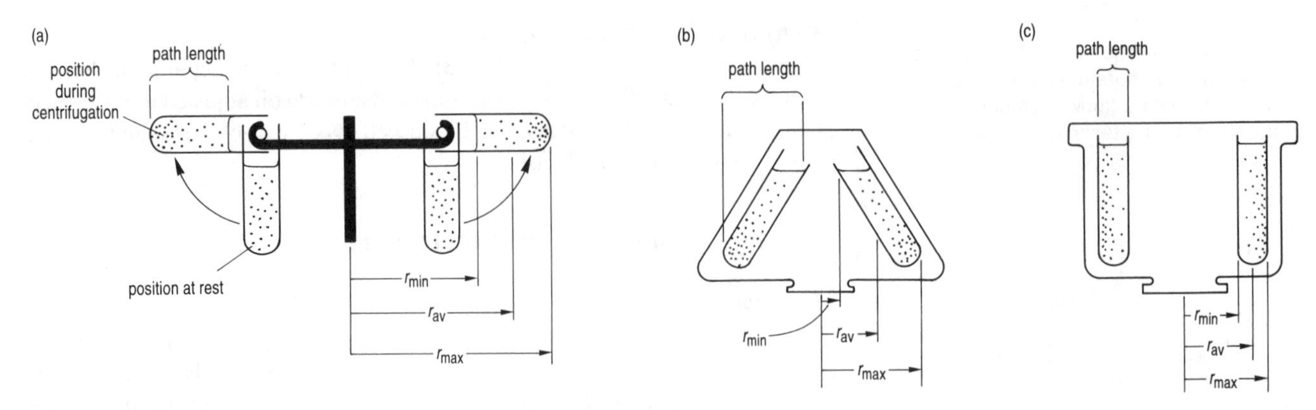

Fig. 38.5 Rotors: (a) swing-out rotor; (b) fixed angle rotor; (c) vertical tube rotor.

Changing a rotor — if you ever have to change a rotor, make sure that you carry it properly (don't knock/drop it), that you fit it correctly (don't cross-thread it, and tighten to the correct setting using a torque wrench) and that you store it correctly (clean it after use and don't leave it lying around).

speed centrifuges: their major drawback is their extended path length and the resuspension of pellets due to currents created during deceleration.

- Fixed-angle rotors: used in many high-speed centrifuges and microcentrifuges (Fig. 38.5b). With their shorter path length, fixed rotors are more effective at pelleting particles than swing-out rotors.
- Vertical tube rotors: used for isopycnic density gradient centrifugation in high-speed centrifuges and ultracentrifuges (Fig. 38.5c). They cannot be used to harvest particles in suspension as a pellet is not formed.

Centrifuge tubes

These are manufactured in a range of sizes (from 1.5 ml up to 1 000 ml) and materials. *Never* be tempted to use a tube or bottle which was not designed to fit the machine you are using (e.g. a general-purpose glass test-tube, or a screw-capped bottle), or you may damage the centrifuge and cause an accident. The following aspects may influence your choice:

- Capacity. This is obviously governed by the volume of your sample. Note that centrifuge tubes must be completely full for certain applications, e.g. for high-speed work.
- Shape. Conical bottomed centrifuge tubes retain pellets more effectively than round-bottomed tubes while the latter may be more useful for density gradient work.
- Maximum centrifugal force. Detailed information is supplied by the manufacturers. Standard Pyrex® glass tubes can only be used at low centrifugal force (up to 2 000*g*).
- Solvent resistance. Glass tubes are inert, polycarbonate tubes are particularly sensitive to organic solvents (e.g. ethanol, acetone), while polypropylene tubes are more resistant. See manufacturer's guidelines for detailed information.
- Sterilization. Disposable plastic centrifuge tubes are often supplied in sterile form. Glass and polypropylene tubes can be repeatedly sterilized. Cellulose ester tubes should *not* be autoclaved. Repeated autoclaving of polycarbonate tubes may lead to cracking/stress damage.
- Opacity. Glass and polycarbonate tubes are clear, while polypropylene tubes are more opaque.
- Ability to be pierced. If you intend to harvest your sample by puncturing the tube wall, cellulose acetate and polypropylene tubes are readily punctured using a syringe needle.

Using microcentrifuge tubes — the integral push-on caps of microcentrifuge tubes must be correctly pushed home before use or they may come off during centrifugation.

- Caps. Most fixed-angle and vertical tube rotors require tubes to be capped, to prevent leakage during use and to provide support to the tube during centrifugation. For low-speed centrifugation, caps must be used for any hazardous samples. Make sure you use the correct caps for your tubes.

Balancing the rotor

Balancing tubes — *never* balance centrifuge tubes 'by eye' — use a balance. Note that a 35 ml tube full of liquid at an RCF of 3 000 *g* has an effective weight greater than a large adult man.

For the safe use of centrifuges, the rotor must be balanced during use, or the spindle and rotor assembly may be permanently damaged; in severe cases, the rotor may fail and cause a serious accident. *It is vital that you balance your loaded centrifuge tubes before use.* As a general rule, *balance all sample tubes to within 1%* using a top-pan balance, or scales. Box 38.1 gives details for a low-speed bench centrifuge. Place balanced tubes opposite each other.

Safe practice

Given their speed of rotation and the extremely high forces generated, centrifuges have the potential to be extremely dangerous, if used incorrectly. For safety reasons, all centrifuges are manufactured with an armoured casing that should contain any fragments in cases of rotor failure. Machines manufactured to BS 4402 have a safety lock to prevent the motor from being switched on unless the lid is closed and to stop the lid from being opened while the rotor is moving. However, some older centrifuges may not have a safety lock, so be particularly careful to avoid opening the lid of such machines during use and make sure that hair and clothing are kept well away from moving parts.

Box 38.1 How to use a low-speed bench centrifuge

1. **Choose the appropriate tube size and material for your application**, with caps where necessary. Most low-speed machines have four-place or six-place rotors — use the correct number of samples to *fill* the rotor assembly whenever possible.
2. **Fill the containers to the appropriate level**: do not overfill, or the sample may spill during centrifugation.
3. **It is vital that the rotor is balanced during use**. Therefore, *identical* tubes must be prepared, to be placed opposite each other in the rotor assembly. This is particularly important for density gradient samples, or for samples containing materials of widely differing densities, e.g. soil samples, since the density profile of the tube will change during a run. However, for low-speed work using small amounts of particulate matter in aqueous solution, it may be sufficient to counterbalance a sample with a second tube filled with water, or a saline solution of similar density to the sample.
4. **Balance each pair of sample tubes** (plus the corresponding caps, where necessary) to within 0.1 g using a top-pan balance; add liquid dropwise to the lighter tube, until the desired weight is reached. Alternatively, use a set of scales. For small sample volumes (up to 10 ml) added to disposable, lightweight plastic tubes, accurate pipetting of your solution may be sufficient for low-speed use.
5. **For centrifuges with swing-out rotors**, check that each holder/ bucket is correctly positioned in its locating slots on the rotor and that it is able to swing freely. All buckets must be in position on a swing-out rotor, even if they do not contain sample tubes — buckets are an integral part of the rotor assembly.
6. **Load the sample tubes into the centrifuge**. Make sure that the outside of the centrifuge tubes, the sample holders and sample chambers are dry: any liquid present will cause an imbalance during centrifugation, in addition to the corrosive damage it may cause to the rotor. For sample holders where rubber cushions are provided, make sure that these are correctly located. Balanced tubes must be placed opposite each other — use a simple code if necessary, to prevent mix-ups.
7. **Bring the centrifuge up to operating speed** by gentle acceleration. Do not exceed the maximum speed for the rotor and tubes used.
8. **If the centrifuge vibrates at any time during use, switch off** and find the source of the problem.
9. **Once the rotor has stopped spinning, release the lid and remove all tubes**. If any sample has spilled, make sure you clean it up thoroughly, so that it is ready for the next user.
10. **Close the lid (to prevent the entry of dust) and return all controls to zero**.

39 Spectroscopic techniques

The absorption and emission of electromagnetic radiation of specific energy (wavelength) is a characteristic feature of many molecules, involving the movement of electrons between different energy states, in accordance with the laws of quantum mechanics. Spectroscopic techniques make use of such changes to:

- identify compounds, by determining their absorption or emission spectra;
- quantify substances, either singly or in the presence of other compounds, by measuring the signal strength at an appropriate wavelength;
- follow reactions, by measuring the rate of disappearance of a substance, or the appearance of a product as a function of time.

Spectrophotometry

This is a widely used technique for measuring the absorption of radiation in the visible and UV regions of the spectrum. A spectrophotometer is an instrument designed to allow precise measurement at a particular wavelength, while a colorimeter is a simpler instrument, using filters to measure broader wavebands (e.g. light in the green, red or blue regions of the visible spectrum).

Principles of light absorption

Two fundamental principles govern the absorption of light by a solution:

- The absorption of light passing through a solution is exponentially related to the number of molecules of the absorbing solute, i.e. the solute concentration C.
- The absorption of light passing through a solution is exponentially related to the length of the absorbing solution, l.

These two principles are combined in the Beer–Lambert relationship, which is usually expressed in terms of the energy of the incident light (I_0) and the energy of the emergent light (I):

$$\log_{10} (I_0/I) = \epsilon \, l \, C \qquad [39.1]$$

where ϵ is a constant for the absorbing substance and the wavelength, termed the absorption coefficient or extinction coefficient, and C is expressed either as mol l^{-1} or g l^{-1} (see p. 197) and l is given in cm. This relationship is extremely useful, since most spectrophotometers are constructed to give a direct measurement of $\log_{10} (I_0/I)$, termed the absorbance (A), or extinction (E), of a solution (older texts may use the outdated term optical density). Note that for substances obeying the Beer–Lambert relationship, A is linearly related to C. Absorbance at a particular wavelength is often shown as a subscript, e.g. A_{550} represents the absorbance at 550 nm. The proportion of light passing through the solution is known as the transmittance (T), and is calculated as the ratio of the emergent and incident light.

Some instruments have two scales:

- an exponential scale from zero to infinity, measuring absorbance;
- a linear scale from 0 to 100, measuring (per cent) transmittance.

For most practical purposes, the Beer–Lambert relationship applies and you should use the absorbance scale.

Definitions

Absorbance (A) — this is given by: $A = \log_{10} (I_0/I)$.

Transmittance (T) — this is usually expressed as a percentage, where $T = (I/I_0) \times 100$ (%).

Colorimeter

This can be used with solutions where the substance of interest is highly coloured and present as the major constituent, e.g. haemoglobin in blood, or where a substance is assayed by adding a reagent which gives a coloured product (a chromophore), e.g. amino acid assay using ninhydrin reagent. Quantification of a particular substance requires a calibration curve to be constructed using known amounts of the compound measured at the same time as the test samples, rather than using the Beer–Lambert relationship directly.

The light source is usually a tungsten filament bulb, focused by a condenser lens to give a parallel beam of light which passes through a glass sample tube or cuvette containing the solution, then through a filter to a photocell detector, which develops an electrical potential in direct proportion to the intensity of the light falling on it. The signal from the photocell is then amplified and passed to a galvanometer, or digital readout, calibrated on a logarithmic scale (see Box 39.1).

The broad bandwidth of most filters means that colorimetry cannot be used to identify a particular compound, nor to distinguish between two compounds with closely related absorption characteristics, e.g. in mixed solution. The photocells used in colorimeters have coefficients of variation (p. 232) of around 0.5%, so they are not suitable for work requiring a high degree of precision or accuracy. In the simplest instruments, the logarithmic measurement scale has arbitrary units, adjusted via sensitivity/scale zero controls, and values

Box 39.1 How to use a colorimeter

1. **Switch on and stabilize** allowing at least 5 min for the lamp to warm up before use.
2. **Choose a filter which is complementary** to the colour of the substance to be measured. Thus haemoglobin is red because it absorbs light within the blue/green regions of the spectrum and should be measured using a blue filter. Similarly, a blue substance should be measured using a red filter.
3. **Set the scale zero** using an appropriate solution blank.
4. **Adjust the sensitivity** to give a reasonable deflection on the galvanometer/readout device, e.g. for a standard solution containing a known amount of the test substance. You would normally adjust the sensitivity so that your highest standard gave a reading close to the maximum on the readout device.
5. **Analyse your samples and standard solutions**, making sure you take all the measurements you need in a single 'run', otherwise they will not be directly comparable.
6. **Use the same tube in the same orientation** in the sample holder to improve precision, since individual tubes may differ in their light absorbing characteristics, wall thickness, etc.
7. **Rinse the tube between samples,** making sure that the rinsing solution does not dilute the next sample and that the outside of the tube is dry before it is loaded into the instrument.
8. **Make frequent checks on reproducibility** by repeat measurements of the same solution (e.g. after every six to eight samples). You should also prepare test and standard solutions in duplicate.
9. **Plot a calibration curve for your standard solutions** and draw the line of best fit through the points. Do not worry if it is a curve, rather than a straight line, or if it does not pass through the origin. Do not extrapolate the calibration curve. If a sample has a reading greater than the highest standard it should be diluted and reassayed.

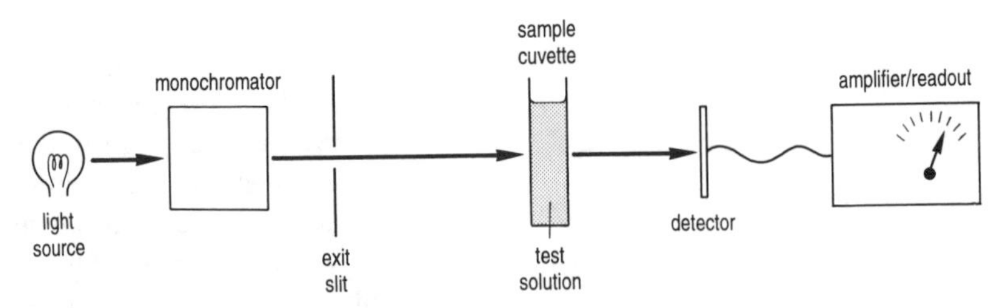

Fig. 39.1 Components of a spectrophotometer.

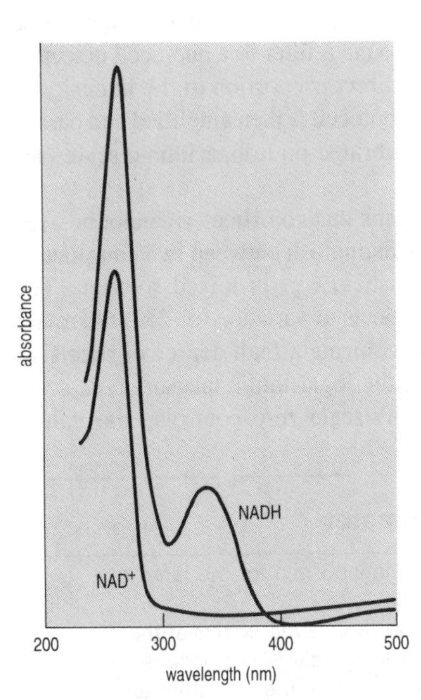

Fig. 39.2 Absorption spectra of nicotinamide adenine dinucleotide in oxidized (NAD^+) and reduced (NADH) form. Note the 340 nm absorption peak, used for quantitative work (p. 197)

Using plastic disposable cuvettes — these are adequate for work in the near-UV region, e.g. for enzyme studies using nicotinamide coenzymes, at 340 nm (p. 161), as well as the visible range.

obtained on one instrument will not be directly comparable to other instruments, or to the same instrument on different settings. In addition, a colorimeter is not suitable for quantitative work at a particular wavelength.

Spectrophotometer

The principal components of a spectrophotometer are shown in Fig. 39.1. High intensity tungsten bulbs are used as the light source in basic instruments, capable of operating in the visible region (i.e. 400–700 nm). Deuterium or tungsten–halogen lamps are used for UV spectrophotometry (200–400 nm); these lamps are fitted with quartz envelopes, since glass does not transmit UV radiation.

A major improvement over the simple colorimeter is the use of a monochromator prism or diffraction grating to produce a parallel beam of monochromatic light from the (polychromatic) light source. In practice the light emerging from such a monochromator does not have a single wavelength, but is a narrow band of wavelengths. This bandwidth is an important characteristic, since it determines the wavelengths used in absorption measurements — the bandwidth of basic spectrophotometers is around 5–10 nm while research instruments have bandwidths of less than 1 nm.

Bandwidth is affected by the width of the exit slit (the slith width), since the bandwidth will be reduced by decreasing the slit width. To obtain accurate data at a particular wavelength setting, the narrowest possible slit width should be used. However, decreasing the slit width also reduces the amount of light reaching the detector, decreasing the signal-to-noise ratio. The extent to which the slit width can be reduced depends upon the sensitivity and stability of the detection/amplification system and the presence of stray light.

Most spectrophotometers are designed to take cuvettes with an optical path length of 10 mm. Disposable plastic cuvettes are suitable for routine work in the visible range using aqueous and alcohol-based solvents, while glass cuvettes are useful for other organic solvents. Glass cuvettes are manufactured to more exacting standards, so use optically matched glass cuvettes for accurate work, especially at low absorbances (< 0.1), where any differences in the optical properties of cuvettes for reference and test samples will be pronounced. Glass and plastic absorb UV light and quartz cuvettes must be used at wavelengths below 300 nm.

Before taking a measurement, make sure that cuvettes are clean, unscratched, dry on the outside, filled to the correct level and in the correct position in their sample holders. Proteins and nucleic acids in biological samples can accumulate on the inside faces of glass/quartz cuvettes, so remove any deposits using acetone on a cotton bud, or soak overnight in 1 mol l^{-1} nitric acid. Corrosive and hazardous solutions must be used in cuvettes with tightly fitting lids, to prevent damage to the instrument and to reduce the risk of accidental spillage.

Basic instruments use photocells similar to those used in colorimeters. In many cases, a different photocell must be used at wavelengths above and below 550–600 nm, due to differences in the sensitivity of such detectors over the visible waveband. Photomultiplier tubes are used in more sophisticated instruments, giving increased sensitivity, accuracy and stability when compared to photocells.

Digital displays are increasingly used in preference to needle-type meters, as they are not prone to parallax errors and misreading of the absorbance scale. Some digital instruments can be calibrated to give a direct readout of the concentration of the test substance.

Examples

The molar extinction coefficient of NADH is 6.333×10^3 l mol^{-1} cm^{-1} at 340 nm. For a test solution giving an absorbance of 0.21 in a cuvette with a light path of 5 mm, using eqn [39.1] this is equal to a concentration of:

$0.21 = 6.333 \times 10^3 \times 0.5 \times C$
$C = 0.000\,066\,3$ mol l^{-1} (or 66.3 μmol l^{-1}).

The specific extinction coefficient (10 g l^{-1}) of double-stranded DNA is 200 at 260 nm, therefore a solution containing 1 g l^{-1} will have an absorbance of 200/10 = 20. For a DNA solution, giving an absorbance of 0.35 in a cuvette with a light path of 1.0 cm, using eqn [39.1] this is equal to a concentration of:

$0.35 = 20 \times 1.0\ C$
$C = 0.0175$ g l^{-1} (equivalent to 17.5 μg ml^{-1}).

Chlorophylls a and b in vascular plants and green algae can be extracted in 90% v/v acetone/water by measuring the absorbance of the mixed solution at 2 wavelengths, according to the formulae:

Chlorophyll a (mg l^{-1}) = $11.93\ A_{664} - 1.93\ A_{647}$

Chlorophyll b (mg l^{-1}) = $20.36\ A_{647} - 5.5\ A_{664}$

Note: different equations are required for other solvents.

The amount of chlorophylls a, b and c can be quantified for phytoplankton extracted in 80% v/v acetone/water by measuring the absorbance of the mixed solution at 3 wavelengths, according to the following formulae:

Chlorophyll a (mg l^{-1}) = $11.85\ A_{664} - 1.54\ A_{647} - 0.08\ A_{630}$

Chlorophyll b (mg l^{-1}) = $21.03\ A_{647} - 5.43\ A_{664} - 2.26\ A_{630}$

Chlorophyll c (mg l^{-1}) = $24.52\ A_{630} - 1.67\ A_{664} - 7.60\ A_{647}$

(See Geider and Osborne 1992).

Types of spectrophotometer

Basic instruments are single beam spectrophotometers in which there is only one light path. The instrument is set to zero absorbance using a blank solution, which is then replaced by the test solution, to obtain an absorbance reading. An alternative approach is used in double beam spectrophotometers, where the light beam from the monochromator is split into two separate beams, one beam passing through the test solution and the other through a reference blank. Absorbance is then measured by an electronic circuit which compares the output from the reference (blank) and sample cuvettes. Double beam spectrophotometry reduces measurement errors caused by fluctuations in output from the light source or changes in the sensitivity of the detection system, since reference and test solutions are measured at the same time (Box 39.2). Recording spectrophotometers are double beam instruments, designed for use with a chart recorder, either by recording the difference in absorbance between reference and test solutions across a predetermined waveband to give an absorption spectrum (Fig. 39.2), or by recording the change in absorbance at a particular wavelength as a function of time (e.g. in an enzyme assay, see Chapter 33).

Quantitative spectrophotometric analysis

A single substance in solution can be quantified using the Beer–Lambert relationship (eqn [39.1]), provided its extinction coefficient is known at a particular wavelength (usually the absorption maximum for the substance, since this will give the greatest sensitivity). The molar extinction coefficient is the absorbance given by a solution with a concentration of 1 mol l^{-1} (= 1 kmol m^{-3}) of the compound in a light path of 1 cm. The appropriate value may be available from tabulated spectral data (e.g. Anon., 1963), or it can be determined experimentally by measuring the absorbance of known concentrations of the substance (Box 39.2) and plotting a standard curve. This should confirm that the relationship is linear over the desired concentration range and the slope of the line will give the molar extinction coefficient.

The specific extinction coefficient is the absorbance given by a solution containing 10 g l^{-1} (i.e. 1% w/v) of the compound in a light path of 1 cm. This is useful for substances of unknown molecular weight, e.g. proteins or nucleic acids, where the amount of substance in solution is expressed in terms of its mass, rather than as a molar concentration. For use in eqn [39.1], the specific extinction coefficient should be divided by 10 to give the solute concentration in g l^{-1}.

This simple approach cannot be used for mixed samples where several substances have a significant absorption at a particular wavelength. In such cases, it may be possible to estimate the amount of each substance by measuring the absorbance at several wavelengths, e.g. protein estimation in the presence of nucleic acids (p. 160).

Atomic spectroscopy

Atoms of certain metals will absorb and emit radiation of specific wavelengths when heated in a flame, in direct proportion to the number of atoms present. Atomic spectrophotometric techniques measure the absorption or emission of particular wavelengths of UV and visible light, to identify and quantify such metals.

Flame emission spectrophotometry (or flame photometry)

The principal components of a flame photometer are shown in Fig. 39.3. A liquid sample is converted into an aerosol in a nebulizer (atomizer) before being introduced into the flame, where a small proportion (typically less than 1 in 10 000) of the atoms will be raised to a higher energy level, releasing this energy as light of a particular wavelength, which is passed through a filter to a photocell detector. Flame photometry is used to measure the alkali metals K, Na, and Ca in biological fluids.

When using a flame photometer:

- Allow time for the instrument to stabilize. Switch on the instrument, light the flame and wait at least 5 min before analysing your solutions.

Box 39.2 How to use a spectrophotometer

1. **Switch on and select the correct lamp** for your measurements (e.g. deuterium for UV, tungsten for visible light).
2. **Allow at least 15 min for the lamp to warm up** and for the instrument to stabilize before use.
3. **Select the appropriate wavelength:** on older instruments a dial is used to adjust the monochromator, while newer machines have microprocessor-controlled wavelength selection.
4. **Select the appropriate detector:** some instruments choose the correct detector automatically (on the basis of the specified wavelength), while others have manual selection.
5. **Choose the correct slit width** (if available): this may be specified in the protocol you are following, or may be chosen on the manufacturer's recommendations.
6. **Insert appropriate reference blank(s):** single beam instruments use a single cuvette, while double beam instruments use two cuvettes (a matched pair for accurate work). The reference blanks should match the test solution in all respects apart from the substance under test, i.e. they should contain all reagents apart from this substance. *Make sure that the cuvettes are positioned correctly, with their polished (transparent) faces in the light path, and that they are accurately located in the cuvette holder(s).*
7. **Check/adjust the 0% transmittance:** most instruments have a control which allows you to zero the detector output in the absence of any light (dark current correction). Some microprocessor-controlled instruments carry out this step automatically.
8. **Set the absorbance reading to zero:** usually via a dial, or digital readout.
9. **Analyse your samples:** replace the appropriate reference blank with a test sample, allow the absorbance reading to stabilize (5 – 10 s) and read the absorbance value from the meter/readout device. For absorbance readings greater than 1 (i.e. $<10\%$ transmission), the signal-to-noise ratio is too low for accurate results, so dilute such samples and measure their absorbance within the range 0 – 1 (don't forget to correct for the dilution in your subsequent calculations). Your analysis may require a calibration curve, as for colorimetry (Box 39.1), or you may be able to use the Beer – Lambert relationship (eqn [39.1]) to determine the concentration of test substance in your samples.
10. **Check the scale zero at regular intervals** using a reference blank, e.g. after every ten samples.
11. **Check the reproducibility of the instrument:** measure the absorbance of a single solution several times during your analysis. It should give the same value.

Problems (and solutions): inaccurate/unstable readings are most often due to incorrect use of cuvettes, e.g. dirt, fingerprints or test solution on outside of cuvette (wipe the polished faces using a soft tissue before insertion into the cuvette holder), condensation (if cold solutions aren't allowed to reach room temperature before use), air bubbles (which scatter light and increase the absorbance; tap to remove), insufficient solution (causing refraction of light at the meniscus), particulate material in the solution (centrifuge before use, where necessary) or incorrect positioning in light path (locate in correct position).

- Check for impurities in your reagents. For example, if you are measuring K in an acid digest of some biological material, check the K content of a reagent blank, containing everything except the biological material, processed in exactly the same way as the samples. Subtract this value from your sample values to obtain the true K content.
- Quantify your samples using a calibration curve. Calibration standards should cover the expected concentration range for the test solutions — your calibration curve may be non-linear (especially at concentrations above 1 mmol l^{-1}, i.e. 1 mol m^{-3} in SI units).
- Analyse all solutions in duplicate, so that reproducibility can be assessed.
- Check your calibration. Make repeated measurements of a standard solution of known concentration after every six or seven samples, to confirm that the instrument calibration is still valid.
- Consider the possibility of interference. Other metal atoms may emit light which is detected by the photocell, since the filters cover a wider waveband than the emission line of a particular element. This can be a serious problem if you are trying to measure low concentrations of a particular metal in the presence of high concentrations of other metals (e.g. Na in sea water), or other substances which form complexes with the test metal, suppressing the signal (e.g. phosphate).

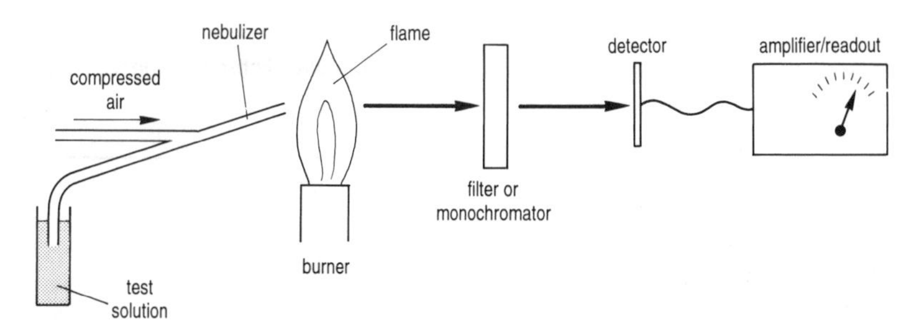

Fig. 39.3 Components of a flame photometer.

Atomic absorption spectrophotometry (or flame absorption spectrophotometry)

This technique is applicable to a broad range of metals, including Pb, Cu, Zn, etc. It relies on the absorption of light of a specific wavelength by atoms dispersed in a flame. The appropriate wavelength is provided by a cathode lamp, coated with the element to be analysed, focused through the flame and onto the detector. When the sample is introduced into the flame, it will *decrease* the light detected in direct proportion to the amount of metal present. Practical advantages over flame photometry include:

- improved sensitivity,
- increased precision,
- decreased interference.

Newer variants of this method include flameless atomic absorption spectrophotometry and atomic fluorescence spectrophotometry, both of which are more sensitive than the flame absorption technique.

Other spectroscopic methods

A wide range of advanced techniques used in biological research are detailed in Table 39.1.

Table 39.1 Advanced spectroscopic techniques in biological research

Technique	Biological applications
Spectrofluorimetry	Quantification of molecules with intrinsic fluorescence, e.g. chlorophyll, porphyrins Enzyme assay, using fluorogenic substrates Membrane structure and function
Infra-red (IR) spectrometry	Gas analysis Identification of molecular structure
Electron spin resonance (ESR) spectrometry	Metalloprotein analysis Free radical behaviour
Mass spectrometry	Biochemical structure/identification Used in conjunction with gas–liquid chromatography (p. 203)
Nuclear magnetic resonance (NMR) spectrometry	Structure of macromolecules (^{1}H and ^{13}C NMR) Metabolic studies (^{13}C and ^{31}P) Intracellular pH (^{31}P NMR)
Circular dichroism spectroscopy	Structure of macromolecules, e.g. proteins and nucleic acids

40 Chromatography

Chromatography is used to separate the individual constituents within a sample on the basis of differences in their physical characteristics, e.g. molecular size, shape, charge, volatility, solubility and/or adsorptivity. The essential components of a chromatographic system are:

- A stationary phase, either a solid, a gel or an immobilized liquid, held by a support matrix.
- A chromatographic bed: the stationary phase may be packed into a glass or metal column, spread as a thin layer on a sheet of glass or plastic, or adsorbed on cellulose fibres (paper).
- A mobile phase, either a liquid or a gas which acts as a solvent, carrying the sample through the stationary phase and eluting from the chromatographic bed.
- A delivery system to pass the mobile phase through the chromatographic bed.
- A detection system to monitor the test substances.

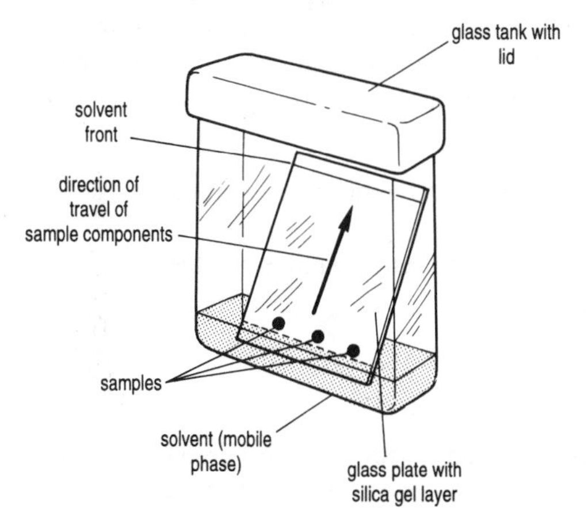

Fig. 40.1 Components of a TLC system.

Using a TLC system — it is essential that you allow the solvent to pre-equilibrate in the chromatography tank for at least 2 h before use, to saturate the atmosphere with vapour.

Individual substances interact with the stationary phase to different extents as they are carried through the system, enabling separation to be achieved. Those substances which interact strongly will be retarded to the greatest extent while those which show little interaction will pass through with minimal delay, leading to differences in distances travelled or elution times.

Types of chromatographic system

Chromatographic systems can be categorized according to the form of chromatographic bed, the nature of the mobile and stationary phases and the method of separation.

Thin-layer and paper chromatography

Here, you apply the sample as a single spot near one end of the sheet, by microsyringe or microcapillary. This sheet is allowed to dry fully, then it is transferred to a glass tank containing a shallow layer of solvent (Fig. 40.1). Remove the sheet when the solvent front has travelled 80–90% of its length.

You can express movement of an individual substance in terms of its relative frontal mobility, or R_F value, where:

$$R_F = \frac{\text{distance moved by substance}}{\text{distance moved by solvent}} \quad [40.1]$$

Alternatively, you may express movement with respect to a standard of known mobility, as R_X, where:

$$R_X = \frac{\text{distance moved by test substance}}{\text{distance moved by standard}} \quad [40.2]$$

The R_F (or R_X) value is a constant for a particular substance and solvent system (under standard conditions) and closely reflects the partitioning of the substance between the stationary and mobile phases. Tabulated values are available for a range of biological molecules and solvents (Stahl, 1965). However, you should analyse one or more reference compounds on the same sheet as your unknown sample, to check their R_F values.

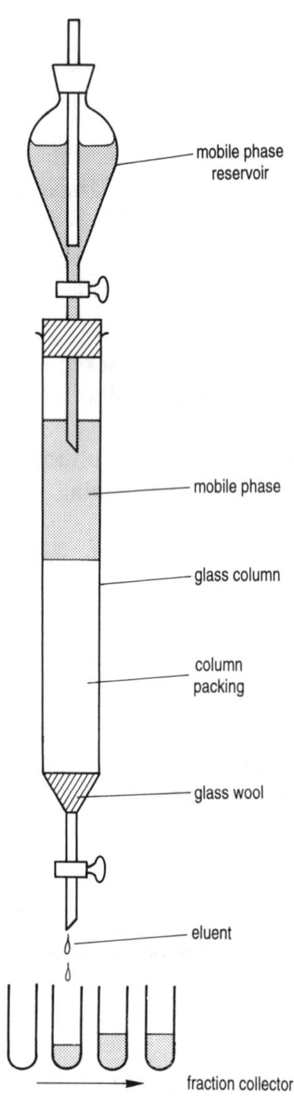

Fig. 40.2 Equipment for column chromatography (gravity feed system).

HPLC is a versatile form of chromatography, used with a wide variety of stationary phases, to separate individual compounds of a particular class of molecules on the basis of size, polarity, solubility or adsorption characteristics.

Problems with peaks — non-symmetrical peaks may result from column overloading, co-elution of solutes, poor packing of the stationary phase, or interactions between the substance and the support material.

Column chromatography

Here, you pack a glass column with the appropriate stationary phase and equilibrate the mobile phase by passage through the column, either by gravity (Fig. 40.2), or using a low pressure peristaltic pump. You can then introduce the sample to the top of the column, to form a discrete band of material. This is then flushed through the column by the mobile phase. If the individual substances have different rates of migration, they will separate within the column, eluting at different times as the mobile phase travels through the column.

You can detect eluted substances by collecting the mobile phase as it elutes from the column in a series of tubes (discontinuous monitoring), either manually or with an automatic fraction collector. Fractions of 2 – 5% of the bed volume are usually collected and analysed, e.g. by chemical assay. You can now construct an elution profile (or chromatogram) by plotting the amount of substance against either time, elution volume or fraction number, which should give a symmetrical peak for each substance (Fig. 40.3).

You can express the migration of a particular substance at a given flow rate in terms of its retention time (t), or elution volume (V_e). The separation efficiency of a column is measured by its ability to distinguish between two similar substances, assessed in terms of:

- selectivity, as measured the difference in retention times of the two peaks (i.e. as $t_a - t_b$);
- resolution (R_s), quantified in terms of the retention time and the base width (W) of each peak:

$$R_s = \frac{2(t_a - t_b)}{W_a + W_b} \qquad [40.3]$$

where the subscripts a and b refer to substances a and b respectively (Fig. 40.3). For most practical purposes, R_s values of 1 or more are satisfactory, corresponding to 98% peak separation for symmetrical peaks.

High performance liquid chromatography (HPLC)

Column chromatography originally used large 'soft' stationary phases that required low pressure flow of the mobile phase to avoid compression; separations were usually time-consuming and of low resolution ('low performance'). Subsequently, the production of small, incompressible, homogeneous particulate support materials and high pressure pumps with reliable, steady flow rates have enabled high performance systems to be developed. These systems operate at pressures up to 10 MPa, forcing the mobile phase through the column at a high flow rate to give rapid separation with reduced band broadening.

HPLC columns are usually made of stainless steel, and all components, valves, etc., are manufactured from materials which can withstand the high pressures involved. The choice of solvent delivery system depends on the type of separation to be performed:

- Isocratic separation: a single solvent (or solvent mixture) is used throughout the analysis.
- Gradient elution separation: the composition of the mobile phase is altered using a microprocessor-controlled gradient programmer, which mixes appropriate amounts of two different substances (using two different pumps) to produce the required gradient.

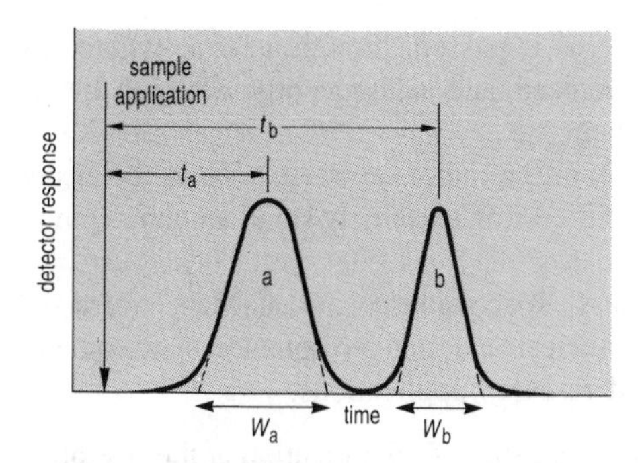

Fig. 40.3 Peak characteristics in a chromatographic separation.

Most HPLC systems are linked to a continuous monitoring detector of high sensitivity, e.g. proteins may be detected spectrophotometrically by monitoring the absorbance of the eluent at 280 nm as it passes through a flow cell (cuvette). Other detectors can be used to measure changes in fluorescence, refractive index, ionization, radioactivity, etc. The detector is linked to a recorder or microcomputer to produce an elution profile. Most detection systems are non-destructive, which means that you can collect eluent with an automatic fraction collector for further study (Fig. 40.4).

The speed, sensitivity and versatility of HPLC makes this the method of choice for the separation of many small molecules of biological interest, often using reverse phase partition chromatography (p. 205). Separation of macromolecules (especially proteins and nucleic acids) usually requires 'biocompatible' systems in which stainless steel components are replaced by titanium, glass or fluoroplastics, using lower pressures to avoid denaturation, e.g. the Pharmacia FPLC® system. Such separations are carried out using ion-exchange, gel permeation and/or hydrophobic interaction chromatography (pp. 205–7).

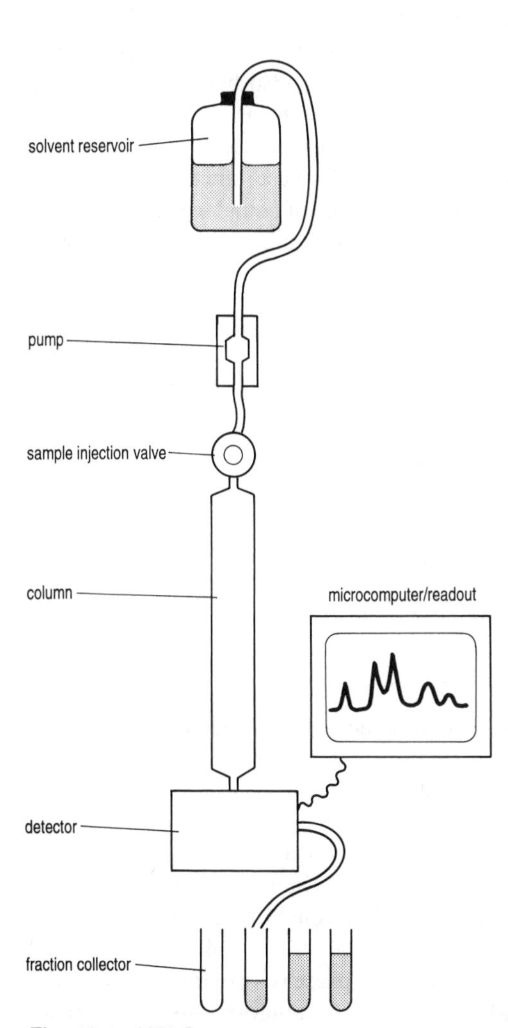

Fig. 40.4 HPLC apparatus.

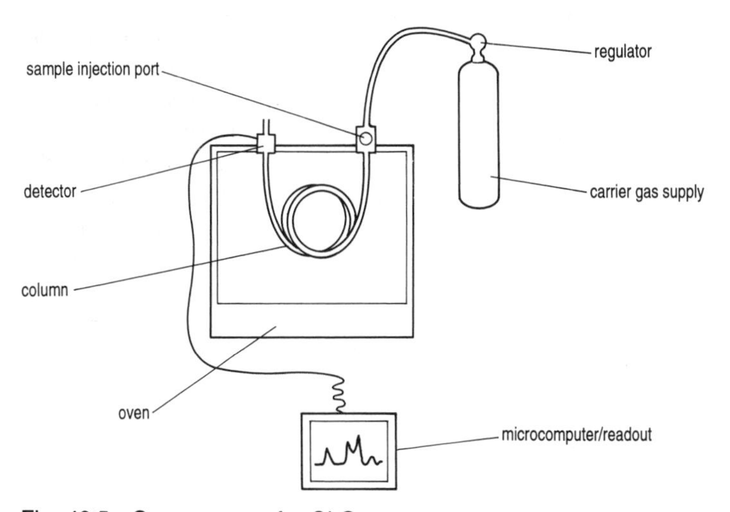

Fig. 40.5 Components of a GLC system.

Gas–liquid chromatography (GLC)

Here, the stationary phase is a liquid, coated onto a column of an inert, particulate support material (Fig. 40.5), and the mobile phase is an inert carrier gas (e.g. nitrogen), driven through the column by pressure supplied by a gas cylinder at up to 0.5 MPa. The liquid stationary phase is usually a silicone gum, which may be:

- selective, where separation is due to the different partitioning of individual compounds between the gas and liquid phases;
- non-selective, where separation is the result of differences in the boiling point of sample components.

The column is maintained at a high temperature (typically 50–250°C) and you inject the sample dissolved in a volatile solvent onto the top of the column through a gas-tight septum using a microsyringe.The output from the column can be monitored by the following methods:

- Thermal conductivity: changes in the composition of the gas at the outflow alter the resistance of a platinum wire.

- Flame ionization: the outflow gas is passed through a flame where any organic compounds will be ionized and subsequently detected by an electrode mounted near the flame tip.
- Electron capture: using a beta-emitting radioisotope (p. 177) as the means of ionization. This is capable of detecting extremely small amounts (pmol) of electrophilic compounds.
- Spectrometry: including mass spectrometry (GLC-MS), infra-red spectrometry (GLC-IR) and nuclear magnetic resonance spectrometry (GLC-NMR) for sophisticated research applications.

Applications of gas–liquid chromatography — GLC is used to separate volatile, non-polar compounds: substances with polar groups must be converted to less polar derivatives prior to analysis, to prevent retention on the column, leading to poor resolution and peak tailing.

GLC can only be used with samples capable of volatilization at the operating temperature of the column, e.g. short chain fatty acids. Other substances may need to be chemically modified to produce more volatile compounds, e.g. long chain saturated fatty acids are usually analysed as methyl esters while monosaccharides are converted to their trimethylsilyl derivatives.

Separation methods

Adsorption chromatography

This is a form of solid – liquid chromatography. The stationary phase is a porous, finely divided solid which adsorbs molecules of the test substance on its surface due to dipole – dipole interactions, hydrogen bonding and/or van der Waals interactions (Fig. 40.6a). The range of adsorbents is limited, e.g. charcoal and polystyrene-based resins (for non-polar molecules), silica, aluminium oxide and calcium phosphate (for polar molecules). Most adsorbents must be activated by heating to 110 – 120°C before use, since their adsorptive capacity is significantly decreased in the presence of bound water. Adsorption chromatography can be carried out in column or thin-layer form, using a wide range of organic solvents.

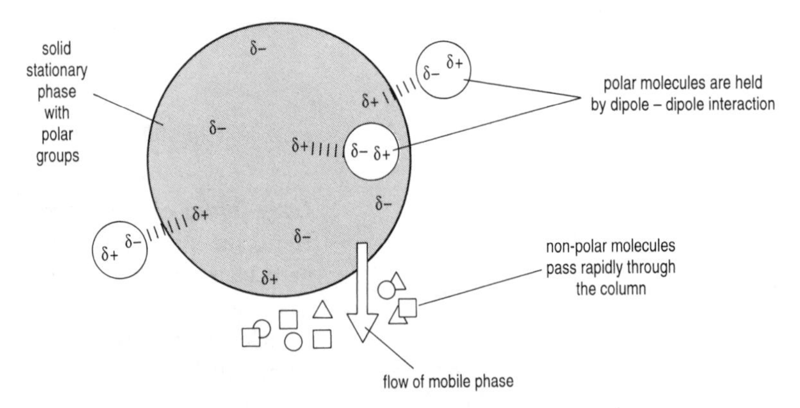

Fig. 40.6(a) Adsorption chromatography (polar stationary phase).

Partition chromatography

This is based on the partitioning of a substance between two liquid phases, in this instance the stationary and mobile phases. Substances which are more soluble in the mobile phase will pass rapidly through the system while those which favour the stationary phase will be retarded (Fig. 40.6b). In normal phase partition chromatography the stationary phase is a polar solvent, usually water, supported by a solid matrix (e.g. cellulose fibres in paper chromatography) and the mobile phase is an immiscible, non-polar organic solvent. For reverse-phase partition chromatography the stationary phase is a non-polar solvent (e.g. a C_{18} hydrocarbon, such as octadecylsilane) which is chemically bonded to the support matrix (e.g. silica), while the mobile phase can be chosen from a wide range of polar solvents. Reverse-phase chromatography is used mainly in HPLC (p. 202), to separate various polar biological molecules, including peptides, oligosaccharides and vitamins.

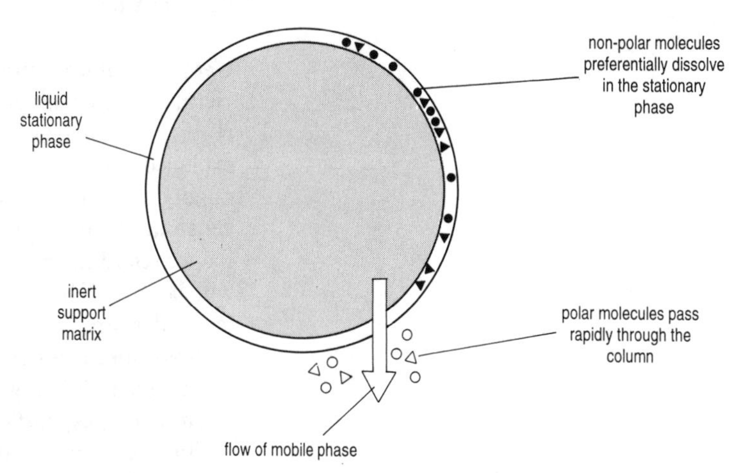

Fig. 40.6(b) Liquid–liquid partition chromatography, e.g. reverse phase HPLC.

Ion-exchange chromatography

Here, separations are carried out using a column packed with a porous matrix which has a large number of ionized groups on its surfaces, i.e. the stationary phase is an ion-exchange resin. The groups may be cation or anion exchangers, depending upon their affinity for positive or negative ions. The net charge on a particular resin depends on the pK_a of the ionizable groups and the pH of the solution, in accordance with the Henderson–Hasselbalch equation (p. 32).

Selecting a separation method — it is often best to select a technique that involves direct interaction between the substance(s) and the stationary phase, (e.g. ion-exchange, affinity or hydrophobic interaction chromatography), due to their increased capacity and resolution compared to other methods, (e.g. partition or gel permeation chromatography).

For most practical applications, you should select the ion-exchange resin and buffer pH so that the test substances are strongly bound by electrostatic attraction to the ion-exchange resin on passage through the system, while the other components of the sample are rapidly eluted (Fig. 40.6c). You can then elute the bound ions by changing the pH (which will alter the affinity of the test substance for the resin) or by raising the salt concentration of the mobile

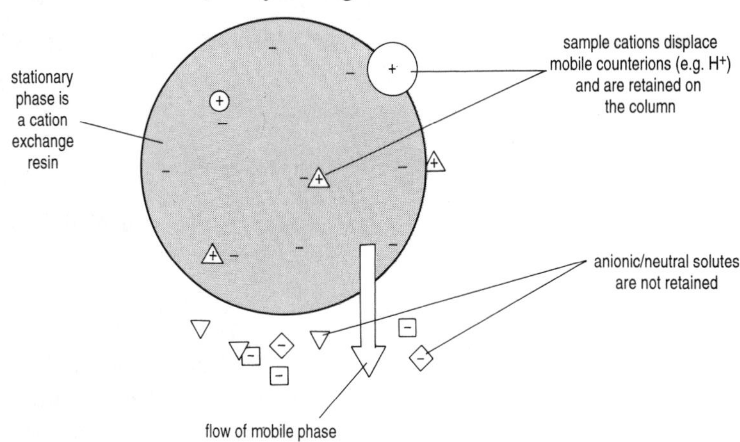

Fig. 40.6(c) Ion exchange chromatography (cation exchanger).

phase (to displace the bound ions); for instance, if you pass a solution of increasing concentration through the system, the weakly bound ions will elute first while the strongly bound ions will be elute at a higher concentration.

Ion-exchange chromatography can be used to separate mixtures of a wide range of ions, including amino acids, peptides, proteins and nucleotides. Electrophoresis is an alternative means of separating charged molecules, as described on p. 166 for DNA.

Gel permeation chromatography (molecular exclusion chromatography)

Here, the stationary phase is a cross-linked gel containing a network of minute pores and channels of various sizes. Large molecules may be completely excluded from these pores, passing through the interstitial spaces and eluting rapidly from the column in the liquid mobile phase. Smaller molecules will penetrate the gel matrix to an extent which will depend upon their size and shape, retarding their progress through the column (Fig. 40.6d). Substances will be eluted from a gel permeation column in decreasing order of their molecular size.

Cross-linked dextrans (e.g. Sephadex®), agarose (e.g. Sepharose®) and polyacrylamide (e.g. Bio-Gel®) are used to separate mixtures of macromolecules, particularly enzymes, antibodies and other globular proteins. Calibration of a gel filtration column using molecules of similar shape and known molecular mass enables the molecular mass of other components to be estimated, since a plot of elution volume (V_e) against $\log_{10}$ molecular mass is approximately linear. A further application of gel permeation chromatography is the general separation of low molecular mass and high molecular mass components, e.g. desalting a protein extract by passage through a Sephadex® G-25 column is faster and more efficient than dialysis.

Using a gel-permeation system — keep your sample volume as small as possible, as the low resolution of this form of chromatography means that the sample will be diluted during passage through the column.

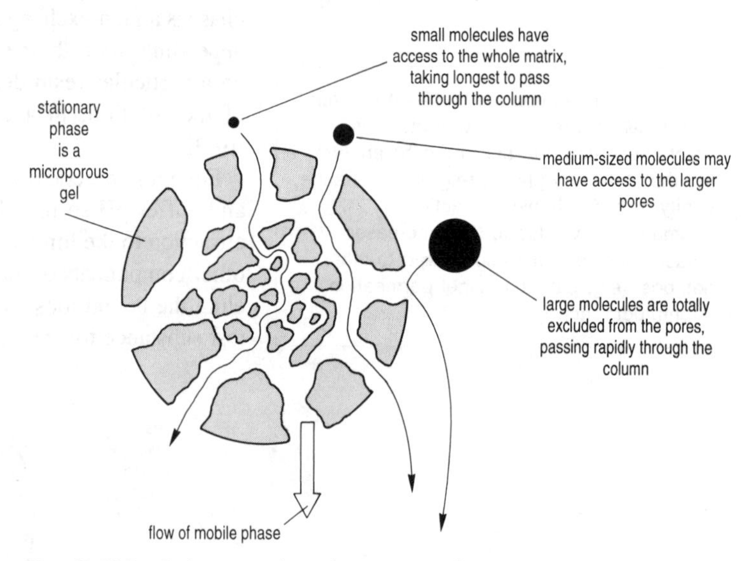

Fig. 40.6(d) Gel permeation chromatography.

Affinity chromatography

This is a highly specific form of adsorption chromatography, where a particular binding molecule (or ligand) is covalently attached to a solid matrix in such a way that the affinity of the ligand for its complementary molecule is unchanged. The immobilized ligand is then packed into a chromatography column where it will selectively adsorb the complementary molecule,

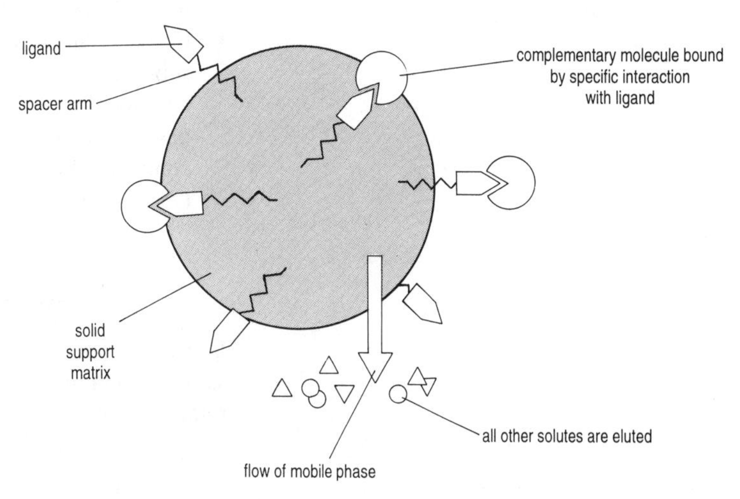

Fig. 40.6(e) Affinity chromatography.

Elution of substances from an affinity system — make sure that your elution conditions do not affect the interaction between the ligand and the stationary phase, or you may elute the ligand from the column.

preventing its passage through the system (Fig. 40.6e). Once the sample has been applied to the column and the contaminating substances washed through with buffer, the purified complementary molecule can be eluted, e.g. by changing the pH or salt concentration, or by adding other substances with a greater affinity for the ligand. Providing a suitable ligand is available, affinity chromatography can be used for single-step purification of small quantities of a particular molecule in the presence of large amounts of contaminating substances. Examples of ligands include:

- triazine dyes, for protein purification;
- enzyme substrates and co-factors, for certain enzymes;
- antibodies, for specific antigens (see p. 153);
- protein A, for IgG antibody purification;
- single-stranded oligonucleotides, for complementary nucleic acids, e.g. mRNA, or particular single-stranded DNA sequences;
- lectins, for specific monosaccharide sub-units.

Hydrophobic interaction chromatography

Learning from experience — if you are unable to separate your test substance(s) using a particular method, do not regard this as a failure, but instead, think about what this tells you about either the substance(s) or your sample.

This technique is used to separate proteins. The stationary phase consists of a non-polar group (e.g. an octyl or phenyl group) bound to a support matrix. This binds proteins differentially, according to the number and position of hydrophobic surface groups. The principle is similar to reverse-phase HPLC (Fig. 40.6b), but samples are analysed under non-denaturing conditions, to retain biological activity.

Quantitative analysis

Most detectors and chemical assay systems give a linear response with increasing amounts of the test substance. Alternative ways of converting the measured response to an amount of substance are:

When using external standardization, samples and standards should be analysed more than once, to confirm the reproducibility of the technique.

- External standardization: this is applicable where the sample volume is sufficiently precise to give reproducible results (e.g. HPLC, column chromatography). You measure the peak areas (or heights) of known amounts of the substance to give a calibration factor or calibration curve which can be used to calculate the amount of test substance in the sample.

- Internal standardization: where you add a known amount of a reference substance (not originally present in the sample) to the sample, to give an additional peak in the elution profile. You determine the response of the detector to the test and reference substances by analysing a standard containing known amounts of both substances, to provide a response factor (r), where

$$r = \frac{\text{peak area (or height) of test substance}}{\text{peak area (or height) of reference substance}} \qquad [40.4]$$

Use this response factor to quantify the amount of test substance (Q_t) in a sample containing a known amount of the reference substance (Q_r), from the relationship:

$$Q_t = \frac{\text{peak area (or height) of test substance}}{\text{peak area (or height) of reference substance}} \times (Q_r/r) \qquad [40.5]$$

Using an internal standard — you should add an internal standard to the sample at the first stage in the extraction procedure, so that any loss or degradation of test substance during purification is accompanied by an equivalent change in the internal standard.

Internal standardization should be the method of choice wherever possible, since it is unaffected by small variations in sample volume (e.g. for GLC microsyringe injection). An additional advantage of an internal standard which is chemically related to the test substance is that it may show up problems due to changes in detector response, incomplete derivatization, etc.

Analysis and presentation of data

41 Manipulating and transforming raw data

Data analysis can be both engrossing and a lot of fun — once experienced, the fascination of discovering meaning from your sets of data can be profound. There are two basic approaches to the investigation of data:

- Exploratory procedures — used to extract the meaning and significance contained within the data.
- Confirmatory procedures — designed to test hypotheses and the quality of the data (see also Chapter 46).

The objective of exploratory procedures is to generate hypotheses for further investigation. Exploratory tools must be quick and easy to use, and to learn. They should be used both early in a study and often during it; they are particularly useful for the scrutiny of 'pilot' data. The purpose of these techniques is to find new and different ways to examine your data, usually in a visual form. Therefore, spreadsheet data manipulation and graphical facilities (see Chapter 47) are extremely useful.

Organizing numbers

Summarizing your results — original, unsummarized data belongs only in your primary record, either in laboratory books or as computer records. You should produce summary tables to organize and condense original data.

In order to organize, manipulate and summarize data, you should follow the steps below:

- Select sub-sets of data for scrutiny: the first step in this process is often to order the data by constructing frequency tables.
- Simplify the numbers, e.g. by rounding or taking means. This avoids the detail becoming overwhelming.
- Rearrange your data in as many ways as possible for comparison.
- Display in graphical form; this provides an immediate visual summary which is relatively easy to interpret.
- Look for an overall pattern in the data; avoid getting lost in the details at this stage.
- Look for any striking exceptions to that pattern (outliers); they often point to special cases of particular interest or to errors in the data produced through mistakes during the acquisition, recording or copying of data.
- Move from graphical interpretations to well-chosen numerical summaries and/or verbal descriptions including where applicable an explanatory hypothesis.

Colour	Tally	Total
Green	///	3
Blue	~~////~~ ///	8
Red	////	4
White	~~////~~ ~~////~~ //	12
Black	/	1
Maroon	///	3
Yellow	//	2
		Σ 33

Fig. 41.1 An example of a tally chart.

Frequency tables

A *neatly* constructed tally chart will double as a rough histogram.

After collecting data, the first step is often to count how often each value occurs and to produce a frequency table (e.g. Table 41.1). The frequency is simply the number of times a value occurs in the data set, and is, therefore, a count. The raw data could be acquired using a tally chart system to provide a simple frequency table. To construct a tally chart (e.g. Fig. 41.1):

- Enter only one tally at a time;
- If working from a data list, cross out each item on the list as you enter it onto the tally chart to prevent double entries;
- Check that all values are crossed out at the end and that the totals agree.

Table 41.1 An example of a frequency table

Size class	Frequency	Relative frequency (%)
0–4.9	7	2.6
5–9.9	23	8.6
10–14.9	56	20.9
15–19.9	98	36.7
20–24.9	50	18.7
25–29.9	30	11.2
30–34.9	3	1.1
Total	267	99.8*

* <100 due to rounding error

stem	leaves
7	23
7	55
7	6
7	9
8	000
8	233
8	45555
8	77
8	888899
9	0000111111
9	2333333
9	44555555555
9	66677777
9	88888999
10	00

Fig. 41.2 A simple 'stem and leaf' plot of a data set. The 'stem' shows the common component of each number, while the 'leaves' show the individual components, e.g. the top line in this example represents the numbers 72 and 73.

Convert the data to a formal table when complete. Because rates or proportions are often more useful than simple totals, the table may contain a column to show the relative frequency of each class (Table 41.1). Relative frequency is usually expressed in decimal form (as a proportion of 1) or as a percentage.

Graphing Data

Graphs are an effective way to investigate trends in data and can reveal features that are difficult to detect from a table. The construction and use of graphs is described in detail in Chapter 42. When manipulating data the main points are as follows:

- Make your data stand out clearly; attention should focus on the actual data, not the labels, scale markings, etc., (contrast with the requirements for constructing a graph for data presentation, see p. 214).
- Avoid clutter in the graph; leave out grid lines and ask yourself whether the graph is more complicated than is necessary for your purpose.
- Use a computer spreadsheet with graphics options whenever possible: the speed and flexibility of these powerful tools should mean that every aspect of your data can be explored rapidly and with relatively little effort.

Displaying distributions

A visual display of a distribution of values is often useful for variables measured on an interval or ratio scale (p. 38). The distribution of a variable can be displayed by a frequency table for each value or, if the possible values are numerous, groups (classes) of possible values of the variable. Graphically, there are two main ways of viewing such data:

- histograms, (see p. 218), generally used for large samples;
- stem and leaf plots (e.g. Fig. 41.2), often used for samples of less than 100: these retain the actual values and are faster to draw by hand. The main drawback is the limitation imposed by the choice of stem values since this predetermines the class boundaries and may obscure some features of the distribution.

These displays allow you to look at the overall shape of a distribution and to observe any significant deviations from the idealized theoretical ones (see Chapter 45). Where necessary, you can use data transformations to investigate any departures from standard distribution patterns such as the normal distribution or the Poisson distribution.

Transforming data

When a distribution is unimodal but not symmetrical, it is often useful to transform the data in an effort to redistribute the values to form a symmetrical distribution. This can provide information about the nature of the underlying relationship and may allow you to use parametric hypothesis-testing statistics (see Chapter 46). The object of such transformation is to find the function which most nearly fits the data into this standard (normal) form. The most common transformation is to take logarithms of one or more sets of values: if the data then approximate to a normal distribution, the relationship is termed

Using transformations — note that if you wish to conserve the *order* of your data, you will need to take negative values when using a reciprocal function (i.e. $-1/n^n$). This is essential when using a box plot to compare graphically the effects of transformations on the five number summary of a data set.

'log-normal'. Some general points about transformations are:

- They should be made on the raw data, not on derived data values: this is simpler, more valid, and more easily interpreted.
- The transformed data can be analysed like any other numbers.
- Transformed data can be examined for outliers, which may be more important if they remain after transformation.

Table 41.2 presents a ladder of transformations which will help you decide which transformations to try (see also Table 46.1). Percentage and proportion data are usually arc-sine transformed which is more complex than most other transformations; consult Sokal and Rohlf (1981) for details.

Table 41.2 Ladder of transformations (due to J.W. Tukey)

$1/x^2$	$1/x$	$\log x$	$\sqrt{x}$	x	x^2	x^3	antilog x
← increasingly strong correction for positive skew in data set				No shape change	increasingly strong correction for negative skew in data set →		

A quick-and-easy way to choose a transformation is as follows:

1. Calculate the 'five-number summary' for the untransformed data (p. 231).
2. Present the summary graphically as a 'box-and-whisker' plot (p. 231).
3. Decide whether you need to correct for positive or negative skew (p. 227).
4. Apply one of the 'mild' transformations in Table 41.2 *on the five number summary values only*.
5. Draw a new box-and-whisker plot and see whether the skewness has been corrected.
6. If the skewness has been under-corrected, try again with a stronger transformation. If it has been over-corrected, try a milder one.
7. When the distribution appears to be acceptable, transform the full data set and recalculate the summary statistics. If necessary, you can use a statistical test to confirm that the transformed data are normally distributed (p. 238).
8. If no simple transformation works well, continue to explore your data, but note that you may need to use non-parametric statistics when comparing data sets.

Transforming by computer — transformation and graphical interpretation is quick and easy to perform with the help of a spreadsheet (Chapter 48), allowing you to use the full data set rather than working with summary statistics.

42 Using graphs

Graphs can be used to show detailed results in an abbreviated form, displaying the maximum amount of information in the minimum space. Graphs and tables present findings in different ways. A graph (figure) gives a visual impression of the content and meaning of your results, while a table provides an accurate numerical record of data values. You must decide whether a graph should be used, e.g. to illustrate a pronounced trend or relationship, or whether a table (p. 219) is more appropriate.

A well-constructed graph will combine simplicity, accuracy and clarity. Planning of graphs is needed at the earliest stage in any write-up as your accompanying text will need to be structured so that each graph delivers the appropriate message. Therefore, it is best to decide on the final form for each of your graphs before you write your text. The text, diagrams, graphs and tables in a laboratory write-up or project report should be complementary, each contributing to the overall message. In a formal scientific communication it is rarely necessary to repeat the same data in more than one place (e.g. as a table and as a graph). However, graphical representation of data collected earlier in tabular format may be applicable in laboratory practical reports.

Selecting a title — it is a common fault to use titles that are grammatically incorrect: a widely applicable format is to state the relationship between the independent and dependent variables within the title, e.g. 'A graph to show the relationship between enzyme activity and external pH'.

Practical aspects of graph drawing

The following comments apply to graphs drawn for laboratory reports. Figures for publication, or similar formal presentation are usually prepared according to specific guidelines, provided by the publisher/organizer.

Remembering which axis is which — a way of remembering the orientation of the *x* axis is that *x* is a 'cross', and it runs 'across' the page (horizontal axis).

- Graphs should be self-contained — they should include all material necessary to convey the appropriate message without reference to the text.
- Every graph must have a concise explanatory title to establish the content. If several graphs are used, they should be numbered, so they can be quoted in the text.
- Consider the layout and scale of the axes carefully. Most graphs are used to illustrate the relationship between two variables (*x* and *y*) and have two axes at right angles (e.g. Fig. 42.1). The horizontal axis is known as the abscissa (*x* axis) and the vertical axis as the ordinate (*y* axis).
- The axis assigned to each variable must be chosen carefully. Usually the *x* axis is used for the independent variable (e.g. treatment) while the dependent variable (e.g. biological response) is plotted on the *y* axis (p. 50). When neither variable is determined by the other, or where the variables are interdependent, the axes may be plotted either way round.
- Each axis must have a descriptive label showing what is represented, together with the appropriate units of measurement, separated from the descriptive label by a solidus (/), as in Fig. 42.1, or brackets as in Fig. 42.2.
- Each axis must have a scale with reference marks on the axis to show clearly the location of all numbers used.
- A figure legend should be used to provide explanatory detail.

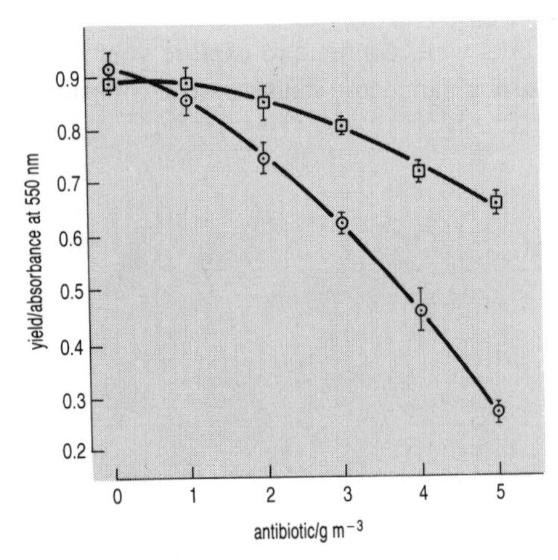

Fig. 42.1 Effect of antibiotic on yield of two bacterial isolates: o, sensitive isolate; □, resistant isolate. Vertical bars show standard errors (n = 6).

Handling very large or very small numbers

To simplify presentation when your experimental data consist of either very large or very small numbers, the plotted values may be the measured numbers multiplied by a power of 10: this multiplying power should be written immediately before the descriptive label on the appropriate axis (as in Fig.

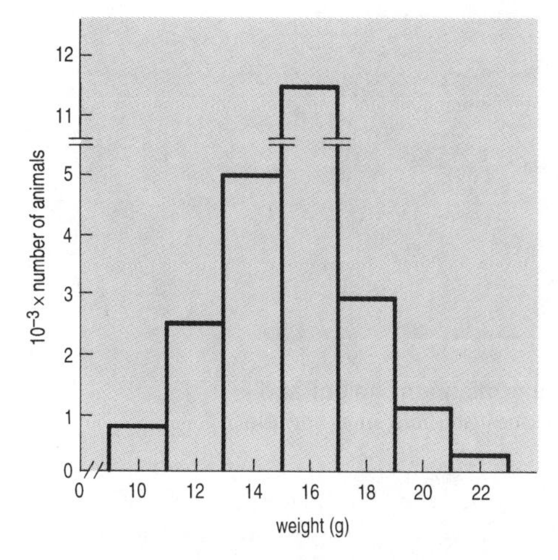

Fig. 42.2 Frequency distribution of weights for a sample of animals (sample size 24 085); the size class interval is 2 g.

Example
For a data set where the smallest number on the log axis is 12 and the largest number is 9 000, three-cycle log-linear paper would be used, covering the range 10 – 10 000.

42.2). However, it is often better to modify the primary unit with an appropriate prefix (p. 41) to avoid any confusion regarding the use of negative powers of 10.

Size

Remember that the purpose of your graph is to communicate information. It must not be too small, so use at least half an A4 page and design your axes and labels to fill the available space without overcrowding any adjacent text. The shape of a graph is determined by your choice of scale for the x and y axes which, in turn, is governed by your experimental data. It may be inappropriate to start the axes at zero (e.g. Fig. 42.1). In such instances, it is particularly important to show the scale clearly, with breaks where necessary, to ensure that the graph does not convey a misleading impression. Note that Fig. 42.1 is drawn with 'floating axes' (i.e. the x and y axes do not meet in the lower left-hand corner), while Fig. 42.2 has clear scale breaks on both x and y axes.

Graph paper

In addition to conventional linear (squared) graph paper, you may need the following:

- Probability graph paper. This is useful when one of the axes is a probability scale (e.g. p. 239)
- Log-linear graph paper. This is appropriate when one of the scales shows a logarithmic progression, e.g. the exponential growth of cells in liquid culture (p. 126). Log-linear paper is defined by the number of logarithmic divisions covered (usually termed 'cycles') so make sure you use a paper with the appropriate number of cycles for your data.
- Log-log graph paper. This is appropriate when both scales show a logarithmic progression, e.g. in allometry (p. 150).

Types of graph

Different graphical forms may be used for different purposes, including:

- Plotted curves — used for data where the relationship between two variables can be represented as a continuum (e.g. Fig. 42.3).
- Scatter diagrams — used to visualize the relationship between individual data values for two interdependent variables, (e.g. Fig. 42.4), often as a preliminary part of a correlation analysis.
- Three-dimensional graphs show the interrelationships of three variables, often one dependent and two independent (e.g. Fig. 42.5). A contour diagram is an alternative method of representing such data.
- Histograms represent frequency distributions of continuous variables (e.g. Fig. 42.6). An alternative is the tally chart (p. 211).
- Frequency polygons emphasize the form of a frequency distribution by joining the coordinates with straight lines, in contrast to a histogram. This is particularly useful when plotting two or more frequency distributions on the same graph (e.g. Fig. 42.7).
- Bar charts represent frequency distributions of a discrete variable (e.g. Fig. 42.8). An alternative representation is the line chart (p. 237).
- Pie charts illustrate portions of a whole (e.g. Fig. 42.9).
- Pictographs give a pictorial representation of data (e.g. Fig. 42.10).

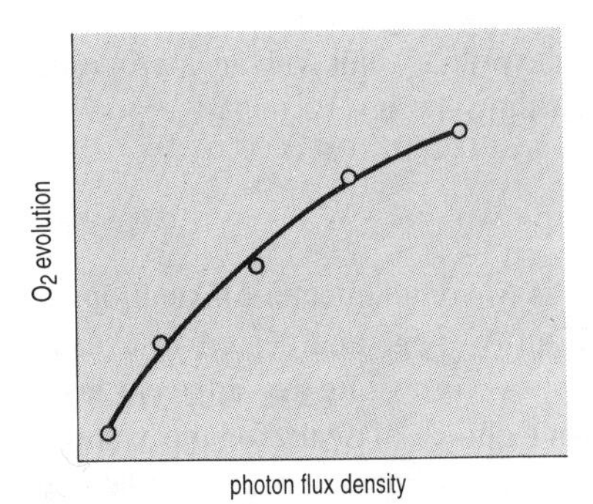

Fig. 42.3 Plotted curve: the rate of photosynthetic O_2 evolution as a function of photon flux density.

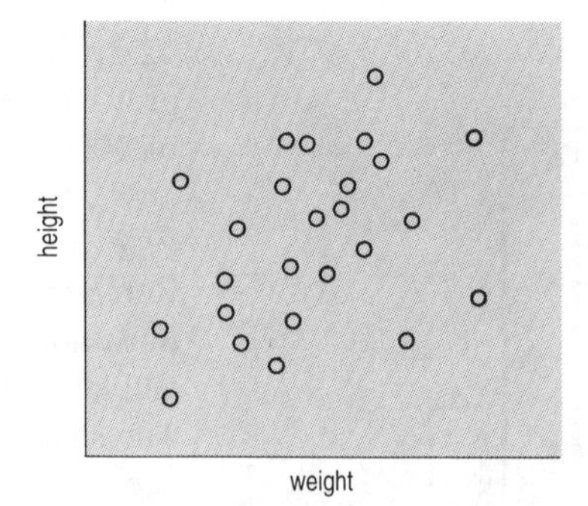

Fig. 42.4 Scatter diagram: height and weight of individual animals in a sample.

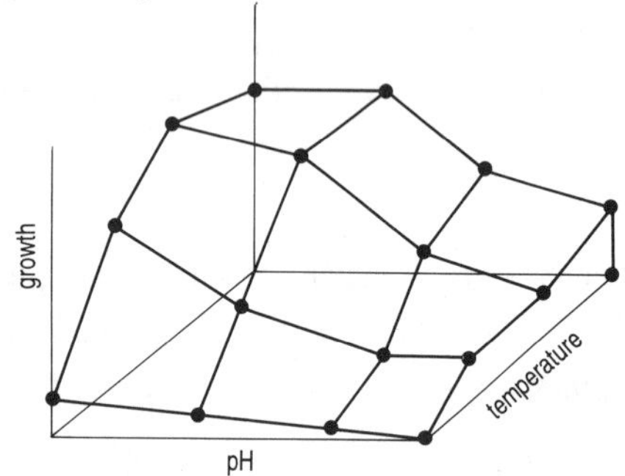

Fig. 42.5 Three-dimensional graph: growth of an organism as a function of temperature and pH.

Fig. 42.6 Histogram: the number of plants within different size classes.

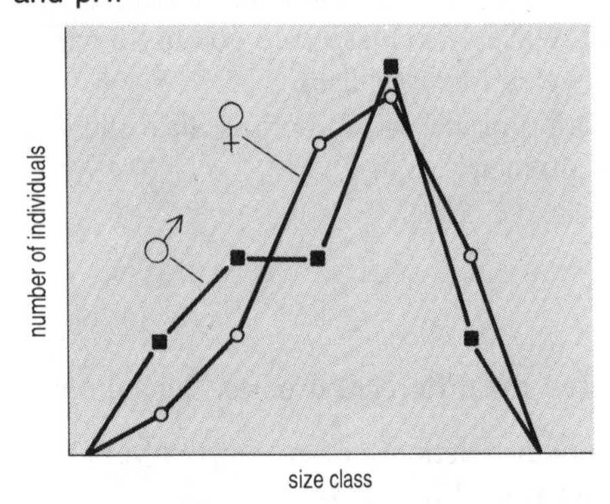

Fig. 42.7 Frequency polygon: frequency distributions of male and female animals according to size.

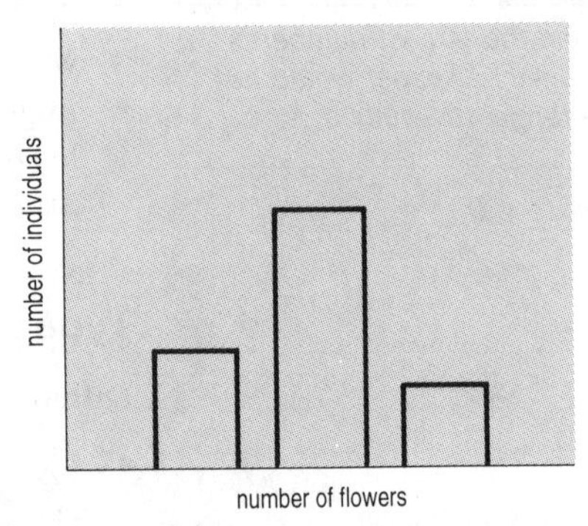

Fig. 42.8 Bar chart: number of flowers per plant.

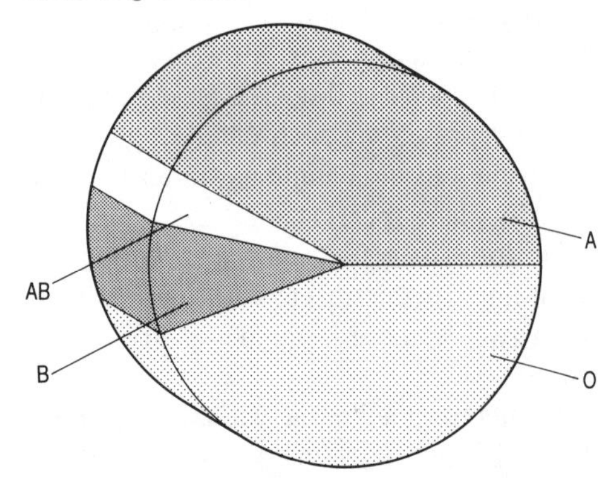

Fig. 42.9 Pie chart: relative abundance of blood groups in man.

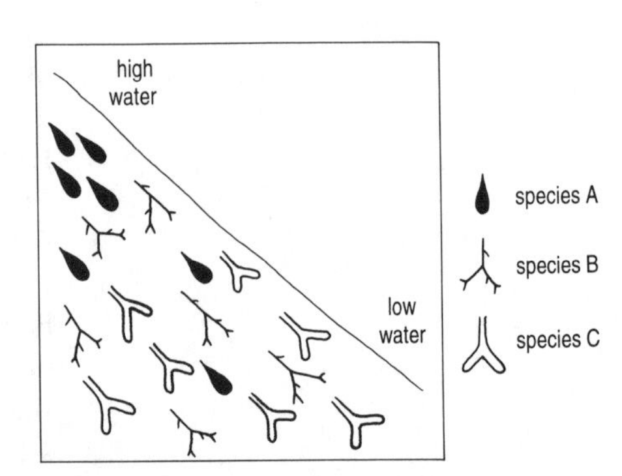

Fig. 42.10 Pictograph: distribution of plants on a rocky shore.

The plotted curve

This is the commonest form of graphical representation used in biology. The key features are outlined below and given in checklist form in Box 42.1.

Data points

Each data point must be shown accurately, so that any reader can determine the exact values of x and y. In addition, the results of each treatment must be readily identifiable. A useful technique is to use a dot for each data point, surrounded by a hollow symbol for each treatment (see Fig. 42.1). An alternative is to use symbols only, though the coordinates of each point are defined less accurately. Use the same symbol for the same entity if it occurs in several graphs and provide a key to all symbols.

Statistical measures

If you are plotting average values for several replicates and if you have the necessary statistical knowledge, you can calculate the standard error, or the 95% confidence limits (p. 240) for each mean value and show these on your graph as a series of vertical bars (see Fig. 42.1). Make it clear in the legend whether the bars refer to standard errors or 95% confidence limits and quote the value of n (the number of replicates per data point). Another approach is to add a least significant difference bar (p. 239) to the graph.

Interpolation

Once you have plotted each point, you must decide whether to link them by straight lines or a smoothed curve. Each of these techniques conveys a different

---

Box 42.1 A checklist for the stages in drawing a graph

The following sequence can be used whenever you need to construct a plotted curve: it will need to be modified if another type of graph is required.

1. **Collect all of the data values and statistical values** (in tabular form, where appropriate).
2. **Decide on the most suitable form of presentation**: this may include transformation of the data (p. 213) to convert the data to linear form.
3. **Compose a concise descriptive title**, together with a reference (figure) number and date, where necessary.
4. **Determine which variable is to be plotted on the x axis and which on the y axis**.
5. **Select appropriate scales for both axes** and make sure that the numbers and their location (scale marks) are clearly shown, together with any scale breaks.
6. **Decide on appropriate descriptive labels for both axes**, with SI units of measurement, where appropriate.
7. **Choose the symbols for each set of data points** and decide on the best means of representation for statistical values.
8. **Plot the points** to show the coordinates of each value with appropriate symbols.
9. **Draw a trend line for each set of points**.
10. **Write a figure legend**, to include a key which identifies all symbols and statistical values and any descriptive footnotes, as required.

---

message to your reader. Joining the points by straight lines may seem the simplest option, but may give the impression that errors are very low or non-existent and that the relationship between the variables is complex. Joining points by straight lines is appropriate in certain graphs involving time sequences (e.g. the number of animals at a particular site each year), or for repeat measurements where measurement error can be assumed to be minimal, (e.g. recording a patient's temperature in a hospital, to emphasize any variation from one time point to the next). However, in most plotted curves the best straight line or curved line should be drawn (according to appropriate mathematical or statistical models, or by eye), to highlight the relationship between the variables – after all, your choice of a plotted curve implies that such a relationship exists! Don't worry if some of your points do not lie on the line: this is caused by errors of measurement and by biological variation (p. 39). Most curves drawn by eye should have an equal number of points lying on either side of the line. You may be guided by 95% confidence limits, in which case your curve should pass within these limits wherever possible.

Conveying the correct message — the golden rule is: 'always draw the simplest line that fits the data reasonably well and is biologically reasonable'.

Curved lines can be drawn using a flexible curve, a set of French curves, or freehand. In the latter case, turn your paper so that you can draw the curve in a single, sweeping stroke by a pivoting movement at the elbow (for larger curves) or wrist (for smaller ones). Do not try to force your hand to make complex, unnatural movements, as the resulting line will not be smooth.

Extrapolation

Be wary of extrapolation beyond the upper or lower limit of your measured values. This is rarely justifiable and may lead to serious errors. Whenever extrapolation is used, a dotted line ensures that the reader is aware of the uncertainty involved. Any assumptions behind an extrapolated curve should also be stated clearly in your text.

Extrapolating plotted curves — try to avoid the need to extrapolate by better experimental or sampling design.

The histogram

While a plotted curve assumes a continuous relationship between the variables by interpolating between individual data points, a histogram involves no such assumptions and is the most appropriate representation if the number of data points is too few to allow a trend line to be drawn. Histograms are also used to represent frequency distributions (p. 211), where the y axis shows the number of times a particular value of x was obtained (e.g. Fig. 42.2). As in a plotted curve, the x axis represents a continuous variable which can take any value within a given range (e.g. plant height), so the scale must be broken down into discrete classes and the scale marks on the x axis should show either the mid-points (mid-values) of each class, or the boundaries between the classes.

In a histogram, each datum is represented by a column with an area proportional to the magnitude of y: in most cases, you should use columns of equal width, so that the height of each column is then directly proportional to y. Shading or stippling may be used to identify individual columns, according to your needs.

The columns are adjacent to each other in a histogram, in contrast to a bar chart, where they are separate because the x axis represents discrete values.

Interpreting graphs

When you look at graphs drawn by other people, make sure you understand the axes before you look at the relationship. It is all too easy to take in the shape of a graph without first considering the scale of the axes, a fact that some advertisers and politicians exploit when curves are used to misrepresent information. Such graphs are often used in newspapers and on television. Examine them critically — many would not pass the stringent requirements of scientific communication and conclusions drawn from them may be flawed.

43 Presenting data in tables

A table is often the most appropriate way to present numerical data in a concise, accurate and structured form. Assignments and project reports should contain tables which have been designed to condense and display results in a meaningful way and to aid numerical comparison. The preparation of tables for recording primary data is discussed on p. 63.

Decide whether you need a table, or whether a graph is more appropriate. Histograms and plotted curves can be used to give a visual impression of the relationships within your data (p. 216). On the other hand, a table gives you the opportunity to make detailed numerical comparisons. Always remember that the primary purpose of your table is to communicate information and allow appropriate comparison, not simply to put down the results on paper!

Preparation of tables

Constructing titles — take care over titles as it is a common mistake in student practical reports to present tables without titles, or to misconstruct the title.

Title

Every table must have a brief descriptive title. If several tables are used, number them consecutively so they can be quoted in your text. The titles within a report should be compared with one another, making sure they are logical and consistent and that they describe accurately the numerical data contained within them.

Saving space in tables — you may be able to omit a column of control data if your results can be expressed as percentages of the corresponding control values.

Structure

Display the components of each table in a way that will help the reader understand your data and grasp the significance of your results. Organize the columns so that each category of like numbers or attributes is listed vertically, while each horizontal row shows a different experimental treatment, organism, sampling site, etc. (as in Table 43.1). Where appropriate, put control values near the beginning of the table. Columns that need to be compared should be set out alongside each other. Use rulings to subdivide your table appropriately, but avoid cluttering it up with too many lines.

Table 43.1 Characteristics of selected photoautotrophs

			Intracellular carbohydrate	
Division	Species	Optimum [NaCl]* (mol m^{-3})	Identity	Quantity† (nmol (g dry wt)$^{-1}$)
Chlorophyta	*Scenedesmus quadruplicatum*	340	Sucrose	49.7
	Chlorella emersonii	780	Sucrose	102.3
	Dunaliella salina	4 700	Glycerol	3 610.7
Cyanobacteria	*Microcystis aeruginosa*	<20‡	None	0.0
	Anabaena variabilis	320	Sucrose	64.2
	Rivularia atra	380	Trehalose	ND

*Determined after 28-day incubation in modified Von Stosch medium.
†Individual samples, analysed by gas–liquid chromatography.
‡Poor growth in all media with added NaCl (minimum NaCl concentration 5 mol m^{-3}).
ND Sample lost: no quantitative data.

Headings and subheadings

These should identify each set of data and show the units of measurement, where necessary. Make sure that each column is wide enough for the headings and for the longest data value.

Examples

If you measured the width of a fungal hypha to the nearest one-tenth of a micrometre, quote the value in the form '52.6 μm'.

Quote the width of a fungal hypha as 52.6 (μm), rather than 0.000 052 6 (m) or 52.6 (10^{-6} m).

Numerical data

Within the table, do not quote values to more significant figures than necessary, as this will imply spurious accuracy (p. 39). By careful choice of appropriate units for each column you should aim to present numerical data within the range 0 to 1 000. As with graphs, it is less ambiguous to use derived SI units, with the appropriate prefixes, in the headings of columns and rows, rather than quoting multiplying factors as powers of 10. Alternatively, include exponents in the main body of the table (see Table 8.1), to avoid any possible confusion regarding the use of negative powers of 10.

Saving further space — in some instances a footnote can be used to replace a whole column of repetitive data.

Other notations

Avoid using dashes in numerical tables, as their meaning is unclear; enter a zero reading as '0' and use 'NT' for not tested or 'ND' if no data value was obtained, with a footnote to explain each abbreviation. Other footnotes, identified by asterisks, superscripts or other symbols in the table, may be used to provide relevant experimental detail (if not given in the text) and an explanation of column headings and individual results, where appropriate. Footnotes should be as condensed as possible. Table 43.1 provides examples.

Statistics

In tables where the dispersion of each data set is shown by an appropriate statistical parameter, you must state whether this is the (sample) standard deviation, the standard error (of the mean) or the 95% confidence limits and you must give the value of n (the number of replicates). Other descriptive statistics should be quoted with similar detail, and hypothesis-testing statistics should be quoted along with the value of P (the probability). Details of any test used should be given in the legend, or in a footnote (see p. 239).

Text

Sometimes a table can be a useful way of presenting textual information in a condensed form (see examples on pp. 178 and 200).

Using microcomputers and word-processing packages — these can be used to prepare high-quality versions of tables for project work (p. 256).

When you have finished compiling your tabulated data, carefully double-check each numerical entry against the original information, to ensure that the final version of your table is free from transcriptional errors. Box 43.1 gives a checklist for the major elements of constructing a table.

Box 43.1 A checklist for preparing a table

Every table should have the following components:

1. A title, plus a reference number and date, where necessary.
2. Headings for each column and row, with appropriate units of measurement.
3. Data values, quoted to the nearest significant figure and with statistical parameters, according to your requirements.
4. Footnotes to explain abbreviations, modifications and details.
5. Rulings to emphasize groupings and distinguish items from each other.

44 Hints for solving numerical problems

Biology often requires a numerical or statistical approach. Not only is mathematical modelling an important aid to understanding, but computations are often needed to turn raw data into meaningful information or to compare them with other data sets. Moreover, calculations are part of laboratory routine, perhaps required for making up solutions of known concentration (see p. 18) or for the calibration of a microscope (see p. 137). In research, 'trial' calculations can reveal what input data are required and where errors in their measurement might be amplified in the final result (see p. 51).

If you have a 'block' about numerical work, practice at problem-solving is especially important because it:

- demystifies the procedures involved, which are normally just the elementary mathematical operations of addition, subtraction, multiplication and division (Table 44.1);

Table 44.1 Sets of numbers and operations

Sets of numbers	
Whole numbers:	0, 1, 2, 3, ...
Natural numbers:	1, 2, 3, ...
Integers:	... −3, −2, −1, 0, 1, 2, 3, ...
Real numbers:	integers and anything between (e.g. −5, 4.376, 3/16, π, $\sqrt{5}$)
Prime numbers:	subset of natural numbers divisible by 1 and themselves only (i.e. 1, 3, 5, 7, 11, 13, ...)
Rational numbers:	p/q where p (integer) and q (natural) have no common factor (e.g. 3/4)
Fractions:	p/q where p is an integer and q is natural (e.g. −6/8)
Irrational numbers:	real numbers with no exact value (e.g. π)
Infinity:	(symbol ∞) is larger than any number (technically not a number as it does not obey the laws of algebra)
Operations and symbols	
Basic operators:	+, −, × and ÷ will not need explanation; however, / may substitute for ÷. * may substitute for × or this operator may be omitted
Powers:	a^n, i.e. 'a to the power n', means a multiplied by itself n times (e.g. $a^2 = a \times a$ = 'a squared', $a^3 = a \times a \times a$ = 'a cubed'). n is said to be the index or exponent. Note $a^0 = 1$ and $a^1 = a$
Logarithms:	the common logarithm (log) of any number x is the power to which 10 would have to be raised to give x (i.e. the log of 100 is 2; $10^2 = 100$); the antilog of x is 10^x. Note that there is no log for 0, so take this into account when drawing log axes by breaking the axis. Natural or Napierian logarithms (ln) use the base e (= 2.718 28...) instead of 10
Reciprocals:	the reciprocal of a real number a is $1/a$ ($a \neq 0$)
Relational operators:	$a > b$ means 'a is greater (more positive) than b', < means less than, ≤ means less-than-or-equal-to and ≥ means greater-than-or-equal-to
Proportionality:	$a \propto b$ means 'a is proportional to b' (i.e. $a = kb$, where k is a constant). If $a \propto 1/b$, a is inversely proportional to b ($a = k/b$)
Sums:	Σx_i is shorthand for the sum of all x values from $i = 0$ to $i = n$ (more correctly the range of the sum is specified under the symbol)
Moduli:	$\lvert x \rvert$ signifies modulus of x, i.e. its absolute value (e.g. $\lvert 4 \rvert = \lvert -4 \rvert = 4$)
Factorials:	$x!$ signifies factorial x, the product of all integers from 1 to x (e.g. $3! = 6$). Note $0! = 1! = 1$

- allows you to gain confidence so that you don't panic when confronted with an unfamiliar or apparently complex form of problem;
- helps you recognize the various forms a problem can take as, for instance, in crossing experiments in classical genetics.

Steps in tackling a numerical problem

The step-by-step approach outlined below may not be the fastest method of arriving at an answer, but most mistakes occur where steps are missing, combined or not made obvious, so a logical approach is often better. Error tracing is distinctly easier when all stages in a calculation are laid out.

Have the right tools ready

Scientific calculators (p. 5) greatly simplify the numerical part of problem-solving. However, the seeming infallibility of the calculator may lead you to accept an absurd result which could have arisen because of faulty key-pressing or faulty logic. Make sure you know how to use all the features on your calculator, especially:

A computer spreadsheet may be very useful in repetitive work or for 'what if?' case studies (see p. 250).

- how the memory works;
- how to obtain an exponent. (Note that the 'exp' button on most calculators gives you 10^x, not 1^x or y^x; so 1×10^6 would be entered as [1] [exp] [6], *not* [10] [exp] [6]).
- how to introduce a constant multiplier or divider.

Approach the problem thoughtfully

If the individual steps have been laid out on a worksheet, the 'tactics' will already have been decided. It is more difficult when you have to adopt a strategy on your own because the problem is presented as a story and it isn't obvious which equations or rules need to be applied.

- Read the problem carefully as the text may give clues as to how it should be tackled. Be certain of what is required as an answer before starting.
- Analyse what kind of problem it is, which effectively means deciding which equation(s) or approach will be applicable. If this is not obvious, consider the dimensions/units of the information available and think how they could be fitted to a relevant formula. In examinations, a favourite ploy of examiners is to present a problem such that the familiar form of an equation must be rearranged (see Table 44.2 and Box 44.1). Another is to make you use two or more equations in series (see Box 44.2).

Checking a formula — if you are unsure whether a formula is correct, a dimensional analysis can help: write in all the units for the variables and make sure that they cancel out to give the expected answer.

- Check that you have, or can derive, all of the information required to use your chosen equation(s). It is unusual but not unknown for examiners to supply redundant information. So, if you decide not to use some of the information given, be sure why you do not require it.
- Decide on what format and units the answer should be presented in. This is sometimes suggested to you. If the problem requires many changes in the prefixes to units, it is a good idea to convert all data to base SI units (multiplied by a power of 10) at the outset.
- If a problem appears complex, break it down into component parts.

Show the steps in your calculations — most markers will only penalize a mistake once and part marks will be given if the remaining operations are performed correctly. This can only be done if those operations are visible!

Present your answer clearly

The way you present your answer obviously needs to fit the individual problem. The example shown in Box 44.2 has been chosen to illustrate several important points, but this format would not fit all situations. Guidelines for presenting an answer include:

(a) Make your assumptions explicit. Most mathematical models of biological phenomena require that certain criteria are met before they can be legitimately applied (e.g. 'assuming the tissue is homogeneous . . .'), while some approaches involve approximations which should be clearly stated (e.g. 'to estimate the mouse's skin area, its body was approximated to a cylinder with radius x and height y . . .').
(b) Explain your strategy for answering, perhaps giving the applicable formula or definitions which suit the approach to be taken. Give details of what the symbols mean (and their units) at this point.
(c) Rearrange the formula to the required form with the desired unknown on the left-hand side (see Table 44.2).
(d) Substitute the relevant values into the right-hand side of the formula, using the units and prefixes as given (it may be convenient to convert values to SI beforehand). Convert prefixes to appropriate powers of 10 as soon as possible.
(e) Convert to the desired units step-by-step, i.e. taking each variable in turn.
(f) When you have the answer in the desired units, rewrite the left-hand side and underline the answer. Make sure that the result is presented with an appropriate number of significant figures (see p. 39).

Units — never write any number without its unit(s) unless it is truly dimensionless.

Rounding off — do not round off numbers until you arrive at the final answer.

Check your answer

Having written out your answer, you should check it methodically, answering the following questions:

- Is the answer of a realistic magnitude? You should be alerted to an error if an answer is absurdly large or small. In repeated calculations, a result standing out from others in the same series should be double-checked.
- Do the units make sense and match up with the answer required? Don't, for example, present a volume in units of m^2.
- Do you get the same answer if you recalculate in a different way? If you have time, recalculate the answer using a different 'route', entering the numbers into your calculator in a different form and/or carrying out the operations in a different order.

Table 44.2 Simple algebra — rules for manipulating equations

If $a = b + c$, then $b = a - c$ and $c = a - b$
If $a = b \times c$, then $b = a/c$ and $c = a/b$
If $a = b^c$, then $b = a^{1/c}$ and $c = \log a/\log b$
$a^{1/n} = \sqrt[n]{a}$
$a^{-n} = 1/a^n$
$a^b \times a^c = a^{(b+c)}$ and $a^b/a^c = a^{(b-c)}$
$(a^b)^c = a^{(b \times c)}$
$a \times b = \text{antilog}(\log a + \log b)$

Some reminders of basic mathematics

Errors in calculations sometimes appear because of faults in mathematics rather than computational errors. For reference purposes, Tables 44.1–44.3 give some basic mathematical principles that may be useful. Eason *et al.* (1980) should be consulted for more advanced needs.

Genetics problems

These can cause problems for many students. The key is to practice and to adopt a logical approach (see Box 44.3).

Box 44.1 Example of using the rules of Table 44.2

Problem: if $a = (b - c)/(d + e^n)$, find e

1. Multiply both sides by $(d + e^n)$; formula becomes:

$$a(d + e^n) = (b - c)$$

2. Divide both sides by a; formula becomes: $d + e^n = \frac{b - c}{a}$

3. Subtract d from both sides; formula becomes: $e^n = \frac{b - c}{a} - d$

4. Raise each side to the power $1/n$; formula becomes:

$$e = \left\{\frac{b - c}{a} - d\right\}^{1/n}$$

Box 44.2 Model answer to a typical biological problem

Problem

Estimate the total length and surface area of the fibrous roots on a maize seedling from measurements of their total fresh weight and mean diameter. Give your answers in m and cm^2 respectively.

Measurements

Fresh weight[a] = 5.00 g, mean diameter[b] = 0.5 mm.

Answer

Assumptions: (1) the roots are cylinders with constant radius[c] and the 'ends' have negligible area; (2) the root system has a density of 1 000 kg m^{-3} (i.e. that of water[d]).

Strategy: from assumption (1), the applicable equations are those concerned with the volume and surface area of a cylinder (Table 44.3), namely:

$$V = \pi r^2 h \quad [44.1]$$

$$A = 2\pi rh \text{ (ignoring ends)} \quad [44.2]$$

where V is volume (m^3), A is surface area (m^2), $\pi \approx$ 3.141 59, h is height (m) and r is radius (m). The total length of the root system is given by h and its surface area by A. We can find h by rearranging eqn [44.1] and then substitute its value in eqn [44.2] to get A.

To calculate total root length: rearranging eqn [44.1], we have $h = V/\pi r^2$. From measurements[e], r = 0.25 mm = 0.25×10^{-3} m.

From density = weight/volume,

$$V = \text{fresh weight/density}$$
$$= 5 \text{ g}/1000 \text{ kg m}^{-3}$$
$$= 0.005 \text{ kg}/1000 \text{ kg m}^{-3}$$
$$= 5 \times 10^{-6} \text{ m}^3$$

Total root length,

$$h = V/\pi r^2$$
$$= 5 \times 10^{-6} \text{ m}^3/3.141\,59 \times (0.25 \times 10^{-3} \text{ m})^2$$

Total root length = 25.46 m

To calculate surface area of roots: substituting value for h obtained above into eqn [44.2], we have:

Root surface area

$$= 2 \times 3.141\,59 \times 0.25 \times 10^{-3} \text{ m} \times 25.46 \text{ m}$$
$$= 0.04 \text{ m}^2$$
$$= 0.04 \times 10^4 \text{ cm}^2$$

(there being $100 \times 100 = 10^4$ cm^2 per m^2)

Root surface area = 400 cm^2

Notes

(a) The fresh weight of roots would normally be obtained by washing the roots free of soil, blotting them dry and weighing.
(b) In a real answer you might show the replicate measurements giving rise to the mean diameter.
(c) In reality, the roots will differ considerably in diameter and each root will not have a constant diameter throughout its length.
(d) This will not be wildly inaccurate as about 95% of the fresh weight will be water, but the volume could also be estimated from water displacement measurements.
(e) Note conversion of measurements into base SI units at this stage and on line 3 of the root volume calculation. Forgetting to halve diameter measurements where radii are required is a common error.

Table 44.3 Geometry and trigonometry — analysing shapes

Shape/object	Diagram	Perimeter	Area
Two-dimensional shapes			
Square	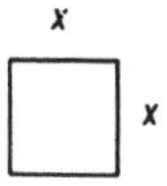	$4x$	x^2
Rectangle	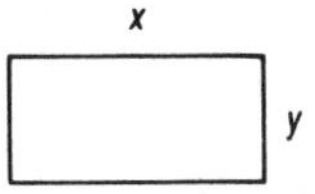	$2(x + y)$	xy
Circle		$2\pi r$	πr^2
Ellipse	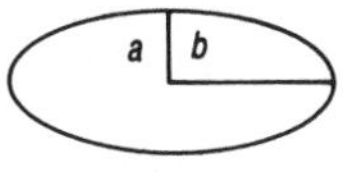	$\pi[1.5(a+b) - \sqrt{a*b}]$ (approx.)	πab
Triangle (general)	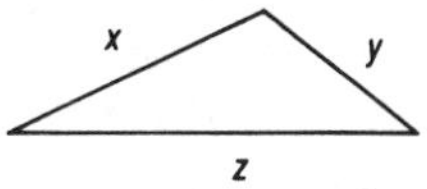	$x+y+z$	$0.5zh$
(right-angled)	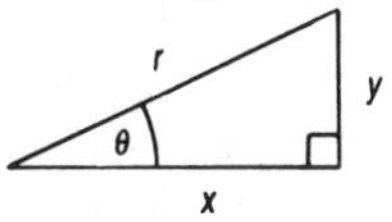	$x+y+r$; $\sin\theta = y/r$, $\cos\theta = x/r$, $\tan\theta = y/x$; $r^2 = x^2 + y^2$	$0.5xy$

Shape/object	Diagram	Surface area	Volume
Three-dimensional objects			
Cube	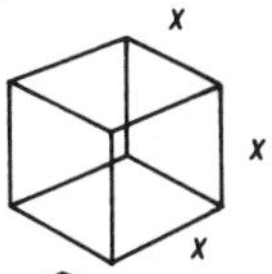	$6x^2$	x^3
Cuboid	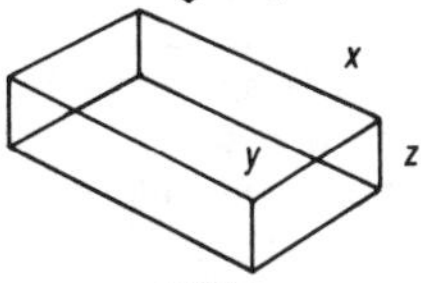	$2xy + 2xz + 2yz$	xyz
Sphere	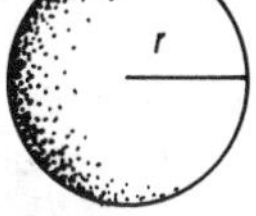	$4\pi r^2$	$4\pi r^3/3$
Ellipsoid	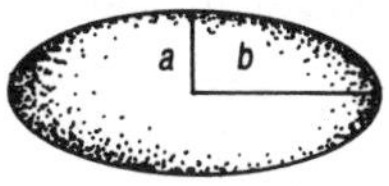	no simple formula	$\pi rab/3$
Cylinder	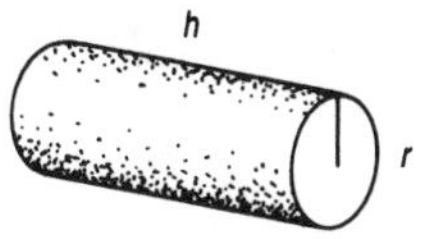	$2\pi rh + 2\pi r^2$	$\pi r^2 h$
Cone and pyramid	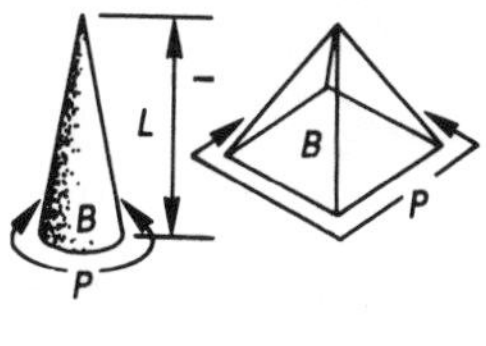	$0.5PL + B$	$BL/3$

x, y, z = sides;
a, b = half minimum and maximum axes;
r = radius or hypotenuse;
h = height;
B = base area;
L = perpendicular height;
P = perimeter of base.

Box 44.3 Hints for answering problems in Mendelian genetics

1. **Learn the terminology and symbols.** You should know what the following mean:

   (a) chromosome, autosome, sex chromosome;
   (b) gene, allele;
   (c) dominant, recessive, lethal;
   (d) homozygous, heterozygous;
   (e) genotype, phenotype;
   (f) P, F_1, F_2, ♀, ♂.

2. **Read the problem carefully:**

   (a) Decide which category it falls into (see below).
   (b) Decide which alleles are dominant and which recessive; sex-linked genes are sometimes obvious from differences in the frequencies of male and female phenotypes.
   (c) Assign symbols to the alleles (upper case letter for dominant, e.g. A, lower case for recessive, e.g. a) if this has not been done for you.

3. **If the problem involves the outcome of a cross, decide how to tackle it:**

   (a) Punnett squares provide a good visual indication of potential combinations of gametes for a given cross. Group together like genotypes to obtain the genotypic ratio then work out the phenotypic ratio. Lay out Punnett squares consistently as shown in Fig. 44.1.

	male gametes A	male gametes a
female gametes A	AA	Aa
female gametes a	Aa	aa

Fig. 44.1 Layout for a simple Punnett square. The genotypic ratios from this cross are 1:2:1 for AA:Aa:aa, and the phenotypic ratio would be 3:1 for characteristic A to characteristic a.

   (b) Probability calculations can be simpler and faster than Punnett squares when more than two genes are considered. The key thing to remember is that the chance of a number of independent events occurring together is equal to the probabilities of each event occurring multiplied together.

   Example: What is the probability of a homozygous blue-eyed mother and a heterozygous brown-eyed father having two sons, both of which are blue eyed?

   P(son) = 0.5; P(blue-eyed child) = 0.5.
   Therefore, P(blue-eyed son) = 0.5 × 0.5 = 0.25.
   Hence, P(two blue-eyed sons) = 0.25 × 0.25 = 0.0625.

4. **Identify the type of problem involved**

   (a) Monohybrid cross: the simplest type of cross, involving two alleles of one gene, e.g. AA × Aa.
   (b) Dihybrid cross: a cross involving two genes each with two alleles, e.g. AaBB × AAbb.
   (c) Test cross: a cross of an unknown genotype with a homozygous recessive, e.g. AaBB × aabb. The genotype is revealed by the phenotypic ratio in the F_1 generation.
   (d) Sex-linked cross: involves a gene carried on the X chromosome; this can be designated as X^A (dominant) or X^a (recessive) $X^aY \times X^AX^a$. Only one copy of the gene will be present in normal (XY) male mammals.
   (e) Crosses with linked genes: genes are linked if they are on the same chromosome. They may become separated when crossing-over occurs between homologous chromosomes during synapsis during meiosis. Linkage can be detected from a test cross, e.g. AaBb × aabb. If the genes A and B are on different chromosomes we expect AaBb, aabb, Aabb and aaBb to occur in the ratio 1:1:1:1; if they are closely linked, the last two genotypes would be rare.
   (f) Chromosome mapping: uses the frequency of crossing-over of linked genes as above to estimate their distance apart from the following formulae:

$$\frac{\text{no. of crossing-over progeny}}{\text{total no. of progeny}} \times 100 = \%\ \text{crossing-over} \qquad [44.3]$$

$$1\%\ \text{crossing-over} = 1\ \text{map unit} \qquad [44.4]$$

45 Descriptive statistics

Whether obtained from observation or experimentation, most data in biology exhibit variability. This can be displayed as a frequency distribution (Fig. 42.6). Descriptive (or summary) statistics quantify aspects of the frequency distribution of a sample. You can use them to:

- condense a large data set for presentation in figures or tables;
- provide estimates of parameters of the frequency distribution of the population being sampled (p. 236).

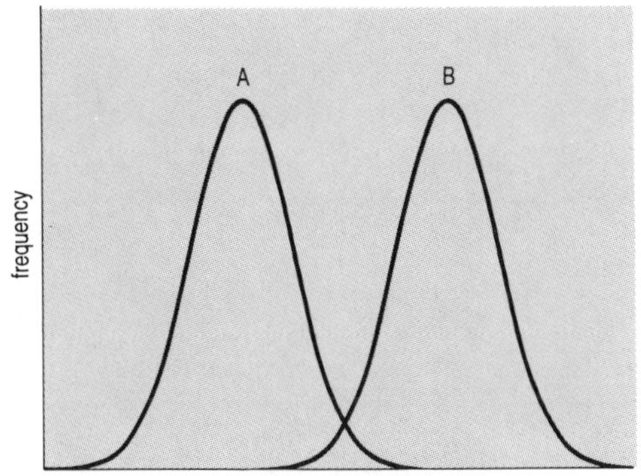

The appropriate descriptive statistics to choose depend on both the type of data, i.e. quantitative, ranked, or qualitative (see p. 37), and the nature of the underlying population frequency distribution. If you have no clear theoretical grounds for assuming what this is like, graph one or more sample frequency distributions, ideally with a sample size >100. The tally system for recording data (see p. 211) can give an immediate visual indication of the frequency distribution as data are collected.

Fig. 45.1 Two distributions with different locations but the same dispersion. The data set labelled B could have been obtained by adding a constant to each datum in the data set labelled A.

The methods used to calculate descriptive statistics depend on whether data have been grouped into classes. You should use the original data set if it is still available, because grouping into classes loses information and accuracy. However, large data sets may make calculations unwieldy, and are best handled by computer programs.

Three important features of a frequency distribution that can be summarized by descriptive statistics are:

- the sample's location, i.e. its position along a given dimension representing the dependent (measured) variable (Fig. 45.1);
- the dispersion of the data, i.e. how spread out the values are (Fig. 45.2);
- the shape of the distribution, i.e. whether symmetrical, skewed, U-shaped, etc. (Fig. 45.3).

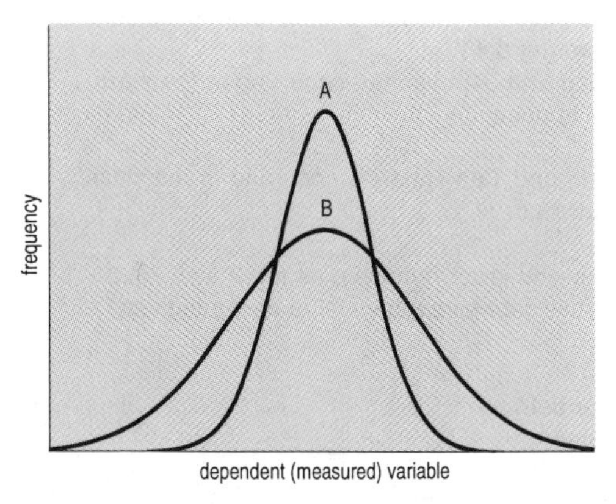

Fig. 45.2 Two distributions with different dispersions but the same location. The data set labelled A covers a relatively narrow range of values of the dependent (measured) variable while that labelled B covers a wider range.

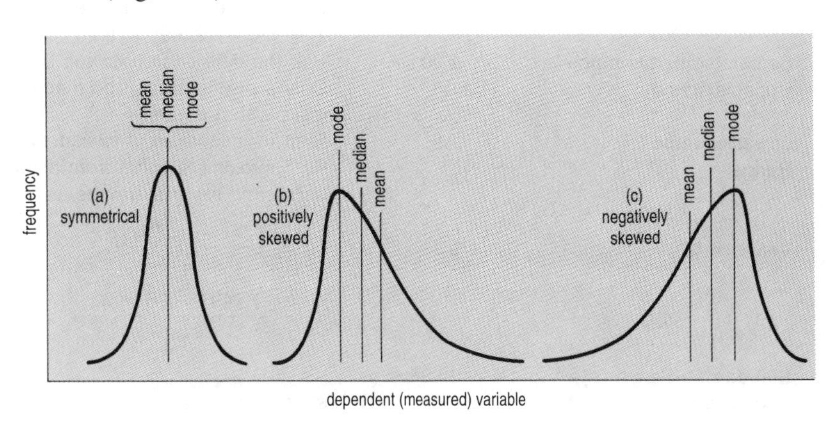

Fig. 45.3 Symmetrical and skewed frequency distributions, showing relative positions of mean, median and mode.

Box 45.1 shows a set of data and the calculated values of the measures of location, dispersion and shape for which methods of calculation are outlined below. Check your understanding by calculating the statistics yourself and confirming that you arrive at the same answers.

Measuring location

Here, the objective is to pinpoint the 'centre' of the frequency distribution, i.e. the value about which most of the data are grouped. The chief measures of location are the mean, median and mode. Fig. 45.4 shows how to choose among these for a given data set.

Box 45.1 Descriptive statistics for a sample of grouped data

Class mid-value* (Y)	Frequency (f)	Cumulative frequency	fY	fY^2
0.5	0	0	0	0
1.5	1	1	1.5	2.25
2.5	1	2	2.5	6.25
3.5	2	4	7	24.5
4.5	5	9	22.5	101.25
5.5	9	18	49.5	272.25
6.5	10	28	65	422.5
7.5	8	36	60	450
8.5	4	40	34	289
9.5	3	43	28.5	270.75
10.5	1	44	10.5	110.25
11.5	0	44	0	0
Totals	$44 = \Sigma f$		$281 = \Sigma fY$	$1\,949 = \Sigma fY^2$

* note that class limits in this example are 0, 1, 2 etc.

Statistic	Value*	How calculated
Mean	6.39	$\Sigma fY/n$, i.e. 281/44
Modal class	6.5	Mid-value of most frequent class
Median	6.45	Value of the $(n + 1)/2$ variate, i.e. half-way between the 22nd and 23rd. From the cumulative frequency data, this appears in the class with mid-value = 6.5, half-way between the 4th and 5th of the 10 values in that class. Assuming the data to be evenly spread, it occurs at (4.5)/10 of the class interval = 0.45. Add this to the lower class limit 6.0 and we get 6.45
Upper quartile	7.69	The upper quartile is between the 33rd and 34th variate, occurring in the class with mid-value = 7.5. It is estimated to occur at $Y = 7.0 + (5.5/8) = 7.69$
Lower quartile	5.28	The lower quartile is between the 11th and 12th variates, occurring in the class with mid-value = 6. It is estimated to occur at $Y = 5.0 + (2.5/9) = 5.28$
Semi-interquartile range	1.20	Half the difference between the upper and lower quartiles, i.e (7.69 − 5.28)/2
Upper extreme	11.5	Only an estimate can be made from the data given: mid-value of the highest class with a datum
Lower extreme	1.5	Ditto for mid-value of lowest class
Range	10	An approximate value from difference between upper and lower extremes as estimated
Variance (s^2)	3.59	$s^2 = \dfrac{\Sigma fY^2 - (\Sigma fY)^2/n}{n - 1} = \dfrac{1\,949 - (281)^2/44}{43}$
Standard deviation (s)	1.895	$\sqrt{s^2}$
Standard error (SE)	0.286	$s/\sqrt{n}$
Coefficient of variation (cov)	29.7%	$100s/\bar{Y}$

* Rounded to appropriate significant figures

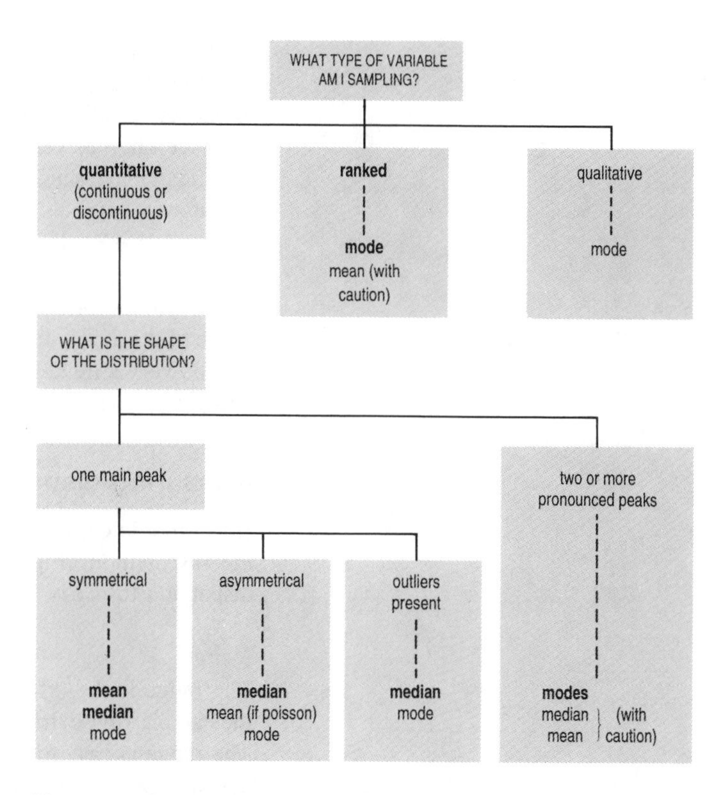

Fig. 45.4 Choosing a statistic for characterizing a distribution's location. Statistics written in bold are the preferred option(s).

Mean

The mean (denoted $\bar{Y}$ and also referred to as the arithmetic mean) is the average value of the observations. It is obtained from the sum of all the observations divided by the number of observations (in symbolic terms, $\Sigma Y/n$). The mean is a good measure of the centre of symmetrical frequency distributions. It uses the numerical values of the sample to describe their location and therefore incorporates all of the information content of the data. However, the value of a mean is greatly affected by the presence of outliers. The arithmetic mean is a very popular statistic in biology, but there are situations when you should be careful about using it (see Box 45.2 for examples).

Definition
An outlier — any datum which has a value much smaller or bigger than most of the data.

Median

The median is the mid-point of the observations when ranked in increasing order. For odd-sized samples, the median is the middle observation; for even-sized samples it is the mean of the middle pair of observations. Where data are grouped into classes, the median must be estimated. This is most simply done from a graph of the cumulative frequency distribution, but can also be worked out by assuming the data to be evenly spread within the class (see Box 45.1). The median may represent the location of the main body of data better than the mean when the distribution is asymmetric or when there are outliers in the sample.

Definition
Rank — the position of a data value when all the data are placed in order of ascending magnitude. If ties occur, an average rank of the tied variates is used. Thus, the rank of the datum 6 in the sequence 1,3,5,6,8,8,10 is 4; the rank of each datum with value 8 is 5.5.

Mode

The mode is the most common value in the sample. The mode is easily found from a tabulated frequency distribution as the most frequent value. If data have been grouped into classes then the term modal class is used for the class containing most values. The mode provides a rapidly and easily found estimate of sample location and is unaffected by outliers. However, the mode is affected by chance variation in the shape of a sample's distribution and it may lie distant from the obvious centre of the distribution. Note that the mode is the only statistic to make sense of qualitative data, e.g. 'the modal (most frequent) egg colour was blue'.

The mean, median and mode have the same units as the variable under discussion. However, whether these statistics of location have the same or similar values for a given frequency distribution depends on the symmetry and shape of the distribution. If it is near-symmetrical with a single peak, all three will be very similar; if it is skewed or has more than one peak, their values will differ to a greater degree (see Fig. 45.3).

Measuring dispersion

Here, the objective is to quantify the spread of the data about the centre of the distribution. Figure 45.5 indicates how to decide which measure of dispersion to use.

Range

The range is the difference between the largest and smallest data values in the sample (the extremes) and has the same units as the measured variable. The range is easy to determine, but is greatly affected by outliers. Its value may also depend on sample size: in general, the larger this is, the greater will be the range.

Semi-interquartile range

The semi-interquartile range is an appropriate measure of dispersion when a median is the appropriate statistic to describe location. For this, you need to determine the first and third quartiles, i.e. the medians for the data values ranked below and above the median of the whole sample (see Fig. 45.6). To calculate

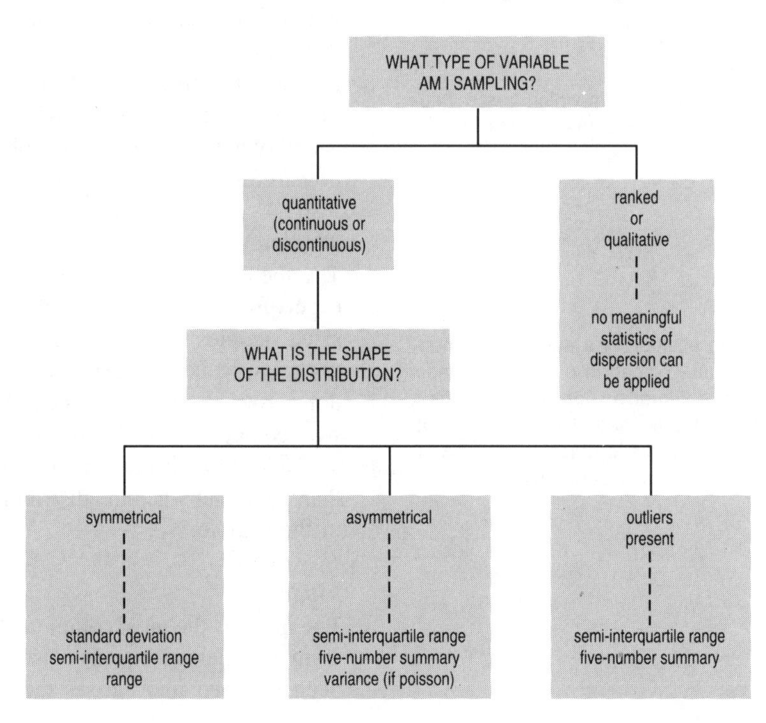

Fig. 45.5 Choosing a statistic for characterizing a distribution's dispersion. Statistics written in bold are the preferred option(s). Note that you should match statistics describing dispersion with those you have used to describe location, i.e. standard deviation with mean, semi-interquartile range with median.

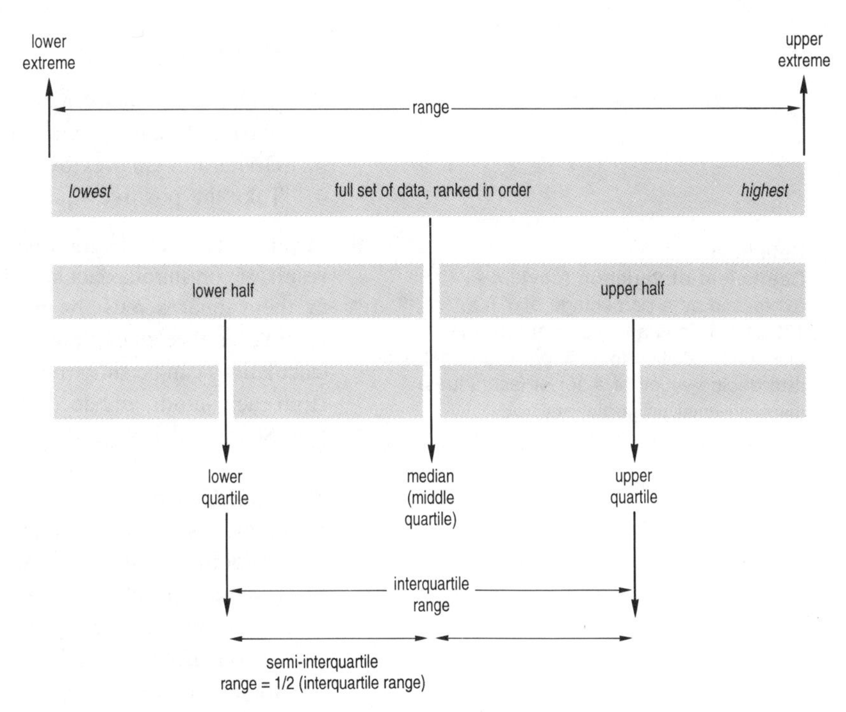

Fig. 45.6 Illustration of median, quartiles, range and semi-interquartile range.

a semi-interquartile range for a data set:

1. Rank the observations in ascending order.
2. Find the values of the first and third quartiles.
3. Subtract the value of the first quartile from the value of the third.
4. Halve this number.

For data grouped in classes, the semi-interquartile range can only be estimated and the calculations are complex (see Box 45.1). Another disadvantage is that it takes no account of the shape of the distribution at its edges. This objection can be countered by using the so-called 'five number summary' of a data set, which consists of the three quartiles and the two extreme values; this can be presented on graphs as a box and whisker plot (see Fig. 45.7) and is particularly useful for summarizing skewed frequency distributions. The corresponding 'six number summary' includes the sample's size.

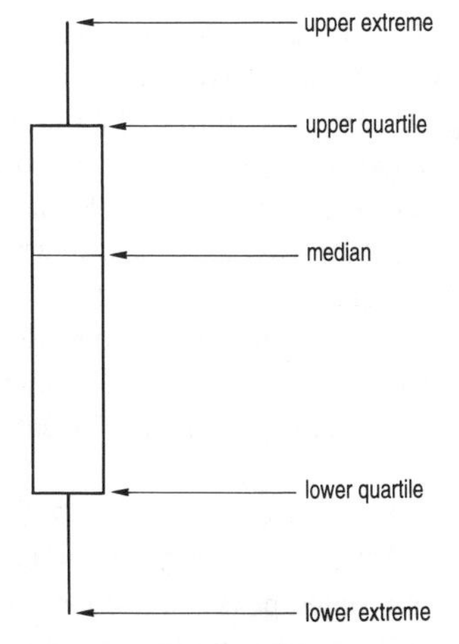

Fig. 45.7 A box and whisker plot, showing the 'five number summary' of a sample as it might be used on a graph.

Variance and standard deviation

For symmetrical frequency distributions, an ideal measure of dispersion would take into account each datum's deviation from the mean and provide a measure of the average deviation from the mean. Two such sample statistics are the variance, which is the sum of squares ($\Sigma(Y - \bar{Y})^2$) divided by $n - 1$ (where n is the sample size), and the standard deviation, which is the positive square root of the variance.

The variance (s^2) has units which are the square of the original units, while the standard deviation (s) is expressed in the original units, one reason s is often preferred as a measure of dispersion. Calculating s or s^2 longhand is a tedious job and is best done with the help of a calculator or computer. If you don't have a calculator that calculates s for you, an alternative formula that simplifies calculations is:

$$s = +\sqrt{\frac{\Sigma Y^2 - (\Sigma Y)^2/n}{n - 1}} \qquad [45.1]$$

To calculate s:

1. Obtain ΣY, square it, divide by n and store in memory.
2. Square Y values, obtain ΣY^2, subtract memory value from this.
3. Divide this answer by $n - 1$.
4. Take the positive square root of this value.

Definition
Coefficient of variation (cov) = s expressed as a percentage of $\bar{Y}$ (i.e. $100s/\bar{Y}$). This is a useful measure of dispersion relative to location; you might use it to decide which of a set of techniques involved least proportional error.

Take care to retain significant figures, or errors in the final value of s will result. If continuous data have been grouped into classes, the class mid-values or their squares must be multiplied by the appropriate frequencies before summation (see example in Box 45.1). When data values are large, longhand calculations can be simplified by coding the data, e.g. by subtracting a constant from each datum, and decoding when the simplified calculations are complete (see Sokal and Rohlf, 1981).

The standard error

You may wish to use a sample mean as an estimate of the population mean. The reliability of this estimate can be indicated by the standard error (SE). For a sample from a normal population, this is given by $s/\sqrt{n}$. The SE is an estimate of the standard deviation of the means of n-sized samples from the population. Data should be quoted as $\bar{Y}$ and SE, with the SE being given to one significant figure more than the mean. The SE is useful because it allows calculation of the confidence limits for the sample mean (see p. 240).

Describing the 'shape' of frequency distributions

Frequency distributions may differ in the following characteristics:

- number of peaks,
- skewness or asymmetry,
- kurtosis or pointedness.

The shape of a frequency distribution of a small sample is affected by chance variation and may not be a fair reflection of the underlying population frequency distribution: check this by comparing repeated samples from the same population or by increasing the sample size. If the original shape were due to random events, it should not appear consistently in repeated samples and should become less obvious as sample size increases.

Genuinely bimodal or polymodal distributions may result from the combination of two or more unimodal distributions, indicating that more than one underlying population is being sampled (Fig. 45.9). An example of a bimodal distribution is the height of adult humans (females and males combined).

A distribution is skewed if it is not symmetrical, a symptom being that the mean, median and mode are not equal (Fig. 45.3). Positive skewness is where the longer 'tail' of the distribution occurs for higher values of the measured variable; negative skewness where the longer tail occurs for lower values. Some biological examples of characteristics distributed in a skewed fashion are volumes of plant protoplasts, insulin levels in human plasma and bacterial colony counts.

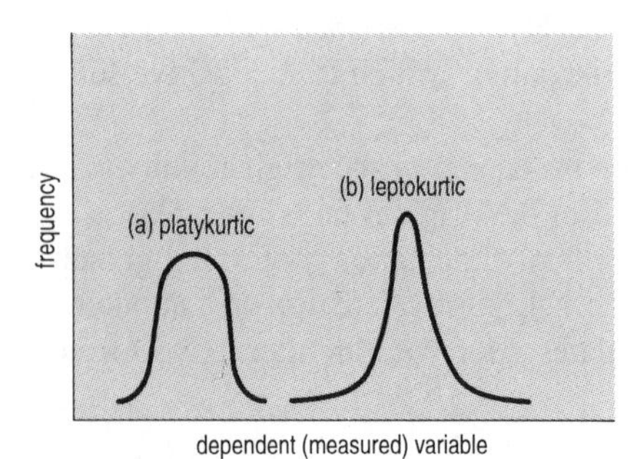

Fig. 45.8 Examples of the two types of kurtosis.

Kurtosis is the name given to the 'pointedness' of a frequency distribution. A platykurtic frequency distribution is one with a flattened peak, while a leptokurtic frequency distribution is one with a pointed peak (Fig. 45.8). The degree of skewness and kurtosis can be quantified and statistical tests exist to testing the 'significance' of observed values (see Sokal and Rohlf, 1981), but the calculations required are complex and best done with a computer.

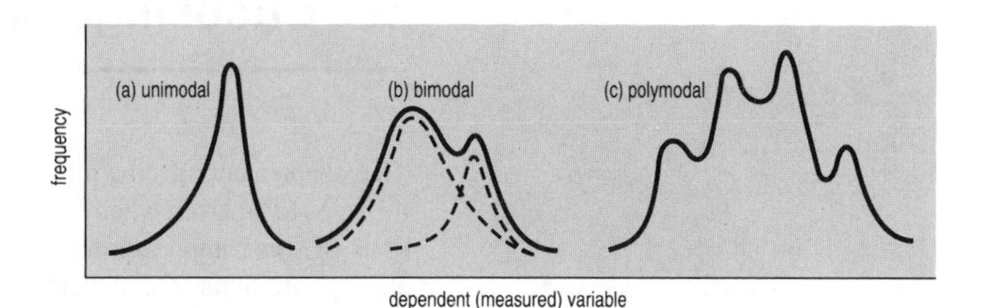

Fig. 45.9 Frequency distributions with different numbers of peaks. A unimodal distribution (a) may be symmetrical or asymmetrical. The dotted lines in (b) indicate how a bimodal distribution could arise from a combination of two underlying unimodal distributions. Note here how the term 'bimodal' is applied to any distribution with two major peaks — their frequencies do not have to be exactly the same.

Box 45.2 Three examples where simple arithmetic means are inappropriate

Mean	*n*
6	4
7	7
8	1

1. If means of samples are themselves meaned, an error can arise if the samples are of different size. For example, the arithmetic mean of the means in the table shown left is 7, but this does not take account of the different 'reliabilities' of each mean. The correct weighted mean is obtained by multiplying each mean by its sample size *n* (a 'weight') and dividing the sum of these values by the total number of observations, i.e. in the case shown, (24 + 49 + 8)/12 = 6.75.

2. When making a mean of ratios (e.g. percentages) for several groups of different sizes, the ratio for the combined total of all the groups is not the mean of the proportions for the individual groups. For example, if 20 rats from a batch of 50 are male, this implies 40% are male. If 60 rats from a batch of 120 are male, this implies 50% are male. The mean percentage of males (50 + 40)/2 = 45% is *not* the percentage of males in the two groups combined, because there are 20 + 60 = 80 males in a total of 170 rats = 47.1% approx.

pH value	[H] (kmol m^{-3})
6	1×10^{-6}
7	1×10^{-7}
8	1×10^{-8}
mean	3.7×10^{-7}
$-\log_{10}$ mean	6.43

3. If the measurement scale is not linear, arithmetic means may give a false value. For example, if three media had pH values 6, 7 and 8, the appropriate mean pH is not 7 because the pH scale is logarithmic. The definition of pH is $-\log_{10}$[H], where [H] is expressed in kmol m^{-3} ('molar'); therefore, to obtain the true mean, convert data into [H] values (i.e. put them on a linear scale) by calculating $10^{(-\text{pH value})}$ as shown. Now calculate the mean of these values and convert the answer back into pH units. Thus, the appropriate answer is pH 6.43 rather than 7. Note that a similar procedure is necessary when calculating statistics of dispersion in such cases, so you will find these almost certainly asymmetric about the mean.

46 Choosing and using statistical tests

This chapter outlines the philosophy of hypothesis-testing statistics, indicates the steps to be taken when choosing a test, and discusses features and assumptions of some important tests. For details of the mechanics of tests, consult appropriate texts (e.g. Sokal and Rohlf, 1981; Wardlaw, 1985). Most tests are now available in statistical packages for computers (see p. 258).

To carry out a statistical test:

1. Decide what it is you wish to test (create a null hypothesis and its alternative).
2. Determine whether your data fit a standard distribution pattern.
3. Select a test and apply it to your data.

Setting up a null hypothesis

Hypothesis-testing statistics are used to compare the properties of samples either with other samples or with some theory about them. For instance, you may be interested in whether two samples can be regarded as having different means, whether the counts of an organism in different quadrats can be regarded as randomly distributed, or whether property A of an organism is linearly related to property B. You can't use statistics to *prove* any of these things, but they can be used to assess *how likely* you are to be wrong.

Statistical testing operates in what at first seems a rather perverse manner. Suppose you think a treatment has an effect. The theory you actually test is that it has no effect; the test tells you how improbable your data would be if this theory were true. This 'no effect' theory is the null hypothesis (NH). If your data are very improbable under the NH, then you may suppose it to be wrong, and this would support your original idea (the 'alternative hypothesis'). The concept can be illustrated by an example. Suppose two groups of subjects were treated in different ways, and you observed a difference in the mean value of the measured variable for the two groups. Can this be regarded as a 'true' difference? As Fig. 46.1 shows, it could have arisen in two ways:

- Because of the way the subjects were allocated to treatments, i.e. all the subjects liable to have high values might, by chance, have been assigned to one group and those with low values to the other (Fig. 46.1a).
- Because of a genuine effect of the treatments, i.e. each group came from a distinct frequency distribution (Fig. 46.1b).

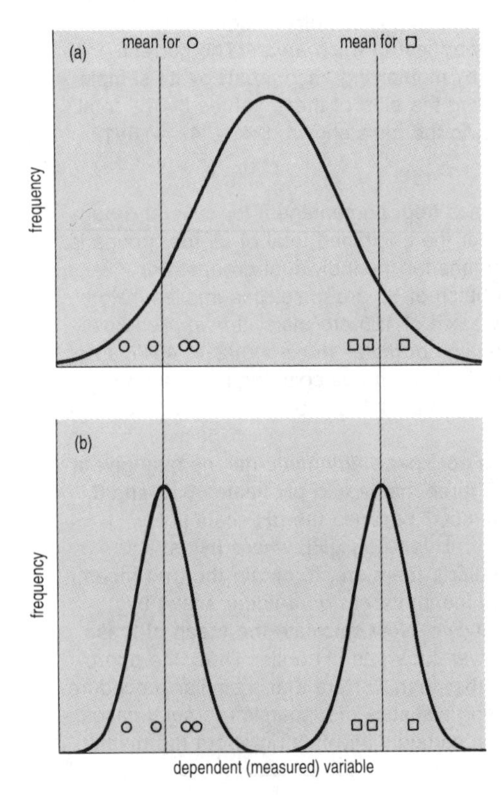

Fig. 46.1 Two explanations for the difference between two means. In case (a) the two samples happen by chance to have come from opposite ends of the same frequency distribution, i.e. there is no true difference between the samples. In case (b) the two samples come from different frequency distributions, i.e. there is a true difference between the samples. In both cases, the means of the two samples are the same.

A statistical test will indicate the probabilities of these options. The NH states that the two groups come from the same population (i.e. the treatment effects are negligible in the context of random variation). To test this, you calculate a test statistic from the data, and compare it with tabulated critical values giving the probability of obtaining the observed or a more extreme result by chance (see Box 46.1). This probability is sometimes called the significance of the test.

Note that you must take into account the degrees of freedom (d.f.) when looking up critical values of most test statistics. The d.f. is related to the size(s) of the samples studied; formulae for calculating it depend on the test being used. Biologists normally use two-tailed tests, i.e. we have no certainty beforehand that the treatment will have a positive or negative effect compared to the control (in a one-tailed test we expect one treatment to be bigger or

Box 46.1 How to interpret critical values for Student's *t* statistic

Suppose the following data were obtained in an experiment (the units are not relevant):

Control: 6.6, 5.5, 6.8, 5.8, 6.1, 5.9 (n = 6, $\bar{Y}$ = 6.12, s = 0.496)

Treatment: 6.3, 7.2, 6.5, 7.1, 7.5, 7.3 (n = 6, $\bar{Y}$ = 6.98, s = 0.475)

To test whether the treatment had a significant effect, we first set up a null hypothesis (NH) that there is no treatment effect. The alternative hypothesis is that there *is* a treatment effect. As we have no reason to suspect that the treatment mean will be larger or smaller than the control, we use a two-tailed test.

We assume that the data are normally distributed in both cases and that their variances are not significantly different, so decide that Student's *t*-test is appropriate (p. 239).

The calculated value of Student's *t* statistic for these data is −3.09. The degrees of freedom are $n_1 + n_2 - 2 = 10$.

Looking at the table below, we see that the modulus of this value exceeds the critical value for $P = 0.05$ at 10 degrees of freedom. We can therefore reject the NH, and conclude that the means are different at the 5% level of significance. If the modulus of the calculated value of *t* had been less than 2.23, we would have accepted the NH.

Selected critical values for Student's *t* statistic (two-tailed test)

Degrees of freedom	$P = 0.05$	$P = 0.01$	$P = 0.001$
5	2.57	4.03	6.87
10	2.23	3.17	4.59
15	2.13	2.95	4.07
20	2.09	2.85	3.85

Note from the table that the critical value that must be exceeded to reject the NH decreases with increasing sample size; but for a given sample size, the critical value increases the lower the probability. If the sample sizes and thus degrees of freedom (d.f.) had been bigger (say giving 15 d.f.), then a *t* value of 3.09 would have been significant at $P < 0.01$. If, alternatively, we had obtained a *t* value of 3.20 for the original 10 d.f., we could conclude that the treatments were significantly different at $P < 0.01$.

smaller than the other). Be sure to use critical values for the correct type of test.

By convention, the critical probability for rejecting the NH is 5% (i.e. $P = 0.05$). This means we reject the NH if the observed result would have come up less than one time in twenty by chance. If the modulus of the test statistic is less than the tabulated critical value for $P = 0.05$, then we accept the NH and the result is said to be 'not significant' (NS for short). If the modulus of the test statistic is greater than the tabulated value for $P = 0.05$, then we reject the NH in favour of the alternative hypothesis that the treatments had different effects.

Two types of error are possible when making a conclusion on the basis of a statistical test. The first occurs if you reject the NH when it is true and the second if you accept the NH when it is false. To limit the chance of the first type of error, choose a lower probability, e.g. $P = 0.01$, but note that the critical value of the test statistic increases when you do this and results in the probability of the second error increasing (see, for example, Box 46.1). The conventional significance levels given in statistical tables (usually 0.05, 0.01, 0.001) are arbitrary. Increasing use of statistical computer programs is likely to lead to the actual probability of obtaining the calculated value of the test statistic being quoted (e.g. $P = 0.037$).

Note that if the NH is rejected, this does not tell you which of many alternative hypotheses is true. Also, it is important to distinguish between statistical and practical significance. Just because you can identify a statistically significant difference between two samples doesn't mean that this will carry any biological importance.

Comparing data with parametric distributions

A parametric test is one which makes particular assumptions about the mathematical nature of the population distribution from which the samples were taken. If these assumptions are not true, then the test is obviously invalid, even though it might give the answer we expect! A non-parametric test does not assume that the data fit a particular pattern, but it may assume some things about the distributions. Used in appropriate circumstances, parametric tests are better able to distinguish between true but marginal differences between samples than their non-parametric equivalents (i.e. they have greater 'power').

The distribution pattern of a set of data values may be biologically relevant, but it is also of practical importance because it defines the type of statistical tests that can be used. The properties of the main distribution types found in biology are given below with both rules-of-thumb and more rigorous tests for deciding whether data fit these distributions.

Choosing between parametric and non-parametric tests — always plot your data graphically when determining whether they are suitable for parametric tests as this may save a lot of unnecessary effort later.

Binomial distributions

These apply to samples of any size from populations when objects occur independently in only two mutually exclusive classes (e.g. type A or type B). They describe the probability of finding the different possible combinations of the attribute for a specified sample size k (e.g. out of 10 specimens, what is the chance of 8 being type A). If p is the probability of the attribute being of type A and q the probability of it being type B, then the expected mean sample number of type A is kp and the standard deviation is $\sqrt{kpq}$. Expected frequencies can be calculated using mathematical expressions (see Sokal and Rohlf, 1981). Examples of the shapes of some binomial distributions are shown in Fig. 46.2. Note that they are symmetrical in shape for the special case $p = q = 0.5$ and the greater the disparity between p and q, the more skewed the distribution.

Some biological examples of data likely to be distributed in binomial fashion are: possession of two alleles for seed coat morphology (e.g. smooth and wrinkly); whether an organism is infected with a microbe or not; whether an animal is male or female. Binomial distributions are particularly useful for predicting gene segregation in Mendelian genetics and can be used for testing whether combinations of events have occurred more frequently than predicted (e.g. more siblings being of the same sex than expected). To establish whether a set of data is distributed in binomial fashion: calculate expected frequencies from probability values obtained from theory or observation, then test against observed frequencies using a χ^2-test or a G-test.

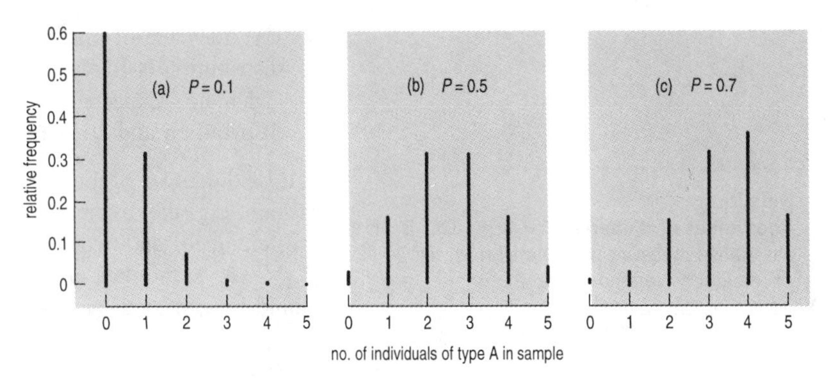

Fig. 46.2 Examples of binomial frequency distributions with different probabilities. The distributions show the expected frequency of obtaining n individuals of type A in a sample of 5. Here P is the probability of an individual being type A rather than type B.

Where samples from a binomial distribution are large (i.e. >15) and p and q are close to 0.5, the proportions can be treated as normally distributed.

For Poisson distributions, if the number of counts recorded in each outcome is greater than about 15, they can be treated as normally distributed.

Poisson distributions

These apply to discrete characteristics which can assume low whole number values, such as counts of events occurring in area, volume or time. The events should be 'rare' in that the mean number observed should be a small proportion of the total that could possibly be found. Also, finding one count should not influence the probability of finding another. The shape of Poisson distributions is described by only one parameter, the mean number of events observed, and has the special characteristic that the variance is equal to the mean. The shape has a pronounced positive skewness at low mean counts, but becomes more and more symmetrical as the mean number of counts increases (Fig. 46.3).

Some examples of characteristics distributed in a Poisson fashion are: number of plants in a quadrat; number of microbes per unit volume of medium; number of animals parasitized per unit time; number of radioactive disintegrations per unit time. One of the main uses for the Poisson distribution is to quantify errors in count data such as estimates of cell densities in dilute suspensions (see p. 129). To decide whether data are Poisson distributed:

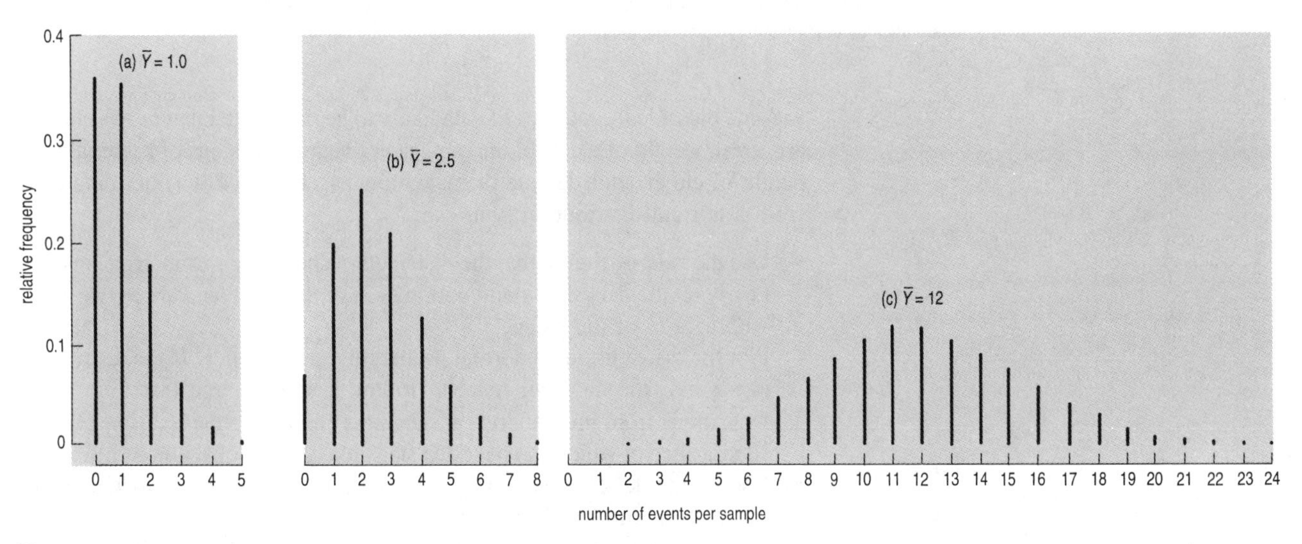

Fig. 46.3 Examples of Poisson frequency distributions differing in mean. The distributions are shown as line charts because the independent variable (events per sample) is discrete.

- Use the rule-of-thumb that if the coefficient of dispersion ≈ 1, the distribution is likely to be Poisson.
- Calculate 'expected' frequencies from the equation for the Poisson distribution and compare with actual values using a χ^2-test or a *G*-test.

Definition
Coefficient of dispersion = $s^2/\bar{Y}$. This is an alternative measure of dispersion to the coefficient of variation (p. 232)

It is sometimes of interest to show that data are *not* distributed in a Poisson fashion, e.g. the distribution of parasite larvae in hosts. If $s^2/\bar{Y} > 1$, the data are 'clumped' and occur together more than would be expected by chance; if $s^2/\bar{Y} < 1$, the data are 'repulsed' and occur together less frequently than would be expected by chance.

Normal distributions

These occur when random events act to produce variability in a continuous characteristic (quantitative variable). This situation occurs frequently in biology, so normal distributions are very useful and much used. The bell-like shape of normal distributions is specified by the population mean and standard deviation (Fig. 46.4): it is symmetrical and configured such that 68.27% of the data will lie within ± 1 standard deviation of the mean, 95.45% within ± 2 standard deviations of the mean, and 99.73% within ± 3 standard deviations of the mean.

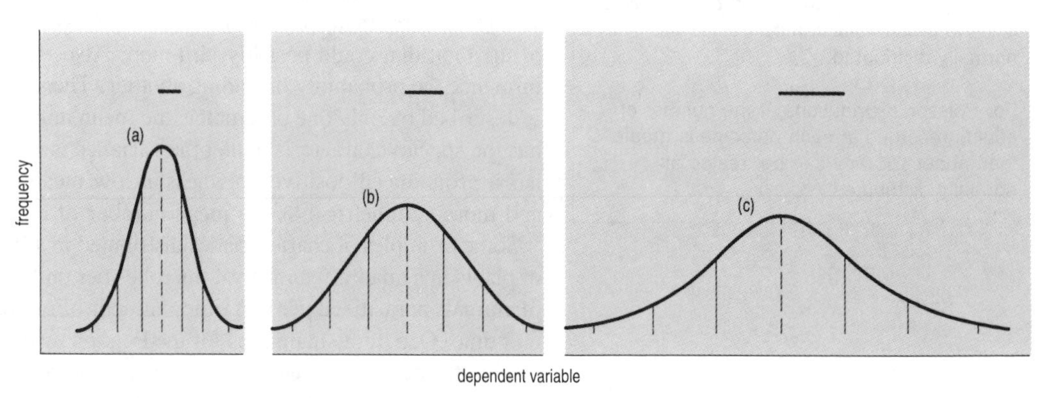

Fig. 46.4 Examples of normal frequency distributions differing in mean and standard deviation. The horizontal bars represent population standard deviations for the curves, increasing from (a) to (c). Vertical dashed lines are population means, while vertical solid lines show positions of values ± 1, 2 and 3 standard deviations from the means.

Some biological examples of data likely to be distributed in a normal fashion are: fresh weight of plants of the same age; linear dimensions of bacterial cells; height of either adult female or male humans. To check whether data come from a normal distribution, you can:

- Use the rule-of-thumb that the distribution should be symmetrical and that nearly all the data should fall within $\pm 3s$ of the mean and about two-thirds within $\pm 1s$ of the mean.
- Plot the distribution on normal probability graph paper. If the distribution is normal, the data will tend to follow a straight line (see Fig. 46.5). Deviations from linearity reveal skewness and/or kurtosis (see p. 232), the significance of which can be tested statistically (see Sokal and Rohlf, 1981).
- Use a suitable statistical computer program to generate predicted normal curves from the $\bar{Y}$ and s values of your sample(s). These can be compared visually with the actual distribution of data and can be used to give 'expected' values for a χ^2-test or a *G*-test.

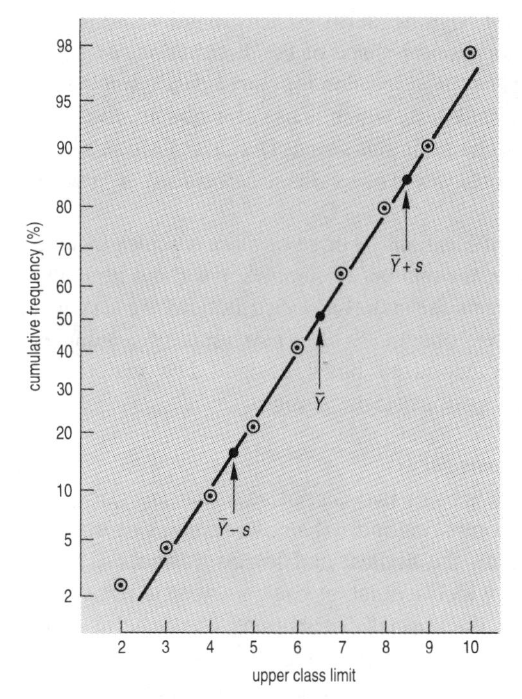

Fig. 46.5 Example of a normal probability plot. The data are those of Box 45.1 (p. 228). For each class, the cumulative frequency as a percentage is plotted against the upper class limit on normal probability graph paper (open symbols). Note that the class limits corresponding to 0 and 100% cannot be used. The straight line is that predicted for a normal distribution with mean = 6.39 and standard deviation = 1.895. This is plotted by calculating the expected positions of points for the mean ± standard deviation. Since 68.3% of the distribution falls within these bounds, the relevant points on the cumulative frequency scale are 50 ± 34.15%; thus, the line is drawn using the points (4.495, 15.85) and (8.285, 84.15) as shown on the graph.

Checking the assumptions of a test — always acquaint yourself with the assumptions of a test. If necessary, test the assumptions before using the test.

Table 46.1 Suggested transformations altering different types of frequency distribution to the normal type. To use, modify data by the formula shown; then examine effects with the tests described on pp. 236–8.

Type of data; distribution suspected	Suggested transformation(s)
Proportions (including percentages); binomial	arcsine $\sqrt{x}$ (also called the angular transformation)
Scores; Poisson	$\sqrt{x}$ or $\sqrt{(x + ½)}$ if zero values present
Measurements; negatively skewed	x^2, x^3, x^4, etc. (in order of increasing strength)
Measurements; positively skewed	$1/\sqrt{x}$, $\sqrt{x}$, ln x, $1/x$ (in order of increasing strength)

The wide availability of tests based on the normal distribution and their relative simplicity means you may wish to transform your data to make them more like a normal distribution. Table 46.1 provides transformations that can be applied (see also Table 41.2). The transformed data should be tested for normality as described above before proceeding — don't forget that you may need to check that transformed variances are homogeneous for certain tests (see below).

A very important theorem in statistics, the Central Limit Theorem, states that as sample size increases, the distribution of a series of means from any frequency distribution will become normally distributed. This fact can be used to devise an experimental or sampling strategy that ensures that data are normally distributed, i.e. using means of samples as if they were your primary data.

Choosing a suitable statistical test

Comparing location (e.g. means)

If you can assume that your data are normally distributed, the main test for comparing two means from independent samples is Student's *t*-test (see Box 46.1). This assumes that the variances of the data sets are homogeneous. Tests based on the *t*-distribution are also available for comparing paired data or for comparing a sample mean with a chosen value.

When comparing means of two or more samples, analysis of variance (ANOVA) is a very useful technique. This method also assumes data are normally distributed and that the variances of the samples are homogeneous. The samples must also be independent (e.g. not sub-samples). The nested types of ANOVA are useful for letting you know the relative importance of different sources of variability in your data. Two-way and multi-way ANOVAs are useful for studying interactions between treatments.

For data satisfying the ANOVA requirements, the least significant difference (LSD) is useful for making planned comparisons among several means. Any two means that differ by more than the LSD will be significantly different. The LSD is useful for showing on graphs.

The chief non-parametric tests for comparing the locations of two samples are the Mann–Whitney *U*-test and the Kolmogorov–Smirnov test. The former assumes that the frequency distributions of the samples are similar, whereas the latter makes no such assumption. In both cases the sample's size must be ≥ 4 and for the Kolmogorov–Smirnov test the samples must have equal sizes.

In the Kolmogorov–Smirnov test, significant differences found with the test could be due to differences in location or shape of the distribution, or both.

Suitable non-parametric comparisons of location for paired data (sample size ≥ 6) include Wilcoxon's signed rank test, which is used for quantitative data and assumes that the distributions have similar shape. Dixon and Mood's sign test can be used for paired data scores where one variable is recorded as 'greater than' or 'better than' the other.

Non-parametric comparisons of location for three or more samples include the Kruskal–Wallis *H*-test. Here, the number of samples is without limit and they can be unequal in size, but again the underlying distributions are assumed to be similar. The Friedman *S*-test operates with a maximum of 5 samples and data must conform to a randomized block design. The underlying distributions of the samples are assumed to be similar.

Comparing dispersions (e.g. variances)

If you wish to compare the variances of two sets of data that are normally distributed, use the *F*-test. For comparing more than two samples, it may be sufficient to use the F_{max}-test, on the highest and lowest variances. Non-parametric tests exist but are not widely available: you may need to transform the data and use a test based on the normal distribution. The Scheffé–Box (log-anova) test is recommended for testing the significance of differences between several variances; it is not particularly sensitive to departures from normality.

Determining whether frequency observations fit theoretical expectation

The χ^2-test is useful for tests of 'goodness of fit', e.g. comparing expected and observed progeny frequencies in genetical experiments or comparing observed frequency distributions with some theoretical function. One limitation is that simple formulae for calculating χ^2 assume that no expected number is less than 5. The *G*-test (2*I* test) is used in similar circumstances.

Comparing proportion data

When comparing proportions between two small groups (e.g. whether 3/10 is significantly different from 5/10), you can use probability tables such as those of Finney *et al.*, (1963) or calculate probabilities from formulae; however, this can be tedious for large sample sizes. Certain proportions can be transformed so that their distribution becomes normal.

Placing confidence limits on an estimate of a population parameter

On many occasions, sample statistics are used to provide an estimate of the population parameters. It is extremely useful to indicate the reliability of such estimates. This can be done by putting a confidence limit on the sample statistic. The most common application is to place confidence limits on the mean of a sample from a normally distributed population. This is done by working out the limits as $\bar{Y} - (t_{P[n-1]} \times \text{SE})$ and $\bar{Y} + (t_{P[n-1]} \times \text{SE})$ where $t_{P[n-1]}$ is the tabulated critical value of Student's *t* statistic for a two-tailed test with $n - 1$ degrees of freedom and SE is the standard error of the mean (p. 232). A 95% confidence limit (i.e. $P = 0.05$) tells you that on average, 95 times out of 100, this limit will contain the population mean. Sokal and Rohlf (1981) discuss how to place confidence limits on statistics other than the mean.

Regression and correlation

These methods are used when testing relationships between samples of two

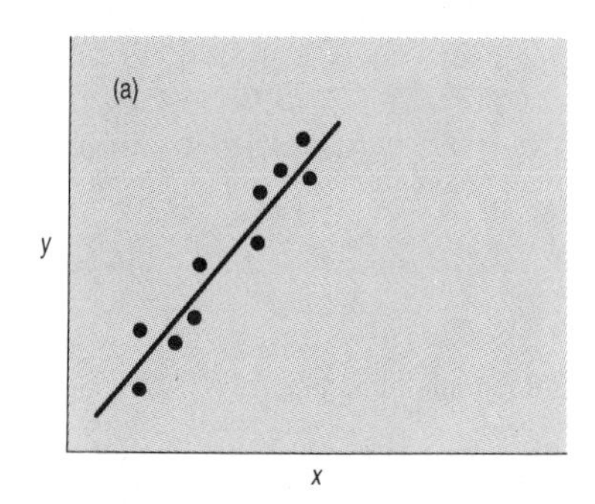

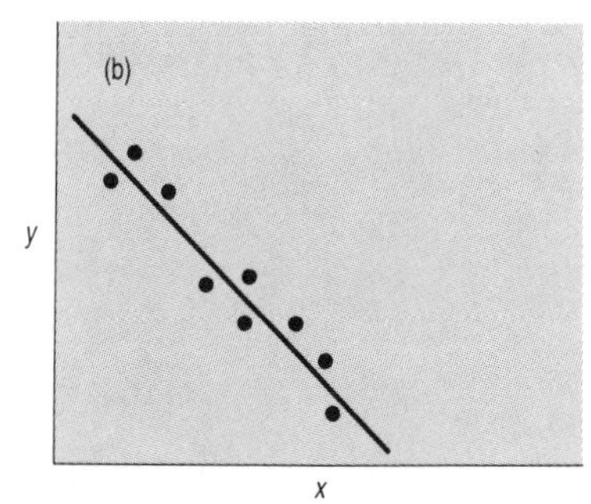

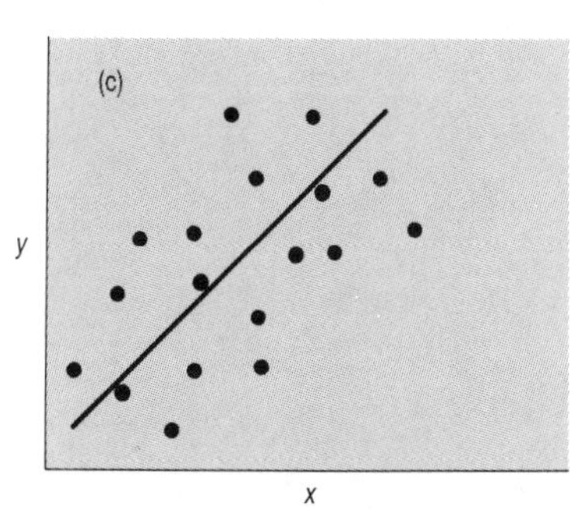

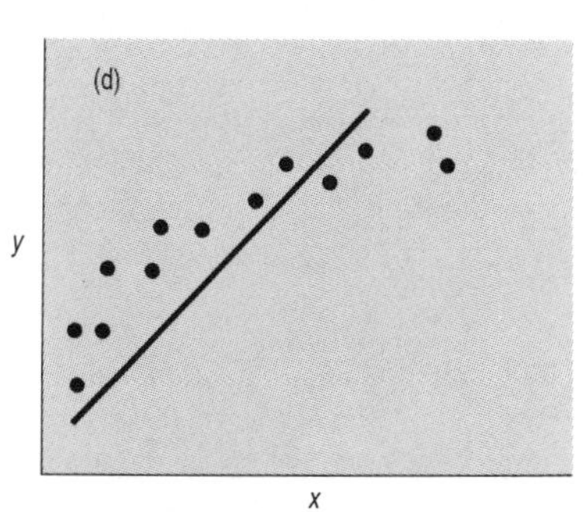

Fig. 46.6 Examples of correlation. The linear regression line is shown. In (a) and (b), the correlation between x and y is good: for (a) there is a positive correlation and the correlation coefficient would be close to 1; for (b) there is a negative correlation and the correlation coefficient would be close to -1. In (c) there is a weak positive correlation and r would be close to 0. In (d) the correlation coefficient may be quite large, but the choice of linear regression is clearly inappropriate.

variables. If one variable is assumed to be dependent on the other then regression techniques are used to find the line of best fit for your data. This does not tell you how well the data fit the line: for this, a correlation coefficient must be calculated. If there is no a priori reason to assume dependency between variables, correlation methods alone are appropriate.

If graphs or theory indicate a linear relationship between a dependent and an independent variable, linear regression can be used to estimate the equation that links them. If the relationship is not linear, try a transformation (see p. 213). For example, this is commonly done in analysis of enzyme kinetics (Fig. 33.3). However, 'linearizations' can lead to errors when carrying out regression analysis: take care to ensure (a) that the data are evenly distributed throughout the range of the independent variable and (b) that the variances of the dependent variable are homogeneous. If these criteria cannot be met, weighting methods may reduce errors. In this situation, it may be better to use non-linear regression but this involves complex calculations best handled by a suitable computer program.

Model I linear regression is suitable for experiments where a dependent variable Y varies with an *error-free* independent variable X and the mean (expected) value of Y is given by $a + bX$. This might occur where you have carefully controlled the independent variable and it can therefore be assumed to have zero error (e.g. a calibration curve). Errors can be calculated for estimates of a and b and predicted values of Y. The Y values should be normally distributed and the variance of Y constant at all values of X.

Model II linear regression is suitable for experiments where a dependent variable Y varies with an independent variable X which has an error associated with it and the mean (expected) value of Y is given by $a + bX$. This might occur where the experimenter is measuring two variables and believes there to be a causal relationship between them; both variables will be subject to errors in this case. The exact method to use depends on whether your aim is to estimate the functional relationship or to estimate one variable from the other.

A correlation coefficient measures the strength of relationships but does not describe the relationship. These coefficients are expressed as a number between -1 and 1. A positive coefficient indicates a positive relationship while a negative coefficient indicates a negative relationship (Fig. 46.6). The nearer the coefficient is to -1 or 1, the stronger the relationship between the variables, i.e. the less scatter there would be about a line of best fit (note that this does *not* imply that one variable is dependent on the other!). A coefficient of 0 implies that there is no relationship between the variables. The importance of graphing data is shown by the case illustrated in Fig. 46.6d.

Pearson's product-moment correlation coefficient (r) is the most commonly used correlation coefficient. If both variables are normally distributed, then r can be used in statistical tests to test whether the degree of correlation is significant. If one or both variables are not normally distributed you can use Kendall's coefficient of rank correlation (τ) or Spearman's coefficient of rank correlation (r_s). They require that data are ranked separately and calculation can be complex if there are tied ranks. Spearman's coefficient is said to be better if there is uncertainty about the reliability of close ranks.

Using microcomputers

47 Microcomputer basics

Although you may use a mainframe computer, you are increasingly likely to use a personal computer (PC) for many tasks. This section is confined to general features of the PC ('microcomputer') as mainframe operations are more machine specific; however, many of the points made are relevant to mainframe use.

Many books are available to help you to use computer systems effectively but hands-on experience is best. The common reaction to computers is concern — fear of doing something wrong and of damaging the machine. Such concerns are ill-founded: all you need to know is how to carry out a few simple steps, and as long as you follow some basic rules (Box 47.1) you will find that the computer becomes a valuable tool.

There are three precepts that will help you to become computer literate:

- Understand the problem you wish to solve and identify how the computer can help you to solve it.
- Understand how to choose and use applications programs (e.g. word processors, spreadsheets, databases, statistics packages and graphics) to achieve your solution.
- Understand enough about the computer to run the applications program. This means knowing how it stores and retrieves files, how to get information into and out of it and how to look after it. Understand the limitations of computers.

A useful maxim for computing — rubbish in–rubbish out!

The computer system

A computer system comprises a number of interacting parts, like a hi-fi system. The central component is the computer cabinet (which contains the CPU or central processing unit) and connected to it are a number of peripherals used to get data into and out of the computer and to store data and programs.

The system comprises the following:

- Computer: performs all the calculations and processes the data (both numeric and textual).
- Input devices: used to enter data for processing in various ways.
- Output devices: used to retrieve data from the computer for analysis. This may be as a screen display or as 'hard copy' (printed material).
- External storage devices: used to store programs and data not currently being processed by the computer; storage is as files.

Definition
Peripheral — any device connected to the core computing circuitry of the CPU and memory, usually taken to mean components outside the main computer cabinet such as printers and disk drives.

Definition
File — a self-contained body of information stored on disk to be retrieved at a later date. It can be, for example, a program, a document from a word processor, data from a database or spreadsheet, or a graphics image.

The operating system

The heart of the computer is the CPU but its operation is automatic and you need have no concern for its workings. The heart of the computer's software is the operating system which controls the operation of the programs: an example is Microsoft® Disk Operating System (MS-DOS). Except when using icon-based systems (e.g. AppleMac®), you will need to know at least a little about the operating system in order to take full advantage of what it offers.

In particular, the operating system provides commands for preparing (formatting) disks, saving and retrieving files, and for manipulating them in various ways (copying, renaming, erasing, making back-up copies, etc.). These

Definition
CPU — central processing unit — the chip with ultimate control of your computer. It is told what to do by your program.

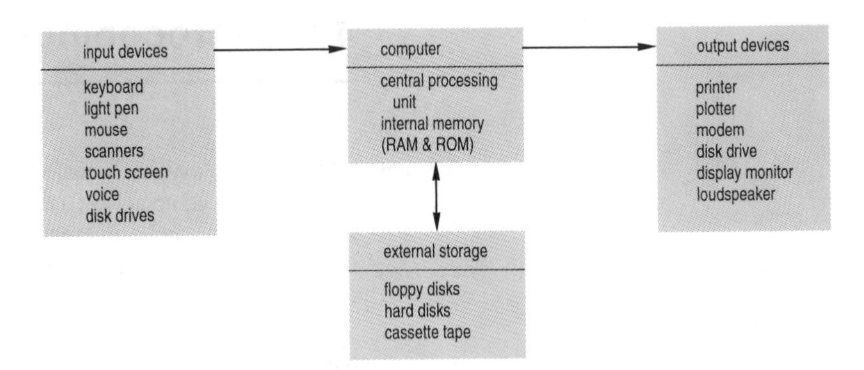

Fig. 47.1 Flowchart illustrating a typical computer system.

commands are specific to each particular operating system but they always have a precise syntax which means that information must be entered in a very specific sequence, including the use of spaces and punctuation. Software written for one particular operating system usually will not run on another. Icon-based operating systems avoid the need to use commands by using graphical images (icons) to represent the functions of the operating system — usually controlled via a mouse.

Memory

The computer's internal memory used for processing is divided into two different types, ROM (read only memory) and RAM (random access memory). ROM remains unchanged when the machine is switched off (non-volatile memory) and cannot be altered by you: it is used by hardware manufacturers to incorporate programs and data that must be permanently available. RAM is that part of the system handling all current operations but, if the machine is switched off, the contents are lost (i.e. it is volatile memory). The type of operations and the amount of information handled are limited by available RAM memory: the bigger the memory, the more the computer can handle and the faster it will be. Memory is usually described in units called bytes, one byte being equivalent to one keyboard character. Make sure that your computer has the required amount of RAM for the program you wish to run. Be aware that the size of RAM may place limits on the amount of data you can enter: this can be important when analysing large data sets or large files.

Interfaces

Your computer needs to communicate with any input or output devices and does so through interfaces which are internal devices operating through sockets (ports) mounted on the computer cabinet. There are two main types, serial (RS-232) and parallel (Centronics), and peripheral equipment must be connected to the right type of port by the correct sort of cable. The parallel port is a socket on the back of your computer used to exchange information with other devices that also have a parallel port, usually printers. It communicates faster than a serial port, which is another socket that can be used to connect to another computer or to devices such as a mouse, modem or plotter that also have a serial port.

Input devices

The most familiar input device is the keyboard but there are various other useful input devices. These include mouses, scanners, joysticks, touch screens, light pens, modems and CD-ROM reading devices. A disk drive is an input device, returning previously stored data and programs to the computer.

Keyboards vary in layout but all have the same types of keys. Alphabetic keys are arranged as on a typewriter using what is known as the 'qwerty' format, after the top line of alphabetic characters. Shift keys produce capitals and there is a CAPS LOCK key for entering only capital letters. Numeric keys are located above the alphabetic keys as are the various symbols produced with the shift key. Many keyboards also have a separate numeric keypad to facilitate entering large quantities of numeric data. Cursor movement keys manoeuvre the cursor around the screen. Function keys provide special operations through the control of the programs — these are used extensively in applications programs and special overlays called templates are usually provided to tell you the role(s) of these keys. Other special keys are also to be found on the keyboard such as delete (DEL) and backspace. It is important to become familiar with the keyboard operations since it is normally through the keyboard that you will communicate with the computer!

Definitions
Mouse — a device that can be pushed around your desk to control an on-screen pointer. Buttons are used to select menu items and perform actions.
Scanner — a device which scans photographs or other illustrations to create either greyscale (black and white) or colour images of them for use in an appropriate program.
Modem — MOdulator/DEModulator — a hardware device that connects the computer to other computers through the telephone system.

Output devices

The most common output devices are the video display monitor (screen) and the printer. Computer programs are usually set up to make a 'bleep' sound when an operation has gone wrong; this is just a warning to let you know you've made an error!

Printers come in a wide variety of types with different qualities of print. The simplest are the dot-matrix printers while the most sophisticated are the laser printers. Ensure that the printer that is available is appropriate for the job: a cheap dot-matrix printer is fine for a draft copy but finished text should have at least an NLQ (near-letter quality) standard. It is also possible to output to colour printers and to plotters. Plotters are used for high resolution graphics while colour printers are suitable for business-type graphics.

Definitions
Dot-matrix printer — a printer where the characters and graphics are formed from a grid of dots produced by wire pins firing against an inked ribbon. They come as 9-pin printers or 24-pin printers: the more pins the better the printing.
Ink-jet/bubble-jet printer — a printer where the characters and graphics are formed from a grid of dots formed by sprays of ink.
Laser printer — a printer which uses photocopier technology to output high-quality pages of text and graphics.
Daisywheel printer — a printer where characters are formed by a technique similar to that used by a typewriter. The hammers are mounted radially on a wheel (the printwheel) which is rotated to bring the correct character into line. Cannot produce graphics!

External storage

Data and program information are stored as files on floppy disks or hard disks, each of which is read by a disk drive. Hard disks are faster and can store large amounts of information but are usually internal devices and not portable. The floppy disk (Fig. 47.2) has more limited storage and is slower in operation but is portable. With this portability comes risk of damage and loss of information. Floppy disks come in two common sizes, 5¼ inch and 3½ inch diameter, the smaller size being encased in a rigid plastic sleeve and less liable to damage. Disks have to be formatted before use, i.e. prepared to receive information which is stored in very specific ways and very specific amounts: only disks formatted in the appropriate way for your computer can be read or copied to. Care of your floppy disks is critical as they are easily damaged, physically or by magnetic fields.

Disk drives are allocated letters by the computer system and you must ensure that you know which is drive A, etc. One will be called the default drive, which is the one the computer examines if you do not tell it to look elsewhere in the command which accesses the disk.

Formatting disks — never format a floppy disk unless you are very sure of what you are doing — you will lose all information on the disk when it is re-formatted. Never re-format a hard disk unless it is your own machine!

Protecting the information on your disks — use the write protect notch/switch on your floppy to protect valuable data or programs.

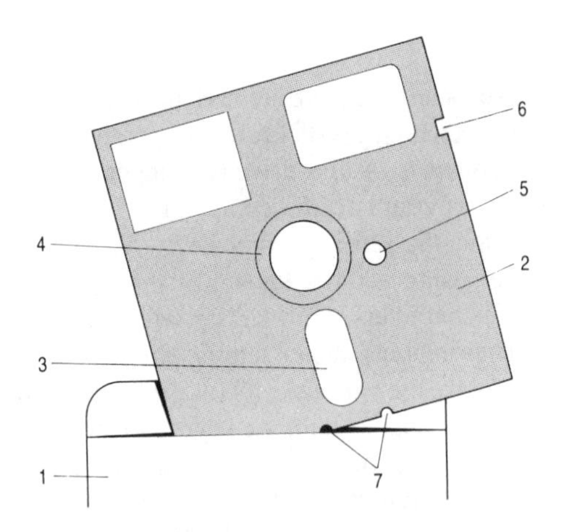

Fig. 47.2(a) Parts of a standard $5\frac{1}{4}''$ floppy disk: 1. paper sleeve/envelope; 2. plastic jacket; 3. read/write slot where the disk drive head contacts the disk surface; 4. hub around central hole used by the disk drive to clamp onto the disk; 5. the sector (index) hole used by the disk drive as a timing mechanism; 6. the write-protect notch; 7. allignment notches used by the disk drive to locate the disk correctly.

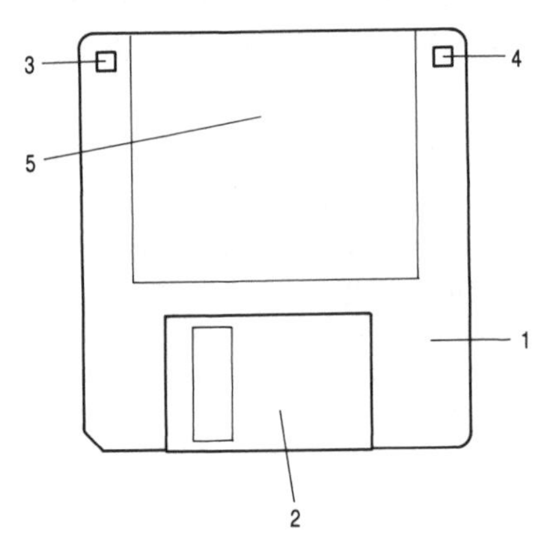

Fig. 47.2(b) Parts of a $3\frac{1}{2}''$ disk: 1. rigid plastic cover; 2. sliding metal cover protecting magnetic disk surface (do not open); 3. locating slot; 4. write protect device — when hole visible, disk is write-protected; 5. label giving information on number of sides, density of magnetic coating and capacity of disk.

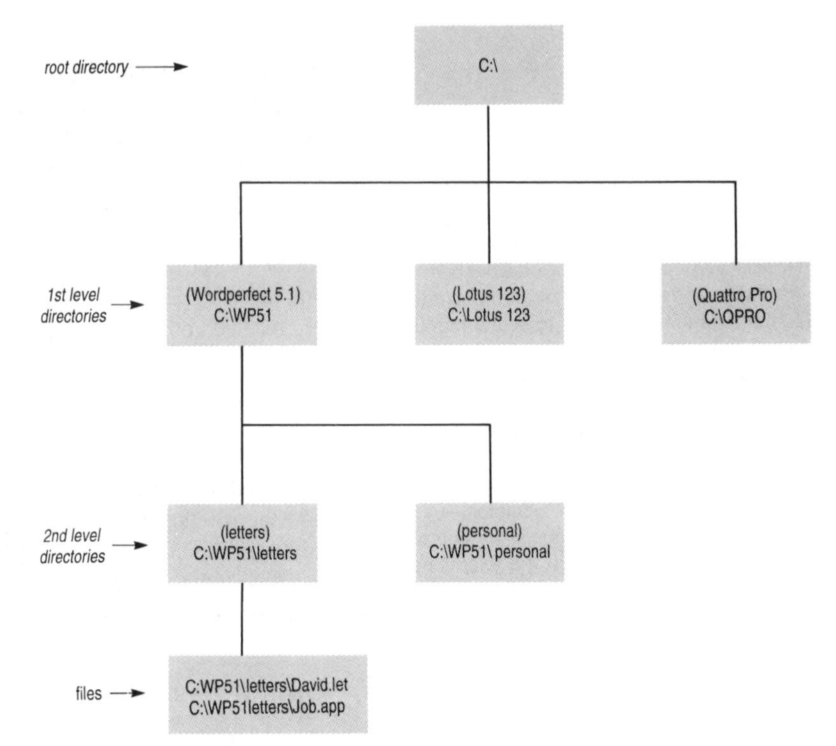

Fig. 47.3 An example of a hierarchical directory structure in MS-DOS.

Files

Information is stored as files, which are grouped in a directory (Fig. 47.3): you can have numerous directories (create ones with appropriate names for your own file collections) which operate in a strict hierarchical manner, the default one being the root directory which cannot be named but is represented by the backslash symbol (\) in MS-DOS. Other directories are called sub-directories. You should learn how to negotiate your way through this system at an early stage. File management is the most important general skill required when using computers. Files are given names, often with limits on the number and type of characters that can be used: try to ensure that names are both unique (to avoid overwriting old files) and meaningful, as it can otherwise be difficult some months later to remember what the file contains.

File extensions are a useful means of indicating the type of document, e.g. .LET for letters, .DFT for drafts, or to indicate the origin of files, e.g. .WP for a WordPerfect file. Sometimes filename extensions are added automatically by a program to indicate the format of a file.

Software

This is the generic term for the sets of instructions (programs) required to direct the computer to perform tasks. It is unlikely that you will need to write your own programs except perhaps in very specialized areas of biology later in your career. Generally you will use programs purchased as 'applications packages' or prepared by your instructors. Make sure you read the instructions for the use of such programs: don't be afraid to learn by experimentation, but reading the manual does pay dividends!

Box 47.1 Do's and don'ts for using microcomputers

- **Do**
  Keep floppy disks in their protective sleeves and boxes.
  Keep disks away from dampness, excess heat or cold.
  Keep disks at least 2 feet away from magnets: be aware of their presence in televisions, loudspeakers, etc.
- **Do not**
  Use disks from others unless first checked for viruses.
  Drink or smoke around the computer.
  Turn off the computer more than is necessary.
  Turn off the electricity supply to the machine while in use.
  Bend, fold or flex floppy disks in any way.
  Touch the floppy disk's recording surface.
  Use a hard-tipped pen to write on the label on the disk.
  Insert or remove a disk from the drive when it is operating (drive light on).
  Leave a disk in the drive when you switch the computer off.
  Attempt to re-format the hard disk except in special circumstances.

The golden rule
Make backup copies of important disks/files and store them well away from your working copies. Be sure that the same accident cannot happen to both copies.

Problem-solving using a computer — before considering writing your own program to tackle a specific problem, see if you can solve it using a spreadsheet.

Although programming is not discussed here, the principles behind writing programs are useful for training of thought processes, since the main requirement is to be able to break down any problem into a logical series of discrete operations (algorithms) which will lead to the required solution.

The whole process of solving a problem by computer involves four stages:

1. Analyse the problem.
2. Design a way to solve the problem: break it down into sequential steps.
3. Use a pre-written program or write a program by coding the solution using a computer language.
4. Thoroughly test the program with data for which the correct solution is known.

Every program requires at least an input section, a calculation section and an output section and should be structured as a flow diagram during the planning stage.

48 Using spreadsheets

The spreadsheet is one of the most powerful and wide-ranging of all microcomputer applications. It is the electronic equivalent of a huge sheet of paper with calculating powers and provides a dynamic method of storing and manipulating data sets. Statistical calculations and graphical presentations are available in many versions and most have scientific functions. Spreadsheets can be used to:

- manipulate raw data by removing the drudgery of repeated calculations, allowing easy transformation of data and calculation of descriptive statistics;
- graph out your data rapidly: they are particularly useful for getting an instant evaluation of results. Print-out can be used in practical and project reports;
- carry out limited statistical analysis by built-in procedures or by allowing construction of formulae for specific tasks;
- model 'what if' situations where the consequences of changes in data can be seen and evaluated;
- store data sets with or without statistical and graphical analysis. This is now common practice in ecology.

Definition
Spreadsheet — a display of a grid of cells into which numbers, text or formulae can be typed to form a worksheet. Each cell is uniquely identifiable by its column and row number combination (i.e. its 2-D coordinates) and can contain a formula which makes it possible for an entry to one cell to alter the contents of one or more other cells.

The spreadsheet (Fig. 48.1) is divided into rows (identified by numbers) and columns (identified by alphabetic characters). Each individual combination of column and row forms a cell which can contain either a data item, a formula, or a piece of text called a label. Formulae can include scientific and/or statistical functions and/or a reference to other cells or groups of cells (often called a range). Complex systems of data input and analysis can be constructed (models). The analysis, in part or complete, can be printed out. New data can be added at any time and the sheet recalculated automatically or at your instigation. You can construct templates, pre-designed spreadsheets without data but containing all the formulae required for repeated data analysis.

The power a spreadsheet offers is directly related to your ability to create models that are accurate and templates that are easy to use. The sequence of

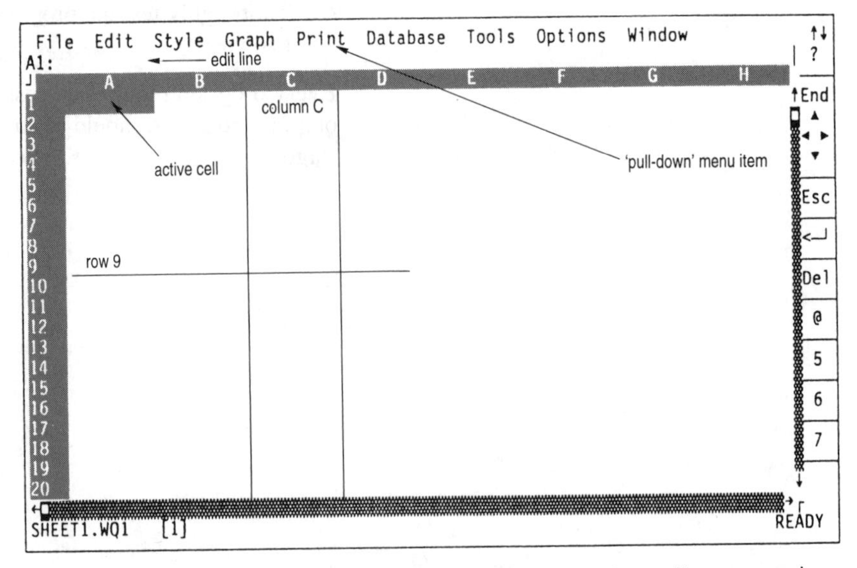

Fig. 48.1 A screen dump of a typical spreadsheet, showing cells, rows and columns.

Key point
Templates should contain:
- **a data input section,**
- **data transformation and/or calculation sections,**
- **a results section, which can include graphics,**
- **text in the form of headings and annotations,**
- **a summary section.**

Constructing a spreadsheet — start with a simple model and extend it gradually, checking for correct operation as you go.

Save your work frequently — otherwise, a power failure or serious mistake could lose hours of work!

operations required is:

1. Determine what information/statistics you want to produce.
2. Identify the variables you will need to use, both for original data that will be entered and for any intermediate calculations that might be required.
3. Set up areas of the spreadsheet for data entry, calculation of intermediate values (statistical values such as sums of squares, etc.), calculation of final parameters/statistics and, if necessary, a summary area.
4. Establish the format of the numeric data if it is different from the default values. This can be done globally (affecting the entire spreadsheet) or locally (affecting only a specified part of the spreadsheet).
5. Establish the column widths required for the various activities.
6. Enter labels: use extensively for annotation.
7. Enter a test set of values to use during formula entry: use a fully worked example to check that formulae are working correctly.
8. Enter the formulae required to make all the calculations, both intermediate and final. Check that results are correct using the test data.

The spreadsheet is then ready for use. Delete all the test data values and you have created your template. Save the template to a disk and it is then available for repeated operations.

Data entry

Spreadsheets have built-in commands which allow you to control the layout of data in the cells. These include number format, the number of decimal places to be shown (the spreadsheet always calculates using eight or more places), the cell width and the location of the entry within the cell (left, right or centre). An auto-entry facility assists greatly in entering large amounts of data by moving the entry cursor either vertically or horizontally as data is entered. Recalculation default is usually automatic so that when a new data value is entered the entire sheet is recalculated immediately. This can dramatically slow down data entry so select manual recalculation mode before entering new data sets if the spreadsheet is large with many calculations.

The parts of a spreadsheet

Labels

These identify the contents of rows and columns. They are text characters, and cannot be used in calculations. Separate them from the data cells by drawing lines, if this feature is available. Programs make assumptions about the nature of the entry being made: most assume that if the first character is a number, then the entry is a number or formula. If it is a letter, then it will be a label. If you want to start a label with a number, you must override this assumption by typing a designated character before the number to tell the program that this is a label; check your program manual for details.

Using hidden (or zero-width) columns — these are useful for storing intermediate calculations which you do not wish to be displayed on the screen or printout.

Numbers

You can also enter numbers (values) in cells for use in calculations. Many programs let you enter numbers in more than one way and you must decide which method you prefer. The way you enter the number does not affect the way it is displayed on the screen as this is controlled by the cell format at the point of entry. There are usually special ways to enter data for percentages, currency and scientific notation for very large and small numbers.

Definition
Function — a pre-programmed code for the transformation of values (mathematical or statistical functions) or selection of text characters (string functions).

Example
A function in Quattro Pro® is @SIN(A5). This will calculate the sine of the number in cell A5. The precise syntax is program specific.

Formulae

These are the 'power tools' of the spreadsheet because they do the calculations. A cell can be referred to by its alphanumeric code, e.g. A5 (column A, row 5) and the value contained in that cell manipulated within a formula, e.g. (A5 + 10) or (A5 + B22) in another cell. Formulae can include a diverse array of pre-programmed functions which can refer to a cell, so that if the value of that cell is changed, so is the result of the formula calculation. They may also include limited branching options through the use of logical operators.

Functions

A variety of functions is usually offered, but only mathematical and statistical functions will be considered here.

Mathematical functions

Spreadsheets have program-specific sets of predetermined functions but they almost all include trigonometrical functions, angle functions, logarithms and random number functions. Functions are invaluable for transforming sets of data rapidly and can be used in formulae required for more complex analyses. Spreadsheets work with an order of preference of the operators in much the same way as a standard calculator and this must always be taken into account when operators are used in formulae. They also require a very precise syntax — the program should warn you if you break this!

Statistical functions

These tend to be basic since spreadsheets are largely seen as business aids, but this is changing. Statistical functions are valuable for rapid summation of large data sets. The main ones found include the following:

- Addition of data: most spreadsheets have the capability to sum all data present in a column, row or segment from different parts of the sheet. These commands usually sum all cells referred to in a formula and empty cells are assigned the value 0.
- Minima and maxima of a range of data cells: beware of programs which assume that empty cells contain values of 0, rendering the minimum value inappropriate.
- Counting cells: a useful operation where you have an unknown or variable number of data values. Again, check how your program deals with empty cells.
- Means: simple mathematical calculations of data from a specified group of cells.

The more sophisticated programs offer standard deviation and variance of groups of data. Some of the best spreadsheets now provide more advanced statistical options such as creating frequency distribution tables, matrix arithmetic, and regression analysis. Be aware of how these work (see the manual) and be sure that you are satisfied with the assumptions used in the calculations.

Copying

All programs provide a means of copying (replicating) formulae or cell contents when required and this is a very useful feature. When copying, references to cells may be either relative, changing with the row/column as they are copied

Distinguishing between absolute and relative cell references — cells containing fixed values must be designated 'absolute' during entry and before copying. Cells not so designated can be changed by subsequent operations.

or absolute, remaining a fixed cell reference and not changing as the formulae are copied. This distinction between cell references is very important and must be understood; it provides one of the most common forms of error when copying formulae. Be sure to understand how your spreadsheet performs these operations.

Naming blocks

When a group of cells (a block) is carrying out a particular function, it is often easier to give the block a name which can then be used in all formulae referring to that block. This powerful feature also allows the spreadsheet to be more readable.

Graphics display

Most spreadsheets now offer a wide range of graphics facilities which are easy to use and this represents an ideal way to examine your data sets rapidly and comprehensively (Chapter 41). The quality of the final graphics output (to a printer) is variable but is usually perfectly sufficient for data exploration and analysis. Many of the options are business graphics styles but there are usually histogram (bar charts), *X-Y* plotting, line and area graphics options available. Note that computer graphics may not come up to the standards expected for the formal presentation of scientific data.

Printing spreadsheets

This is usually a straightforward menu-controlled procedure, made difficult only by the fact that your spreadsheet may be too big to fit on one piece of paper. Try to develop an area of the sheet which contains only the data that you will be printing, i.e. perhaps a summary area. Remember that columns can usually be hidden for printing purposes and you can control whether the printout is in portrait or landscape mode, and for continuous paper or single sheets (depending on printer capabilities). Use a screen preview option, if available, to check your layout before printing. The more sophisticated versions of spreadsheets are now WYSIWYG (What You See Is What You Get) so that the appearance on the screen is a realistic impression of the printout. A 'print to fit' option is also available in some programs, making the fitting of output to a page much easier.

Using string functions — these allow you to manipulate text within your spreadsheet and include functions such as 'search and replace' and alphabetical or numerical 'sort'.

Use as a database

Many spreadsheets can be used as databases, using rows and columns to represent the fields and records (see Chapter 49). For many biological purposes, this form of database is perfectly adequate and should be seriously considered before using a full-feature database program.

49 Using word processors

The word processor has facilitated writing because of the ease of revising text. Word processing is a transferable skill valuable beyond the immediate requirements of your degree course. Using a word processor should improve your writing skills and speed because you can create, check and change your text on the screen before printing it as 'hard copy' on paper. Once entered and saved, multiple uses can be made of a piece of text with little effort.

When using a word processor you can:

- refine material many times before submission;
- insert material easily, allowing writing to take place in any sequence;
- use a thesaurus when composing your text;
- use a spell-checker to check your text;
- produce high quality final copies;
- reuse part or all of the text in other documents.

The disadvantages of using a word processor include:

- lack of ready access to suitable systems;
- time taken to learn the operational details of the program;
- 'trivial' revisions are a real temptation.

Word processors come as 'packages' comprising the program and a manual, often with a tutorial program. They vary from elementary to state-of-the-art programs such as WordPerfect® and Microsoft® Word. Most word processors have similar general features but differ in operational detail; it is best to pick one and stick to it as far as possible so that you become familiar with it. Learning to use the package is like learning to drive a car — you need only to know how to drive the computer and its program, not to understand how the engine (program) and transmission (data transfer) work, although a little background knowledge is often helpful and will allow you to get the most from the program.

In the best word processors, the appearance of the screen realistically represents what the printout on paper will look like (WYSIWYG). Word processing files actually contain large amounts of code relating to text format, etc., but these clutter the screen if visible, as in non-WYSIWYG programs. Some word processors are menu-driven, others require keyboard entry of codes: menus are easier to start with and the more sophisticated programs allow you to choose between these options.

Using textbooks, manuals and tutorials — the manuals that come with some programs are not very user-friendly and it is often worth investing in one of the textbooks that are available for most word processing programs. Alternatively, use an on-line tutorial, available with the more sophisticated packages.

Because of variation in operational details, only general and strategic information is provided in this chapter: you must learn the details of your word processor through use of the appropriate manual.

Before starting you will need:

- the program (ideally on a hard disk);
- a floppy disk for storage, retrieval and back-up of your own files when created;
- the appropriate manual or textbook giving operational details;
- a draft page layout design: in particular you should have decided on page size, page margins, typeface (font) and size, type of text justification, and format of page numbering;
- an outline of the text content;
- access to a suitable printer: this need not be attached to the computer you

are using since your file can be taken to an office where such a printer is available, providing that it has the same word processing program.

Laying out (formatting) your document

Although you can lay out (format) your text at any time, it is good practice to enter the basic format commands at the start of your document: entering them later can lead to considerable problems due to reorganization of the text. If you use a particular set of layout criteria regularly, e.g. an A4 page with space for a letterhead, make a template containing the appropriate codes that can be called up whenever you start a new document. Be aware that various printers may respond differently to particular codes, so that a draft typed on a dot matrix printer may look quite different when printed on a laser printer.

Typing the text

Think of the screen as a piece of typing paper. The cursor marks the position where your text/data will be entered and can be moved around the screen by use of the cursor-control keys. When you type, don't worry about running out of space on the line because the text will wrap around to the next line automatically. Do not use a carriage return (usually the ENTER key) unless you wish to force a new line, e.g. when a new paragraph is wanted. If you make a mistake when typing, correction is easy! You can usually delete characters or words or lines and the space is closed automatically. You can also insert new text in the middle of a line or word. You can insert special codes to carry out a variety of tasks, including changing text appearance such as underlining, **bolding**, *italics* and redline. Paragraph indentations can be automated using TAB as on a typewriter but you can also indent whole blocks of text. The function keys are usually pre-programmed to assist in many of these operations.

Editing features

Word processors usually have an array of features designed to make editing documents easy. In addition to the simple editing procedures described above, the program usually offers facilities to allow blocks of text to be moved ('cut and paste'), copied or deleted.

Deleting and restoring text — because deletion can sometimes be made in error, there is usually an 'undelete' or 'restore' feature which allows the last deletion to be recovered.

An extremely valuable editing facility is the search procedure: this can rapidly scan through a document looking for a specified word, phrase or punctuation. This is particularly valuable when combined with a replace facility so that, for example, you could replace the word 'test' with 'trial' throughout your document simply and rapidly.

Most WYSIWYG word processors have a command which reveals the normally hidden codes controlling the layout and appearance of the printed text. When editing, this can be a very important feature, since some changes to your text will cause difficulties if these hidden codes are not taken into account; in particular, make sure that the cursor is at the precisely correct point before making changes to text containing hidden code, otherwise your text will sometimes change in apparently mystifying ways.

Fonts and line spacing

Most word processors offer a variety of fonts depending upon the printer being used. Fonts come in a wide variety of types and sizes, but they are defined in particular ways as follows:

- Typeface: the term for a family of characters of a particular design, each of which is given a particular name. The most commonly used for normal text is Times Roman but many others are widely available, particularly for the better quality printers. They fall into three broad groups: serif fonts with curves and flourishes at the ends of the characters (e.g. Times Roman); sans serif fonts without such flourishes, providing a clean, modern appearance (e.g. Helvetica, also known as Swiss); and decorative fonts used for special purposes only, such as the production of newsletters and notices.
- Size: measured in points, a point being the smallest typographical unit of measurement, there being 72 points to the inch (about 28 points per cm). The standard sizes for text are 10, 11 and 12 point, but typefaces are often available up to 72 point or more.
- Appearance: many typefaces are available in a variety of styles and weights such as **boldface**, *italic*, outline and shadow. Many of these are not designed for use in scientific literature but for desk-top publishing.
- Spacing: can be either fixed, where every character is the same width, or proportional, where the width of every character, including spaces, is varied. Typewriter fonts such as Elite and Prestige use fixed spacing and are useful for filling in forms or tables, but proportional fonts make the overall appearance of text more pleasing and readable.
- Pitch: specifies the number of characters per horizontal inch of text. Typewriter fonts are usually 10 or 12 pitch, but proportional fonts are never given a pitch value since it is inherently variable.
- Justification is the term describing the way in which text is aligned vertically. Left justification is normal, but for formal documents, both left and right justification may be used.

Preparing draft documents — use double spacing to allow room for your editing comments on the printed page.

Preparing final documents — for most work, use a 12 point proportional serif typeface with spacing dependent upon the specifications for the work.

You should also consider the vertical spacing of lines in your document. Drafts and manuscripts are frequently double-spaced (3 lines per vertical inch) whereas normal spacing (single spacing) is usually 6 lines per vertical inch. If your document has unusual font sizes, this may well affect line spacing although most word processors will cope with this automatically.

Table construction

Tables can be produced by a variety of methods:

- Using tabs as on a typewriter: this moves the cursor to predetermined positions on the page, equivalent to the start of each tabular column. You can define the positions of these tabs as required at the start of each table.
- Using special table-constructing procedures available in the top of the range processors. Here the table construction is largely done for you and it is much easier than using tabs, providing you enter the correct information when you set up the table.
- Using a spreadsheet to construct the table and then copying it to the word processor in some appropriate manner. This procedure requires considerably more manipulation than using the word processor directly and is best reserved for special circumstances, such as the presentation of a very large or complex table of data, especially if the data are already stored as a spreadsheet.

Graphics and special characters

Many word processors can now incorporate graphics from other programs into the text of a document. Files must be compatible (see your manual) but if this is so, it is a relatively straightforward procedure. For highly professional documents this is a valuable facility, but for most undergraduate work it is probably better to produce and use graphics as a separate operation, e.g. employing a spreadsheet.

You can draw lines and other graphical features directly from within most word processors and special characters may be available dependent upon your printer's capabilities. It is a good idea to print out a full set of characters from your printer so that you know what it is capable of. Laser printers, in particular, often have large additional character sets which include symbols and Greek characters, often useful in biological work.

Tools

Many word processors also offer you special tools, the most important of which are:

- Macros: special sets of files you can create when you have a frequently repeated set of keystrokes to make. You can record these keystrokes as a 'macro' so that it can provide a short-cut for repeated operations.
- Thesaurus: used to look up alternative words of similar or opposite meaning while composing text at the keyboard.
- Spell-check: a very useful facility which will check your spellings against a dictionary provided by the program. This dictionary is often expandable to include specialist words which you use in your work. The danger lies in becoming too dependent upon this facility as they all have limitations: in particular, they will not pick up incorrect words which happen to be correct in a different context (i.e. 'was' typed as 'saw' or 'meter' rather than 'metre'). Beware of American spellings in programs from the USA, e.g. 'color' instead of 'colour'. The rule, therefore, is to use the spell-check first and then carefully read the text for errors which have slipped through.
- Word count: useful when you are writing to a prescribed limit.

Using a spell-check facility — do not rely on this to spot all errors. Remember that spell-check programs do not correct grammatical errors.

Printing from your program

Word processors require you to specify precisely the type of printer you are using. Do this at the start of your document. Laser printers also offer choices as to text and graphics quality, so choose draft (low) quality for all but your final copy since this will save both time and materials.

Use a screen preview option to show the page layout if it is available. Assuming that you have entered appropriate layout and font commands, printing is a straightforward operation carried out by the word processor at your command. Problems usually arise because of some incompatibility between the criteria you have entered and the printer's own capabilities. Make sure that you know what your printer offers before starting to type: although parameters are modifiable at any time, changing the page size, margin size, font size, etc., all cause your text to be re-arranged, and this can be frustrating if you have spent hours carefully laying out the pages!

50 Using databases and other packages

Databases

A database is an electronic filing system whose structure is similar to a manual record card collection. Its collection of records is termed a file. The individual items of information on each record are termed fields. Once the database is constructed, search criteria can be used to view files through various filters according to your requirements. The computerized catalogues in your library are just such a system; you enter the filter requirements in the form of author or subject keywords.

You can use a database to catalogue, search, sort, and relate collections of information. The benefits of a computerized database over a manual card-file system are:

- The information content is easily amended/updated.
- Printout of relevant items can be obtained.
- It is quick and easy to organize through sorting and searching/selection criteria, to produce sub-groups of relevant records.
- Record displays can easily be redesigned, allowing flexible methods of presenting records according to interest.
- Relational databases can be combined, giving the whole system immense flexibility. The older 'flat-file' databases store information in files which can be searched and sorted, but cannot be linked to other databases.

Choosing between a database and a spreadsheet — use a database only after careful consideration. Can the task be done better within a spreadsheet? A database program can be complex to set up and usually needs to be updated regularly.

Relatively simple database files can be constructed within the more advanced spreadsheets using the columns and rows as fields and records respectively. These are capable of limited sorting and searching operations and are probably sufficient for the types of databases you are likely to require as an undergraduate. You may also make use of a bibliography database especially constructed for that purpose.

Statistical analysis packages

Statistical packages vary from small programs designed to carry out very specific statistical tasks to large sophisticated packages (Statgraphics®, Minitab®, etc.) intended to provide statistical assistance from experimental design to the analysis of results. Consider the following features when selecting a package:

- The data entry and editing section should be user-friendly, with options for transforming data.
- Data exploration options should include descriptive statistics and exploratory data analysis techniques.
- Hypothesis testing techniques should include ANOVA, regression analysis, multivariate techniques and parametric and non-parametric statistics.
- The program should provide assistance with experimental design and sampling methods.
- Output facilities should be suitable for graphical and tabular formats.

Some programs have very complex data entry systems, limiting the ease of using data in diverse tests. The data entry and storage system should be based upon a spreadsheet system, so that subsequent editing and transformation operations are straightforward.

You will probably have only a limited choice of available programs. Make sure that you understand the statistical basis for your test and the computational techniques involved *before* using a particular program. Mainframe computers will give you access to even more powerful packages, but there is rarely any need for using these at undergraduate level.

Graphics/presentation packages

Many of these packages are specifically designed for business graphics rather than science. They do, however, have considerable value in the preparation of materials for posters and talks where visual quality is an important factor. There is a variety of packages available for microcomputers such as Freelance Graphics® and Harvard Graphics® which provide numerous templates for the preparation of overhead transparencies, slide transparencies, and paper copy, both black and white and in colour. They usually incorporate a 'freehand' drawing option, allowing you to make your own designs.

Although the facilities offered are often attractive, the learning time required for some of the more complex operations is considerable and they should be considered only for specific purposes: routine graphical presentation of data sets is best done from within a spreadsheet or statistical package. There may be a service provided by your institution for the preparation of such material and this should be seriously considered before trying to learn to use these programs. The most important points regarding their use are:

- Graphics quality: the built-in graphics are sometimes of only moderate quality. Use of annotation facilities can improve graphics considerably. Do not use inappropriate graphics for scientific presentation.
- The production of colour graphics: this requires a good quality colour printer/plotter to be available.
- Importing of graphics files: graphs produced by other statistical programs can usually be imported into graphics programs — this is useful for adding legends, annotations, etc., when statistics/spreadsheet programs are inadequate. Check that the format of files produced by your statistics/spreadsheet program can be recognized by your graphics program. The different types of file are distinguished by the three-character filename extension.

Key point
Computer graphics are not always satisfactory for scientific presentation. While they may be useful for exploratory procedures (p. 211), they may need to be re-drawn by hand for your final report. It may be helpful to use a computer-generated graph as a template for the final version.

Communicating results

51 General aspects of scientific writing

Written communication is an essential component of all sciences. Most courses include writing exercises in which you will learn to describe ideas and results accurately, succinctly and in an appropriate style and format. The following are features common to all forms of scientific writing.

Organizing time

Making a timetable at the outset helps ensure that you give each stage adequate attention and complete the work on time. To create and use a timetable:

1. Break down the task into stages.
2. Decide on the proportion of the total time each stage should take.
3. Set realistic deadlines for completing each stage, allowing some time for slippage.
4. Refer to your timetable frequently as you work: if you fail to meet one of your deadlines, make a serious effort to catch up as soon as possible.

Organizing information and ideas

Before you write, you need to gather and/or think about relevant material. You must then decide:

- what needs to be included and what doesn't;
- in what order it should appear.

Start by jotting down headings for everything of potential relevance to the topic (this is sometimes called 'brainstorming'). A spider diagram (Fig. 51.1) will help you organize these ideas. Use it as the basis for an outline of your text. This will:

- force you to think about and plan the structure;
- provide a checklist so nothing is missed out;
- ensure the content is balanced in content and length;
- help you organize figures and tables by showing where they will be used.

An informal outline can be made simply by indicating the order of parts on a spider diagram (as in Fig. 51.1). Two methods of constructing more formal outlines are shown on Fig. 51.2.

In an essay or review, the structure of your writing should help the reader to assimilate and understand your main points. Sub-divisions of the topic could simply be related to the physical nature of the subject matter (e.g. zones of an ecosystem) and should proceed logically (e.g. low water mark to high water mark).

A chronological approach is good for evaluation of past work (e.g. the development of the concept of evolution), whereas a step-by-step comparison might be best for certain exam questions (e.g. 'Discuss the differences between prokaryotes and eukaryotes'). There is little choice about structure for practical and project reports (see p. 271).

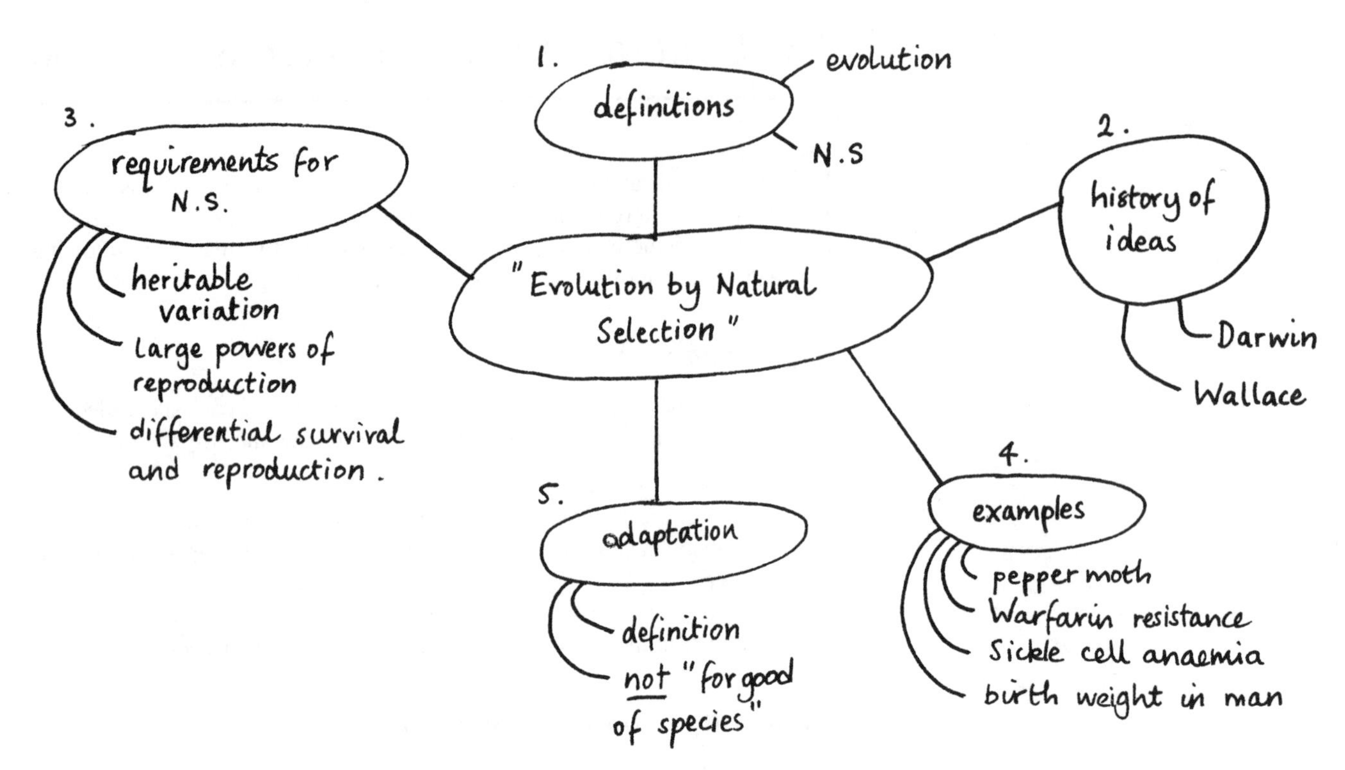

Fig. 51.1 Spider diagram showing how you might 'brainstorm' an essay with the title 'Evolution by Natural Selection'. Write out the essay title in full to form the spider's body, and as you think of possible content, place headings around this to form its legs. Decide which headings are relevant and which are not and use arrows to note connections between subjects. This may influence your choice of order and may help to make your writing flow because the links between paragraphs will be natural. You can make an informal outline directly on a spider diagram by adding numbers indicating a sequence of paragraphs (as shown). This method is best when you must work quickly, as with an essay written under exam conditions.

Writing

Adopting a scientific style

Your main aim in developing a scientific style should be to get your message across directly and unambiguously. While you can try to achieve this through a set of 'rules' (see Box 51.1), you may find other requirements driving your writing in a contradictory direction. For instance, the need to be accurate and complete may result in text littered with technical terms, and the flow may be continually interrupted by references to the literature. The need to be succinct also affects style and readability through the use of, for example, stacked noun-adjectives (e.g. 'restriction fragment length polymorphism') and acronyms (e.g. 'RFLP'). Finally, style is very much a matter of taste and each tutor, examiner, supervisor or editor will have pet loves and hates which you may have to accommodate.

Developing technique

Writing is a skill that can be improved, but not instantly. It takes a long-term commitment involving changes in work habits and some extra reading. An essential component is to build up and *make full use of* a personal reference library (see Box 51.2).

Getting started

A common problem is 'writer's block' — inactivity or stalling brought on by

(a)

```
Evolution by Natural Selection
introduction
        definitions
                evolution
                natural selection
        history of ideas
                Darwin
                Wallace
requirements for natural selection
        heritable variation
        large powers of reproduction
        differential survival and reproduction
examples of selection
        pepper moth
        Warfarin resistance
        sickle cell anaemia
        birthweight in man
consequences of natural selection
        adaptation
        refute 'good of species' notion
```

(b)

```
Evolution by Natural Selection
1. INTRODUCTION
        1.1 definitions
                1.1.1 evolution
                1.1.2 natural selection
        1.2 history of ideas
                1.2.1 Darwin
                1.2.2 Wallace
2. REQUIREMENTS FOR NATURAL SELECTION
        2.1 heritable variation
        2.2 large powers of reproduction
        2.3 differential survival and reproduction
3. EXAMPLES OF SELECTION
        3.1 pepper moth
        3.2 Warfarin resistance
        3.3 sickle cell anaemia
        3.4 birthweight in man
4. CONSEQUENCES OF NATURAL SELECTION
        4.1 adaptation
        4.1 refute 'good of species' notion
```

Fig. 51.2 Formal outlines. These are useful for a long piece of work where you or the reader might otherwise lose track of the structure. The headings for sections and paragraphs are simply written in sequence with the type of lettering and level of indentation indicating their hierarchy. Two different forms of formal outline are shown, a minimal form (a) and a numbered form (b). Note that the headings used in an outline are often repeated within the essay to emphasize its structure. The content of an outline will depend on the time you have available and the nature of the work, but the most detailed hierarchy you should reasonably include is the subject of each paragraph.

a variety of causes. If blocked, ask yourself these questions:

- Are you comfortable with your surroundings? Make sure you are seated comfortably at a reasonably clear desk and have minimized the possibility of interruptions and distractions.
- Are you trying to write too soon? Have you clarified your thoughts on the subject? Have you done enough preliminary reading? Talking to a friend about your topic might bring out ideas or reveal deficiencies in your knowledge.
- Are you happy with the underlying structure of your work? If you haven't made an outline, try this. If you are unhappy because you can't think of a particular detail at the planning stage, just start writing — it is more likely to come to you while you are thinking of something else.
- Are you trying to be too clever? Your first sentence doesn't have to be earth-shattering in content or particularly smart in style. A short statement of fact or a definition is fine. If there will be time for revision, get your ideas down on paper and revise grammar, content and order later.
- Do you really need to start writing at the beginning? Try writing the opening remarks after a more straightforward part. With reports of experimental work, the Materials and Methods section may be the easiest to start at.
- Are you too tired to work? Don't try to 'sweat it out' by writing for long periods at a stretch: stop frequently for a rest.

Box 51.1 How to achieve a clear, readable style

Paragraphs

- Keep short and restricted to a distinct theme.
- Use repeated key words (same subject or verb) or appropriate linking phrases (e.g. 'On the other hand . . .') to connect sentences and emphasize the flow of text.
- The first sentence should introduce the topic of a paragraph and the following sentences explain, illustrate or give examples.

Sentences

- Don't make over-long or complicated.
- Introduce variety in structure and length.
- Make sure you understand how and when to use punctuation.
- If unhappy with the structure of a sentence, try chopping it into a series of shorter sentences.

Words and phrases

- Choose short clear words and phrases rather than long ones: e.g. use 'build' rather than 'fabricate'; 'now' rather than 'at the present time'. At certain times, technical terms must be used for precision, but don't use jargon if you don't have to.
- Don't worry too much about repeating words, especially when to introduce an alternative might subtly alter your meaning.
- Where appropriate, use the first person to describe your actions ('We decided to . . .'; 'I conclude that . . .'), but not if this is specifically discouraged by your supervisor.
- Favour active forms of speech ('the solution was placed in a beaker') rather than the passive voice ('the beaker was filled with the solution').
- Use tenses consistently. Past tense is always used for materials and methods ('samples were taken from . . .') and for reviewing past work ('Smith (1990) concluded that . . .'). The present tense is used when describing data ('Fig. 1 shows . . .'), for generalizations ('Most authorities agree that . . .') and conclusions ('I conclude that . . .').
- Avoid statements in parentheses — they disrupt the reader's attention to your central theme.
- Avoid clichés and colloquialisms — they are usually inappropriate in a scientific context.

Note: If you're not sure what is meant by any of the terms used here, consult a guide on writing (see Box 51.2).

Revising your text

Wholesale revision of your first draft is strongly advised for all writing apart from in exams. If a word processor is available, this can be a simple process. Where possible, schedule your writing so you can leave the first draft to settle for at least a couple of days. When you return to it fresh, you will see more easily where improvements can be made. Try the following structured revision process, each stage being covered in a separate scan of your text:

Box 51.2 Improve your writing ability by consulting a personal reference library

Using dictionaries

We all know that a dictionary helps with spelling and definitions, but how many of us use one effectively? You should:

- Keep a dictionary beside you when writing and always use it if in any doubt about spelling or definitions.
- Use it to prepare a list of words which you have difficulty in spelling: apart from speeding up the checking process, the act of writing out the words helps commit them to memory.
- Use it to write out a personal glossary of terms. This can help you memorize definitions. From time to time, test yourself.

Not all dictionaries are the same! Ask your tutor or supervisor whether he/she has a preference and why. Try out the *Oxford Advanced Learner's Dictionary*, which is particularly useful because it gives examples of use of all words and helps with grammar, e.g. by indicating which prepositions to use. Dictionaries of biology tend to be variable in quality, possibly because the subject is so wide and new terms are continually being coined. *Henderson's Dictionary of Biological Terms* (Longman) is a useful example.

Using a thesaurus

A thesaurus contains lists of words of similar meaning grouped thematically; words of opposite meaning always appear nearby.

- Use a thesaurus to find a more precise and appropriate word to fit your meaning, but check definitions of unfamiliar words with a dictionary.
- Use it to find a word or phrase 'on the tip of your tongue' by looking up a word of similar meaning.
- Use it to increase your vocabulary.

Roget's Thesaurus is the standard. Collins publish a combined dictionary and thesaurus.

Using guides for written English

These provide help with the use of words.

- Use a guide to solve grammatical problems such as when to use 'shall' or 'will', 'which' or 'that', 'effect' or 'affect', etc.
- Use it for help with the paragraph concept and the correct use of punctuation.
- Use it to learn how to structure writing for different tasks.

Recommended guides include the following:

Kane, T.S. (1983) *The Oxford Guide to Writing*. Oxford University Press, New York. This is excellent for the basics of English — it covers grammar, usage and the construction of sentences and paragraphs.

Partridge, E. (1953) *You Have a Point There*. Routledge and Kegan Paul, London. This covers punctuation in a very readable manner.

Tichy, H.J. (1988) *Effective Writing for Engineers, Managers and Scientists*. John Wiley and Sons, New York. This is strong on scientific style and clarity in writing.

1. Examine content. Have you included everything you need to? Is all the material relevant?
2. Check the grammar and spelling. Can you spot any 'howlers'?
3. Focus on clarity. Is the text clear and unambiguous? Does each sentence really say what you want it to say?
4. Try to achieve brevity. What could be missed out without spoiling the essence of your work? It might help to imagine an editor has set you the target of reducing the text by 15%.
5. Improve style. Could the text read better? Consider the sentence and paragraph structure and the way your text develops to its conclusion.

Revising your text — to improve clarity and shorten your text, 'distil' each sentence by taking away unnecessary words and 'condense' words or phrases by choosing a shorter alternative.

52 Writing essays

The function of an essay is to show how much you understand about a topic and how well you can organize and express your knowledge.

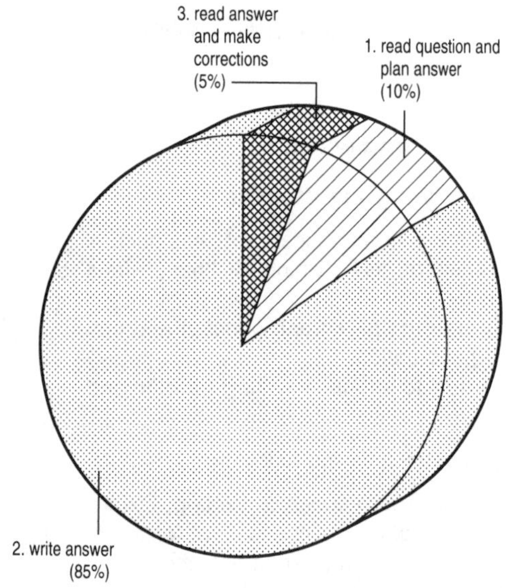

Organizing your time

Most essays have a relatively straightforward structure and it is best to divide your time into three main parts (Fig. 52.1). For exam strategies, see Chapter 57.

3. read answer and make corrections (5%)
1. read question and plan answer (10%)
2. write answer (85%)

Fig. 52.1 Pie chart showing a typical division of time for an essay.

Making a plan for your essay

Dissect the meaning of the essay question or title

Read the title very carefully and think about the topic before starting to write. Consider the definitions of each of the important nouns (this can help in approaching the introductory section). Also think about the meaning of the verb(s) used and try to follow each instruction precisely (see Table 52.1). Don't get side-tracked because you know something about one word or phrase in the title: consider the whole title and all its ramifications. If there are two or more parts to the question, make sure you give adequate attention to each part.

Consider possible content and examples

The spider diagram technique (p. 264) is a speedy way of doing this. If you have time to read several sources, consider their content in relation to the essay title. Can you spot different approaches to the same subject? Which do you

Table 52.1 Instructions often used in essay questions and their meanings. When more than one instruction is given (e.g. compare and contrast; describe and explain), make sure you carry out *both* or you may lose a large proportion of the available marks

Account for:	give the reasons for
Analyse:	examine in depth and describe the main characteristics of
Assess:	weigh up the elements of and arrive at a conclusion about
Comment:	give an opinion on and provide evidence for your views
Compare:	bring out the similarities between
Contrast:	bring out dissimilarities between
Criticize:	judge the worth of (give both positive and negative aspects)
Define:	explain the exact meaning of
Describe:	use words and diagrams to illustrate
Discuss:	provide evidence or opinions about, arriving at a balanced conclusion
Enumerate:	list in outline form
Evaluate:	weigh up or appraise; find a numerical value for
Explain:	make the meaning of something clear
Illustrate:	use diagrams or examples to make clear
Interpret:	express in simple terms, providing a judgement
Justify:	show that an idea or statement is correct
List:	provide an itemized series of statements about
Outline:	describe the essential parts only, stressing the classification
Prove:	establish the truth of
Relate:	show the connection between
Review:	examine critically, perhaps concentrating on the stages in the development of an idea or method
State:	express clearly
Summarize:	without illustrations, provide a brief account of
Trace:	describe a sequence of events from a defined point of origin

prefer as a means of treating the topic in relation to your title? Which examples are most relevant to your case, and why?

Construct an outline

Every essay should have a structure related to its title. Most marks for essays are lost because the written material is badly organized or is irrelevant. An essay plan, by definition, creates order and, if thought about carefully, can ensure relevance. Your plan should be written down (but scored through later if written in an exam book). Think about an essay's content in three parts:

1. The introductory section, in which you should include definitions and some background information on the context of the topic being considered. You should also tell your reader how you plan to approach the subject.
2. The middle of the essay, where you develop your answer and provide relevant examples. Decide whether a broad analytical approach is appropriate or whether the essay should contain more factual detail.
3. The conclusion, which you can make quite short. You should use this part to summarize and draw together the components of the essay, without merely repeating previous phrases. You might mention such things as: the broader significance of the topic; its future; its relevance to other important areas of biology. Always try to mention both sides of any debate you have touched on, but beware of 'sitting on the fence'.

Writing your essay — it is rarely enough simply to lay down facts for the reader — you must analyse them and comment on their significance.

Use paragraphs to make the essay's structure obvious. Emphasize them with headings and sub-headings unless the material beneath the headings would be too short or trivial.

Now start writing!

- Never lose track of the importance of content and its relevance. Repeatedly ask yourself: 'Am I really answering this question?' Never waffle just to increase the length of an essay. Quality rather than quantity is important.
- Illustrate your answer appropriately. Use examples to make your points clear, but remember that too many similar examples can stifle the flow of an essay. Use diagrams where a written description would be difficult or take too long. Use tables to condense a lot of information into a small space.
- Take care with your handwriting. You can't get marks if your writing is illegible! Try to cultivate an open form of handwriting, making the individual letters large and distinct. If there is time, make out a rough draft from which a tidy version can be copied.

Using drawings and diagrams — give a title and legend for each drawing so that it makes sense in isolation and point out in the text when the reader should consult it (e.g. 'as shown in Fig. 1 . . .' or 'as can be seen in the accompanying diagram, . . .').

Reviewing your answer

Don't stop yet!

- Re-read the question to check that you have answered all points.
- Re-read your essay to check for errors in punctuation, spelling and content. Make any corrections obvious. Don't panic if you suddenly realize you've missed a large chunk out as the reader can be redirected to a supplementary paragraph if necessary.
- Learn from lecturers' and tutors' comments. Ask for further explanations if you don't understand a comment or why an essay was less successful than you thought it should be.

53 Reporting practical work

Practical reports, project reports, theses and scientific papers differ greatly in depth, scope and size, but they all have the same basic structure (Box 53.1).

Practical and project reports

These are exercises designed to make you think more deeply about your experiments and to practise and test the skills necessary for writing up research work. Special features are:

- Introductory material is generally short and unless otherwise specified should outline the aims of the experiment(s) with a minimum of background material.
- Materials and methods may be provided by your supervisor for practical reports. With project work, your lab notebook (see p. 64) should provide the basis for writing this section.
- Great attention in assessment will be paid to presentation and analysis of data. Take special care over graphs (see p. 217). Make sure your conclusions are justified by the evidence.

Theses

Theses are submitted as part of the examination for a degree following an extended period of research. They act to place on record full details about your experimental work and will normally only be read by those with a direct interest in it — your examiners or colleagues. Note the following:

- You are allowed scope to expand on your findings and to include detail that might otherwise be omitted in a scientific paper.
- You may have problems with the volume of information that has to be organized. One method of coping with this is to divide your thesis into chapters, each having the standard format (as in Box 53.1). A General Introduction can be given at the start and a General Discussion at the end. Discuss this option with your supervisor as it is not universally favoured.
- There may be an oral exam ('viva') associated with the submission of the thesis. The primary aim of the examiners will be to ensure that you understand what you did and why you did it.

Steps in the production of a practical report or thesis

Choose the experiments you wish to describe and decide how best to present them

Try to start this process before your lab work ends, because at the stage of reviewing your experiments, a gap may become apparent (e.g. a missing control) and you might still have time to rectify the deficiency. Irrelevant material should be ruthlessly eliminated, at the same time bearing in mind that negative results can be extremely important (see p. 49). Use as many different forms of data presentation as are appropriate, but avoid presenting the same data in more than one form. Graphs are generally easier for the reader to assimilate, while tables can be used to condense a lot of data into a small space. Relegate large tables of data to an appendix and summarize the important points. Make sure that the experiments you describe are representative: always state the number of times they were repeated and how consistent your findings were.

Repeating your experiments — remember, if you do an experiment twice, you have repeated it only once!

Box 53.1 The structure of reports of experimental work

Undergraduate practical and project reports are generally modelled on this structure or a close variant of it, because this is the structure used for nearly all research papers and theses. The more common variations include Results and Discussion combined into a single section for convenience and Conclusions appearing separately as a series of points arising from the work. In scientific papers, a list of Key Words (for computer cross-referencing systems) may be included following the Abstract. Regarding variations in positioning, Acknowledgements may appear after the Contents section, rather than near the end. Department or faculty regulations for producing theses and reports may specify a precise format; they often require a title page to be inserted at the start and a list of figures and tables as part of the Contents section. These regulations may also specify declarations and statements to be made by the student and supervisor.

Part (in order)	Contents/purpose	Checklist for reviewing content
Title	Explains what the project was about	Does it explain what the text is about succinctly?
Authors plus their institutions	Explains who did the work and where; also where they can be contacted now	Are all the details correct?
Abstract/Summary	Synopsis of methods, results and conclusion of work described. Allows the reader to grasp quickly the essence of the work	Does it explain why the work was done? Does it outline the whole of your work and your findings?
List of Contents	Shows the organization of the text; directs reader to relevant pages (not required for short papers)	Are all the sections covered? Are the page numbers correct?
Abbreviations	Lists all the abbreviations used (but not those of SI, chemical elements, or standard biochemical terms)	Have they all been explained? Are they all in the accepted form? Are they in alphabetical order?
Introduction	Orientates the reader, explains why the work has been done and its context in the literature, why the methods used were chosen, why the experimental organisms were chosen. Indicates the central hypothesis behind the experiments	Does it provide enough background information and cite all the relevant references? Is it of the correct depth for the readership? Have all the technical terms been defined? Have you explained why you investigated the problem? Have you explained your methodological approach to the problem?
Materials and Methods	Explains how the work was done. Should contain sufficient detail to allow another competent worker to repeat the work	Is each experiment covered and have you avoided unnecessary duplication? Is there sufficient detail to allow repetition of the work? Are proper scientific names and authorities given for all organisms? Have you explained where you got them from? Are the correct names, sources and grades given for all chemicals?
Results	Displays and describes the data obtained. Should be presented in a form which is easily assimilated (graphs rather than tables, small tables rather than large ones)	Is the sequence of experiments logical? Are the parts adequately linked? Are the data presented in the clearest possible way? Have SI units been used properly throughout? Has adequate statistical analysis been carried out? Is all the material relevant? Have appendices been used where appropriate? Are the figures and tables all numbered in the order of their appearance? Are their titles appropriate? Do the figure and table legends provide all the information necessary to interpret the data without reference to the text? Have you presented the same data more than once?
Discussion/ Conclusions	Discusses the results: their meaning, their importance; compares the results with those of others; suggests what to do next	Have you explained the significance of the results? Have you compared your data with other published work? Are your conclusions justified by the data presented?

continued overleaf

Box 53.1 continued

Acknowledgements	Gives credit to those who helped carry out the work	Have you listed everyone that helped including any grant-awarding bodies?
Literature Cited	Lists all references cited in appropriate format: provides enough information to allow the reader to find the reference in a library	Do all the references in the text appear on the list? Do all the listed references appear in the text? Do the years of publication and authors match? Are the journal details complete and in the correct format? Is the list in alphabetical order, or correct numerical order?
Appendices	Contain material that would otherwise disrupt the text (e.g. large tables of raw data)	Are data presented in the most succinct form?

Presenting your results — remember that the order of results presented in a report need not correspond with the order in which you carried out the experiments: you are expected to rearrange them to provide a logical sequence of findings.

Make up plans or outlines for the component parts
The overall plan is well defined (see Box 53.1), but individual parts will need to be organized as with any other form of writing (see Chapter 51).

Write!
The Materials and Methods section is often the easiest to write once you have decided what to report. Remember to use the past tense and do not allow results or discussion to creep in. The Results section is the next easiest as it should only involve description. At this stage, you may benefit from jotting down ideas for the Discussion — this may be the hardest part to compose as you need an overview both of your own work and of the relevant literature. It is also liable to become wordy, so try hard to make it succinct. The Introduction shouldn't be too difficult if you have fully understood the aims of the experiments. Write the Abstract and complete the list of references at the end. To assist with the latter, it is a good idea as you write to jot down the references you use or to pull out their cards from your index system.

Revise the text
Once your first draft is complete, try to answer all the questions given in Box 53.1. Show your work to your supervisors and learn from their comments. Let a friend or colleague who is unfamiliar with your subject read your text; they may be able to pinpoint obscure wording and show where information or explanation is missing. If writing a thesis, double-check that you are adhering to your institution's thesis regulations.

Prepare the final version
Markers appreciate neatly produced work but a well-presented document will not disguise poor science! If using a word processor, print the final version with the best printer available. Make sure figures are clear and in the correct size and format.

Submit your work
Your department will specify when to submit a thesis or project report, so plan your work carefully to meet this deadline or you may lose marks. Tell your supervisor early of any circumstances that may cause delay.

Producing a scientific paper

Scientific papers are the means by which research findings are communicated to others. They are published in journals with a wide circulation among academics and are 'peer reviewed' by one or more referees before being accepted. Each journal covers a well-defined subject area and publishes details of the format they expect. It would be very unusual for an undergraduate to submit a paper on his or her own — this would normally be done in collaboration with your project supervisor and only then if your research has satisfied appropriate criteria. However, it is important to understand the process whereby a paper comes into being (Box 53.2), as this can help you when interpreting the primary literature.

Box 53.2 Producing a scientific paper

Scientific papers are the lifeblood of any science and it is a major landmark in your scientific career to publish your first paper. The major steps in doing this should include the following:

Assessing potential content

The work must be of an appropriate standard to be published and should be 'new, true and meaningful'. Therefore, before starting, the authors need to review their work critically under these headings. The material included in a scientific paper will generally be a sub-set of the total work done during a project, so it must be carefully selected for relevance to a clear central hypothesis — if the authors won't prune, the referees and editors of the journal certainly will!

Choosing a journal

There are thousands of journals covering biology and each covers a specific area (which may change through time). The main factors in deciding on an appropriate journal are the range of subjects it covers, the quality of its content and the number and geographical distribution of its readers. The choice of journal always dictates the format of a paper since authors must follow to the letter the journal's 'Instructions to Authors'.

Deciding on authorship

In multi-author papers, a contentious issue is often who should appear as an author and in what order they should be cited. Where authors make an equal contribution, an alphabetical order of names may be used. Otherwise, each author should have made a substantial contribution to the paper and should be prepared to defend it in public. Ideally, the order of appearance will reflect the amount of work done rather than seniority. This may not happen in practice!

Writing

The paper's format will be similar to that shown in Box 53.1 and the process of writing will include outlining, reviewing, etc., as discussed elsewhere in this chapter. Figures must be finished to an appropriate standard and this may involve preparing photographs of them.

Submitting

When completed, copies of the paper are submitted to the editor of the chosen journal with a simple covering letter. A delay of one to two months usually follows while the manuscript is sent to one or more anonymous referees who will be asked by the editor to check that the paper is novel, scientifically correct and that its length is fully justified.

Responding to referees' comments

The editor will send on the referees' comments and the authors then have a chance to respond. The editor will decide on the basis of the comments and replies to them whether the paper should be published. Sometimes quite heated correspondence can result if the authors and referees disagree!

Checking proofs and waiting for publication

If a paper is accepted, it will be sent off to the typesetters. The next the authors see of it is the proofs (first printed version in style of journal), which have to be corrected carefully for errors and returned. Eventually, the paper will appear in print, but a delay of six months following acceptance is not unusual. Most journals offer the authors reprints which can be sent to other researchers in the field or to those who send in reprint request cards.

54 Writing literature surveys and reviews

The literature survey or review is a specialized form of essay which summarizes and reviews the evidence and concepts concerning a particular area of research. It offers you full scope for exercising your critical faculties — the best reviews are those which analyse information rather than recite it.

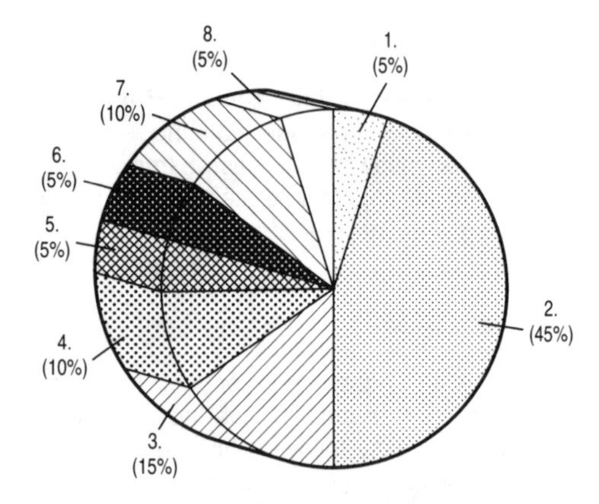

Fig. 54.1 Pie chart showing how you might allocate time for a literature survey: 1. select a topic; 2. scan the literature; 3. plan the review; 4. write first draft; 5. leave to settle; 6. structured review of text; 7. write final draft; 8. produce top copy.

Making up a timetable

Figure 54.1 illustrates how you might divide up your time for writing a literature survey. There are many sub-divisions in this chart because of the size of the task: in general, for lengthy tasks, it is best to divide up the work into manageable chunks. Note also that proportionately less time is allocated to writing itself than with an essay. In a literature survey, make sure that you spend adequate time on research and revision.

Selecting a topic

You may have no choice in the topic to be covered, but if you do, carry out your selection as a three-stage process:

1. Identify a broad subject area that interests you.
2. Find and read relevant literature in that area. Try to gain a broad impression of the field from books and general review articles. Discuss your ideas with your supervisor.
3. Select a precise and concise title. The wording should be considered very carefully as it will define the content expected by the reader. A narrow subject area will cut down on the amount of literature you will be expected to review, but will also restrict the scope of the conclusions you can make (and vice versa for a wide subject area).

Scanning the literature and organizing your references

You will need to carry out a thorough investigation of the literature before you start to write. The key problems are as follows:

- Getting an initial toe-hold in the literature. Seek help from your supervisor, who may be willing to supply a few key papers to get you started. Hints on expanding your collection of references are given on p. 72.
- Assessing the relevance and value of each article. This is the essence of writing a review, but it is difficult unless you already have a good understanding of the field (Catch 22!). Try reading earlier reviews in your area.
- Clarifying your thoughts. Sometimes you can't see the wood for the trees! Sub-dividing the main topic and assigning your references to these smaller subject areas may help you gain a better overview of the literature.

Using index cards (see p. 75) — these can help you organize large numbers of references. Write key points on each card — this helps when considering where the reference fits into the literature. Arrange the cards in subject piles, eliminating irrelevant ones. Order the cards in the sequence you wish to write in.

Deciding on structure and content

The general structure and content of a literature survey is described overleaf.

Introduction

The introduction should give the general background to the research area, concentrating on its development and importance. You should also make a statement about the scope of your survey; as well as defining the subject matter to be discussed, you may wish to restrict the period being considered.

Review

The review itself should discuss the published work in the selected field. Within each portion of a review, the approach is usually chronological, with appropriate linking phrases (e.g. 'Following on from this, . . .'; 'Meanwhile, Bloggs (1980) tackled the problem from a different angle . . .'). However, a good review is much more than a chronological list of work done. It should:

- allow the reader to obtain an overall view of the current state of the research area, identifying the key areas where knowledge is advancing;
- show how techniques are developing and discuss the benefits and disadvantages of using particular organisms or experimental systems;
- assess the relative worth of different types of evidence — this is the most important aspect. Do not be intimidated from taking a critical approach as the conclusions you may read in the primary literature aren't always correct;
- indicate where there is conflict in findings or theories, suggesting if possible which side has the stronger case;
- indicate gaps in current knowledge.

Conclusions

The conclusions should draw together the threads of the preceding parts and point the way forward, perhaps listing areas of ignorance or where the application of new techniques may lead to advances.

References, etc.

The references or literature cited section should provide full details of all papers referred to in the text (see p. 76). The regulations for your department may also specify a format and position for the title page, list of contents, acknowledgements, etc.

Style of literature surveys

The *Annual Review* series (available in most university libraries) provides good examples of the style expected in a biological literature survey. A review of literature poses stylistic problems because of the need to cite large numbers of papers and in the *Annual Review* series this is overcome by using numbered references (see p. 75).

55 Organizing a poster display

A scientific poster is a visual display of the results of an investigation, usually mounted on a rectangular board. Posters are used at scientific meetings, to communicate research findings, and in undergraduate courses, to display project results or assignment work.

In a written report you can include a reasonable amount of specific detail and the reader can go back and re-read difficult passages. However, if a poster is long-winded or contains too much detail, your reader is likely to lose interest. A poster session is like a competition — you are competing for the attention of people in a room. Because you need to attract and hold the attention of your audience, make your poster as interesting as possible. Think of it as an advertisement for your work and you will not go far wrong.

Preliminaries

Before considering the content of your poster, you should establish:

- the dimensions of your poster area, typically 1.5 m wide by 1.0 m high;
- the composition of the poster board and the method of attachment, whether drawing pins, Velcro® tape, or some other form of adhesive;
- whether fixatives will be provided — in any case, it's safer to bring your own;
- the time(s) when the poster should be set up and when you should attend;
- the room where the poster session will be held.

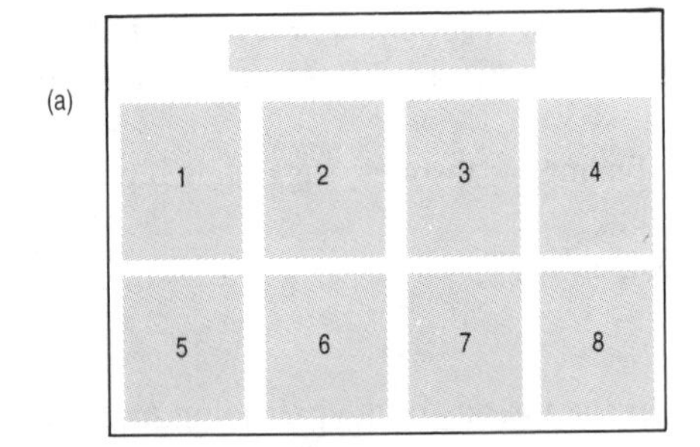

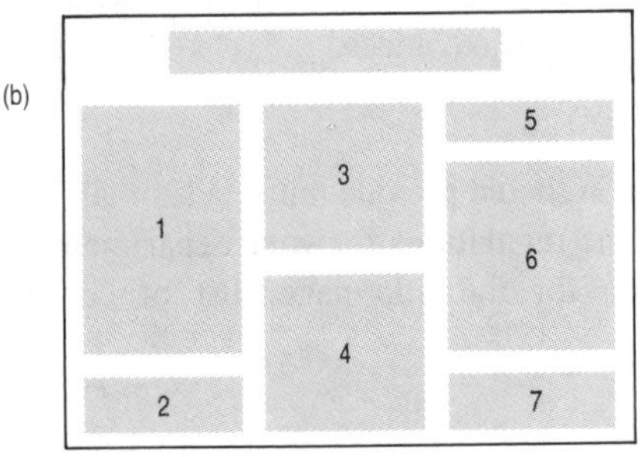

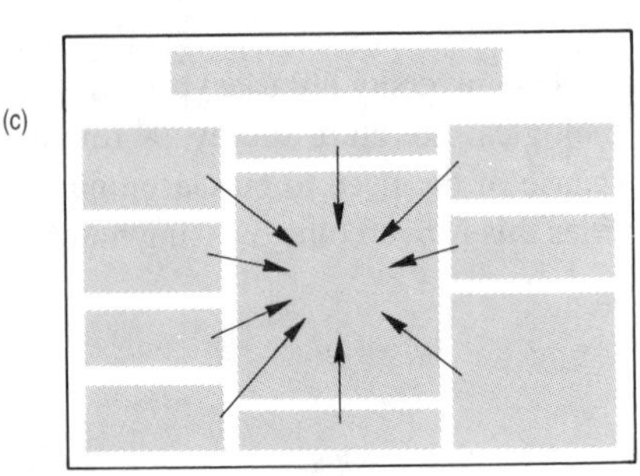

Fig. 55.1 Poster design. (a) An uninspiring design: sub-units of equal area, reading left to right, are not recommended. (b) This design is more interesting and the text will be easier to read (column format). (c) An alternative approach, with a central focus and arrows/tapes to guide the reader.

Design

Plan your poster with your audience in mind, as this will dictate the appropriate level for your presentation. Aim to make your poster as accessible as possible to a broad audience. Since a poster is a visual display, you must pay particular attention to the presentation of information: work that may have taken hours to prepare can be ruined in a few minutes by the ill-considered arrangement of items. Begin by making a draft sketch of the major elements of your poster. It is worth discussing your intended design with someone else, as constructive advice at the draft stage will save a lot of time and effort when you prepare the final version.

Layout

Usually the best approach is to divide the poster into several smaller areas, perhaps six or eight in all, and prepare each as a separate item on a piece of thick card. Some people prefer to produce a single large poster on one sheet of paper or card and store it inside a protective cardboard tube. However, a single large poster will bend and crease, making it difficult to flatten out. In addition, photographs and text attached to the backing sheet often work loose.

Sub-dividing your poster means that each smaller area can be prepared on a separate piece of paper or card, of A4 size or slightly larger, making transport and storage easier. It also breaks the reading matter up into smaller pieces, looking less formidable to a potential reader. By using pieces of card of different colours you can provide emphasis for key aspects, or link textual material with figures or photographs.

You will need to guide your reader through the poster. It is often appropriate to use either a numbering system, with large, clear numbers at the top of each piece of card, or a system of arrows (or thin tapes), to show the relationship of sections within the poster (see Fig. 55.1). Make sure that the relationship is clear and that the arrows or tapes do not cross.

Title

Your chosen title should be concise (no more than eight words), specific and interesting, to encourage people to read the poster. Make the title large and bold — it should run across the top of your poster, in letters at least 4 cm high, so that it can be read from the other side of the room. Coloured spirit-based marker and block capitals drawn with a ruler work well, as long as your writing is readable and neat (the colour can be used to add emphasis). Alternatively, you can use Letraset®, or similar lettering. Details of authors, together with their addresses (if appropriate), should be given, usually in the top right-hand corner in somewhat smaller lettering than the title. At conferences, a passport-sized photograph of the contributor is sometimes useful for identification.

Text

Do not overcrowd your poster with text. Keep it as visual as possible and make effective use of the spaces between the blocks of text. Your final text should be double-spaced and should have a minimum capital letter height of 5 mm, preferably greater, so that the poster can be read at a distance of 1 m. One method of obtaining text of the required size is to photo-enlarge standard typescript (using a good-quality photocopier), or use a high-quality (laser) printer. It is best to avoid continuous use of text in capitals, since it slows reading and makes the text less interesting to the reader. Also avoid italic, 'balloon' or similar styles of lettering.

Key point
Keep text to a minimum — aim to have a maximum of 500 words in your poster. Write in short sentences: avoid verbosity.

Sub-titles and headings

These should have a capital letter height of at least 10 mm, and should be restricted to two or three words. They can be produced by photo-enlargement, by stencilling, Letraset® or by hand, using pencilled guidelines (but make sure that no pencil marks are visible on your finished poster).

Adhesives

Liquid glues are best avoided. Attach the text to its backing card using photographic mountant, Pritt®, or similar adhesive. Using window mounts for photographs stops the corners from curling up and may allow you to highlight an appropriate area without cutting the original photograph.

Colour

Consider the overall visual effect of your chosen display, including the relationship between text, diagrams and the backing board. Colour can be used to highlight key aspects of your poster. However, it is very easy to ruin a poster by the inappropriate choice and application of colour. Careful use of two, or at most three, complementary colours will be easier on the eye and may aid comprehension. Colour can be used to link the text with the visual images (e.g. by picking out a colour in a photograph and using the same colour on the mounting board for the accompanying text). Use coloured inks or water-based paints to provide colour in diagrams and figures, as felt pens rarely give satisfactory results.

Content

The typical format is that of a scientific paper (see Box 53.1), i.e. with the same headings, but with a considerably reduced content. Never be tempted to spend the minimum amount of time converting a piece of scientific writing into poster format. At scientific meetings, the least interesting posters are those where the author simply displays pages from a written communication (e.g. a journal article) on the poster board! Keep references within the text to a minimum — interested parties can always ask you for further information.

Introduction

This should give the reader background information on the broad field of study and the aims of your own work. It is vital that this section is as interesting as possible, to capture the interest of your audience. It is often worth listing your objectives as a series of numbered points.

Designing the materials and methods section — photographs or diagrams of apparatus can help to break up the text of the Materials and Methods section and provide visual interest. It is sometimes worth preparing this section in a smaller typeface.

Materials and Methods

Keep this short, and describe only the principal techniques used. You might mention any special techniques or problems of general interest.

Results

Don't present your raw data: use data reduction wherever possible, i.e. figures and simple statistical comparisons. Graphs, diagrams, histograms and pie charts give clear visual images of trends and relationships and should be used in place of tabulated data (see p. 214). Final copies of all figures should be produced so that the numbers can be read from a distance of 1 m. Each should have a concise title and legend, so that it is self-contained: if appropriate, a series of numbered points can be used to link a diagram with the accompanying text. Where symbols are used, provide a key on each graph (symbol size should be at least 5 mm). Avoid using graphs straight from a written version, e.g. a project report, textbook, or a paper, without considering whether they need modification to meet your requirements.

Key point
Keep your graphs and diagrams as simple as possible. Avoid composite graphs with different scales for the same axis, or with several trend lines (use a maximum of three trend lines per graph).

Conclusions

This is where many readers will begin, and they may go no further unless you make this section sufficiently interesting. This section needs to be the strongest part of your poster. Refer to your figures here to draw the reader into the main part of your poster. A slightly larger or bolder typeface may add emphasis, though too many different typefaces can look messy.

Listing your conclusions — a series of numbered points is a useful approach, if your findings fit this pattern.

The poster session

If you stand at the side of your poster throughout the session, you are likely to discourage some readers, who may not wish to become involved in a detailed conversation about the poster. Stand nearby. Find something to do — talk to someone, or browse among the other posters, but remain aware of people reading your poster and be ready to answer any queries they may raise. Do not be too discouraged if you aren't asked lots of questions: remember, it is meant to be a self-contained, visual story, without need for further explanation.

A poster display will never feel like an oral presentation, where the nervousness beforehand is replaced by a combination of satisfaction and relief as you unwind after the event. However, it can be a very satisfying means of communication, particularly if you follow these guidelines.

Consider providing a handout — this is a useful way to summarize the main points of your poster, so that your readers have a permanent record of the information you have presented.

56 Giving an oral presentation

Most students feel very nervous about giving talks. This is natural, since very few people are sufficiently confident and outgoing that they look forward to speaking in public. Additionally, the technical nature of the topic may give you cause for concern, especially if you feel that some members of the audience have a greater knowledge than you have. However, this is a fundamental method of scientific communication and it therefore forms an important component of many courses.

The comments in this chapter apply equally to informal talks, e.g. those based on assignments and project work, and to more formal conference presentations. It is hoped that the advice and guidance given below will encourage you to make the most of your opportunities for public speaking, but there is no substitute for practice. Do not expect to find all of the answers from this, or any other, book. Rehearse and learn from your own experience.

Preparation

Preliminary information

Begin by marshalling all of the details you need to plan your presentation, including:

- the duration of the talk;
- whether time for questions is included;
- the size and location of the room;
- the projection/lighting facilities provided, and whether pointers or similar aids are available.

Key points
The three 'Rs' of successful public speaking are:
Reflect: give sufficient thought to all aspects of your presentation, particularly at the planning stage.
Rehearse: to improve your delivery.
Rewrite: modify the content and style of your material in response to your own ideas and to the comments of others.

It is especially important to find out whether the room has the necessary equipment for slide projection (slide projector and screen, black-out curtains or blinds, appropriate lighting) or overhead projection before you prepare your audio-visual aids. If you concentrate only on the spoken part of your presentation at this stage, you are inviting trouble later on. Have a look around the room and try out the equipment at the earliest opportunity, so that you are able to use the lights, projector, etc., with confidence.

Audio-visual aids

Find out whether your department has facilities for preparing overhead transparencies and slides, whether these facilities are available for your use and the cost of materials. Adopt the following guidelines:

- Keep text to a minimum: present only the key points, with up to 20 words per slide/transparency.
- Make sure the text is readable: try out your material beforehand.
- Use several simpler figures rather than a single complex graph.
- Avoid too much colour on overhead transparencies: blue and black are easier to read than red or green.
- Don't mix slides and transparencies as this is often distracting.
- Use spirit-based pens for transparencies: use alcohol for corrections.
- Transparencies can be produced from typewritten or printed text using a photocopier, often giving a better product than pens. Note that you must use special heat-resistant acetate sheets for photocopying.

Audience

You should consider your audience at the earliest stage, since they will determine the appropriate level for your presentation. If you are talking to fellow students you may be able to assume a common level of background knowledge. In contrast, a research lecture given to your department, or a paper at a meeting of a scientific society, will be presented to an audience from a broader range of backgrounds. An oral presentation is not the place for a complex discussion of specialized information: build up your talk from a low level. The speed at which this can be done will vary according to your audience. As long as you are not boring or patronizing, you can cover basic information without losing the attention of the more knowledgeable members in your audience. The general rule should be: 'do not overestimate the background knowledge of your audience'. This sometimes happens in student presentations, where fears about the presence of 'experts' can encourage the speaker to include too much detail, overloading the audience with facts.

Content

While the specific details in your talk will be for you to decide, most oral presentations share some common features of structure, as described below.

Introductory remarks

It is vital to capture the interest of your audience at the outset. Consequently, you must make sure your opening comments are strong, otherwise your audience will lose interest before you reach the main message. Remember it takes a sentence or two for an audience to establish a relationship with a new speaker. Your opening sentence should be some form of preamble and should not contain any key information. For a formal lecture, you might begin with 'Mr Chairman, ladies and gentlemen, my talk today is about . . .' then re-state the title and acknowledge other contributors, etc. You might show a transparency or slide with the title printed on it, or an introductory photograph, if appropriate. This should provide the necessary settling-in period.

After these preliminaries, you should introduce your topic. Begin your story on a strong note — there is no room for timid or apologetic phrases. You should:

- explain the structure of your talk;
- set out the aims and objectives of your work;
- explain your approach to the topic.

Opening remarks are unlikely to occupy more than 10% of the talk. However, because of their significance, you might reasonably spend up to 25% of your preparation time on them. Make sure you have practised this section, so that you can deliver the material in a flowing style, with less chance of mistakes.

The main message

This section should include the bulk of your experimental results or literature findings, depending on the type of presentation. Keep details of methods to the minimum needed to explain your data. This is *not* the place for a detailed description of equipment and experimental protocol (unless it is a talk about methodology!). Results should be presented in an easily-digested format.

Do not expect your audience to cope with large amounts of data; use a maximum of six numbers per slide. Present summary statistics rather than individual results. Show the final results of any analyses in terms of the statistics calculated, and their significance (p. 234), rather than dwelling on details of the procedures used. Remember that graphs and diagrams are usually better than tables of raw data, since the audience will be able to see the trends and relationships in your data (p. 214). However, figures should not be crowded with unnecessary detail. Every diagram should have a concise title and the symbols and trend lines should be clearly labelled, with an explanatory key where necessary. When presenting graphical data always 'introduce' each graph by stating the units for each axis and describing the relationship for each trend line or data set. Summary slides can be used at regular intervals, to maintain the flow of the presentation and to emphasize the key points.

Allowing time for slides — as a rough guide you should allow at least two minutes per illustration, although some diagrams may need longer, depending on content.

Take the audience through your story step-by-step at a reasonable pace. Try not to rush the delivery of your main message due to nervousness. Avoid complex, convoluted story-lines — one of the most distracting things you can do is to fumble backwards through slides or overhead transparencies. If you need to use the same diagram or graph more than once then you should make two (or more) copies. In a presentation of experimental results, you should discuss each point as it is raised, in contrast to written text, where the results and discussion may be in separate sections. The main message typically occupies approximately 80% of the time allocated to an oral presentation.

Concluding remarks

Having captured the interest of your audience in the introduction and given them the details of your story in the middle section, you must now bring your talk to a conclusion. At all costs, do not end weakly, e.g. by running out of steam on the last slide. Provide your audience with a clear 'take-home message', by returning to the key points in your presentation. It is often appropriate to prepare a slide or overhead transparency listing your main conclusions as a numbered series.

Final remarks — make sure you give the audience sufficient time to assimilate your final slide: some of them may wish to write down the key points. Alternatively, you might provide a handout, with a brief outline of the aims of your study and the major conclusions.

Signal the end of your talk by saying 'finally . . .', 'in conclusion . . .', or a similar comment and then finish speaking after that sentence. Your audience will lose interest if you extend your closing remarks beyond this point. You may add a simple end phrase (for example, 'thank you') as you put your notes into your folder, but do not say 'that's all folks!', or make any similar offhand remark. Finish as strongly and as clearly as you started.

Hints on presentation

Notes

Many accomplished speakers use abbreviated notes for guidance, rather than reading from a prepared script. When writing your talk:

- Prepare a first draft as a full script: write in spoken English, keeping the text simple and avoiding an impersonal style. Aim to *talk* to your audience, not read to them.
- Use note cards with key phrases and words: it is best to avoid using a full script at the final presentation. As you rehearse and your confidence improves, a set of cards may be a more appropriate format for your notes.
- Consider the structure of your talk: keep it as simple as possible and announce each sub-division, so your audience is aware of the structure.
- Mark the position of slides/key points, etc.: each note card should contain details of structure, as well as content.
- Memorize your introductory/closing remarks: you may prefer to rely on a full written version for these sections, in case your memory fails.

- Use notes: write on only one side of the card/paper, in handwriting large enough to be read easily during the presentation. Each card or sheet must be clearly numbered, so that you do not lose your place.
- Rehearse your presentation: ask a friend to listen and to comment constructively on parts that were difficult to follow.
- Use 'split times' to pace yourself: following rehearsal, note the time at which you should arrive at key points of your talk. These timing marks will help you keep to time during the 'real thing'.

Using slides — check that the lecture theatre has a lectern light, otherwise you may have problems reading your notes when the lights are dimmed.

Image

Ensure that the image you project is appropriate for the occasion:

- Consider what to wear: aim to be respectable without 'dressing up', otherwise your message may be diminished.
- Develop a good posture: it will help your voice projection if you stand upright, rather than slouching, or leaning over the lectern.
- Project your voice: speak towards the back of the room.
- Make eye contact: look at members of the audience in all parts of the room. Avoid talking to your notes, or to only one section of the audience.
- Deliver your material with expression: arm movements and subdued body language will help maintain the interest of your audience. However, you should avoid extreme gestures (it may work for some TV personalities, but it isn't recommended for the beginner!).
- Manage your time: avoid looking at your watch as it gives a negative signal to the audience. Use a wall clock, if one is provided, or take off your watch and put it beside your notes, so you can glance at it without distracting your audience.
- Try to identify and control any distracting repetitive mannerisms, e.g. repeated empty phrases, fidgeting with pens, keys, etc., as this will distract your audience. Practising in front of a mirror may help.
- Practise your delivery: use the comments of your friends to improve your performance.

Questions

Many students are worried by the prospect of questions after their oral presentation. Once again, the best approach is to prepare beforehand:

- Consider what questions you may be asked: prepare brief answers.
- Do not be afraid to say 'I don't know': your audience will appreciate honesty, rather than vacillation, if you don't have an answer for a particular question.
- Avoid arguing with a questioner: suggest a discussion afterwards rather than becoming involved in a debate about specific details.
- If no questions are forthcoming you may pose a question yourself, and then ask for opinions from the audience: if you use this approach you should be prepared to comment briefly if your audience has no suggestions.

After the questions, you can sit down and relax — you've earned it!

Learning from experience — use your own experience of good and bad lecturers to shape your performance. Some of the more common errors include:

- speaking too quickly
- reading to notes and ignoring the audience
- unexpressive, impersonal or indistinct speech
- distracting mannerisms
- poorly structured material with little emphasis on key information
- factual information too complex and detailed
- too few visual aids

57 Sitting exams

You are unlikely to have reached this stage in your education without being exposed to the examination process. However, the following comments should help you to identify and improve on the skills required for exam success.

Preparation

Begin by finding out as much as you can about the exam, including:

- its format and duration;
- the date and location;
- the types of question;
- whether any questions/sections are compulsory;
- whether the questions are internally or externally set or assessed;
- whether calculators are required.

Your course tutor is likely to give you details of exam structure and timing well beforehand, so that you can plan your revision: the course handbook and past papers (if available) can provide further useful details. Check with your tutor that the nature of the exam has not changed before you consult past papers.

Lecture notes, assignments and practical reports

Given their importance as a source of material for revision, you must sort out any deficiencies or omissions in lecture notes/practical reports at an early stage. For example, you may have missed a lecture or practical due to illness, etc., but the exam is likely to assume attendance throughout the year. Make sure you attend classes whenever possible and keep your notes up to date.

Your practical reports and any assignment work will contain specific comments from the teaching staff, indicating where marks were lost, corrections, mistakes, inadequacies, etc. It is always worth reading these comments as soon as your work is returned, to improve the standard of your subsequent reports. If you are unsure about why you lost marks in an assignment, or about some particular aspects of a topic, ask the appropriate member of staff for further explanation. Most lecturers are quite happy to discuss such details with students on a one-to-one basis and this information may provide you with 'clues' to the expectations of individual lecturers that may be useful in exams set by the same members of staff. However, you should *never* 'fish' for specific information on possible exam questions, as this is likely to be counter-productive.

Revision

Begin your revision early, to avoid last-minute panic. Start in earnest about 6 weeks beforehand:

- Prepare a revision timetable — an 'action plan' that gives details of specific topics to be covered. Try to keep to this timetable. Time management during this period is as important as keeping to time during the exam itself.
- Remember, your concentration span is limited to 15–20 min: make sure you have two or three short (5 min) breaks during each hour of revision.

- Make your revision as active and interesting as possible: the least productive approach is simply to read and re-read your notes.
- Include recreation within your schedule: there is little point in punishing yourself with too much revision, as this is unlikely to be profitable.
- Ease back on the revision near the exam: plan your revision, to avoid last-minute cramming and consequent overload fatigue.

Active revision

The following techniques may prove useful in devising an active revision strategy:

- Prepare revision sheets with details for a particular topic on a single sheet of paper, arranged as a numbered checklist. Wall posters are another useful revision aid.
- Memorize definitions and key phrases: definitions can be a useful starting point for many exam answers.
- Use mnemonics and acronyms to commit specific factual information to memory. The dafter they are, the better they work!
- Prepare answers to past or hypothetical questions, e.g. write essays or work through calculations and problems, within appropriate time limits. However, you should not rely on 'question spotting': this is a risky practice!
- Use spider diagrams as a means of testing your powers of recall on a particular topic (p. 264).
- Try recitation as an alternative to written recall.
- Draw diagrams from memory: make sure you can label them fully.
- Form a revision group to share ideas and discuss topics with other students.
- Use a variety of different approaches to avoid boredom during revision (e.g. record information on audio tape, use cartoons, or any other method, as long as it's not just reading notes!).

Preparing for an exam — make a checklist of the items you'll need (e.g. pens, pencils, sharpener and eraser, ruler, calculator, paper tissues, watch)

The evening before your exam should be spent in consolidating your material, and checking through summary lists and plans. Avoid introducing new material at this late stage: your aim should be to boost your confidence, putting yourself in the right frame of mind for the exam itself.

Final preparations — try to get a good night's sleep before an exam. Last minute cramming will be counter-productive if you are too tired during the exam.

The examination

On the day of the exam, give yourself sufficient time to arrive at the correct room, without the risk of being late (e.g. what if your bus breaks down?).

The exam paper

Begin by reading the instructions at the top of the exam paper carefully, so that you do not make any errors based on lack of understanding. Make sure that you know:

- how many questions are set;
- how many must be answered;
- whether the paper is divided into sections;
- whether any parts are compulsory;
- what each question/section is worth, as a proportion of the total mark;
- whether different questions should be answered in different books.

If you are unsure about anything, ask — the easiest way to lose marks in an exam is to answer the wrong number of questions, or to answer a different question from the one set by the examiner. Underline the key phrases in the instructions, to reinforce their message.

Next, read through the set of questions. If there is a choice, decide on those questions to be answered and decide on the order in which you will tackle them. Prepare a timetable which takes into account the amount of time required to complete each question and which reflects the allocation of marks — there is little point in spending one-quarter of the exam period on a question worth only 5% of the total marks! Use the exam paper to mark the sequence in which the questions will be answered and write the finishing times alongside: refer to this timetable during the exam to keep yourself on course.

Using the exam paper — unless this is forbidden, you *should* write on the question paper to plan your strategy, keep to time and organise your answers.

Do not be tempted to spend too long on any one question: the return in terms of marks will not justify the loss of time from other questions (see Fig. 57.1). Take the first 10 min or so to read the paper and plan your strategy, before you begin writing. Do not be put off by those who begin immediately; it is almost certain they are producing unplanned work of a poor standard.

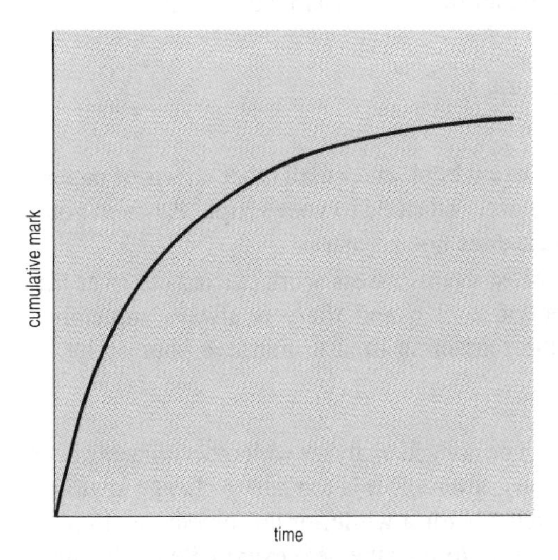

Fig. 57.1 Exam marks as a function of time. The marks awarded in a single answer will follow the law of diminishing returns — it will be far more difficult to achieve the final 25% of the available marks than the initial 25%. Do not spend too long on any one question.

Providing answers

Before you tackle a particular question, you must be sure of what is required in your answer. Ask yourself 'What is the examiner looking for in this particular question?' and then set about providing a *relevant* answer. Consider each individual word in the question and highlight, underline or circle the key words. Make sure you know the meaning of the terms given in Table 52.1 (p. 268)

Box 57.1 Writing under exam conditions

Never go into an exam without a strategy for managing the available time.

- Allocate some time (say 5% of the total) for consideration of which questions to answer and in which order.
- Share the rest of the time among the questions. Aim to optimize the marks obtained. A potentially good answer should be allocated *slightly* more time than one you don't feel so happy about. However, don't concentrate on any one answer (see Fig. 57.1).
- For each question divide the time into planning, writing and revision phases (see p. 268).

Employ time-saving techniques as much as possible.

- Use spider diagrams (p. 264) to organize and plan your answer.
- Use diagrams and tables to save time in making difficult and lengthy explanations.
- Use abbreviations to save time repeating text but *always* explain them at the first point of use.
- Consider speed of writing and neatness when selecting the type of pen to use — ball-point pens are fastest, but they can smudge. You can only gain marks if the examiner can read your script!
- Keep your answer simple and to the point, with clear explanations of your reasoning.

Make sure your answer is relevant.

- Don't include irrelevant facts just because you memorized them during revision as this may do you more harm than good. You must answer the specific question that has been set.
- Time taken to write irrelevant material is time lost from another question.

so that you can provide the appropriate information, where necessary. Refer back to the question as you write, to confirm that you are keeping to the subject matter. Box 57.1 gives advice on writing essays under exam conditions.

It is usually a good idea to begin with the question that you are most confident about. This will reassure you before tackling more difficult parts of the paper. If you run out of time, write in note form. Examiners are usually understanding, as long as the main components of the question have been addressed and the intended structure of the answer is clear.

The final stage

At the end of the exam, you should allow at least 10 min to read through your script, to check for:

- grammatical and spelling errors;
- mathematical errors.

Make sure your name is on each exam book and on all other sheets of paper, including graph paper, even if securely attached to your script, as it is in your interest to ensure that your work does not go astray.

Never leave any exam early. Most exams assess work carried out over the previous year in a time period of 2–3 h and there is always something constructive you can do with the remaining time to improve your script.

After the exam

Try to avoid becoming involved in prolonged analyses with other students over the 'ideal' answers to the questions; after all, it is too late to change anything at this stage. Go for a walk, watch TV for a while, or do something else that helps you relax, so that you are ready to face the next exam with confidence.

Practical exams

The prospect of a practical examination in biology may cause you more concern than a theory exam. This may be due to a limited experience of practical examinations, or the fact that practical and observational skills are tested, as well as recall, description and analysis of factual information. Your first thoughts may be that it is not possible to prepare for a practical exam but, in fact, you can improve your performance by mastering the various practical techniques described in this book. The principal types of question you are likely to encounter include:

- Manipulative exercises, often based on work carried out as part of your practical course (e.g. dissection).
- 'Spot' tests: short answer questions requiring identification, or brief descriptive notes on a specific item (e.g. a prepared slide).
- Numerical exercises, including the preparation of aqueous solutions at particular concentrations (p. 18) and statistical exercises (p. 234). General advice is given in Chapter 44.
- Data analysis, including the preparation and interpretation of graphs (p. 214) and numerical information, from data either obtained during the exam or provided by the examiner.
- Drawing a specimen, where accurate representation and labelling will be important (Chapter 15).
- Preparation of a specimen for examination with a microscope: this will test staining technique (p. 116) and skills in light microscopy (p. 133).

- Interpretation of photographic material: sometimes used when it is not possible to provide living specimens, e.g. in relation to field work, or electron microscopy.

Practical reports

You may be allowed to take your laboratory reports and other texts into the practical exam. Don't assume that this is a soft option, or that revision is unnecessary: you will not have time to read large sections of your reports or to familiarize yourself with basic principles, etc. The main advantage of 'open book' exams is that you can check specific details of methodology, reducing your reliance on memory, provided you know your way around your practical manual. In all other respects, your revision and preparation for such exams should be similar to theory exams. Make sure you are familiar with all of the practical exercises, including any work carried out in class by your partner (since exams are assessed on individual performance). Check with the teaching staff to see whether you can be given access to the laboratory, to complete any exercises that you have missed.

The practical exam

At the outset, determine or decide on the order in which you will tackle the questions. A question in the latter half of the paper may need to be started early on in the exam period (e.g. an enzyme assay requiring 2-h incubation in a 3-h exam). Such questions are included to test your forward-planning and time-management skills. You may need to make additional decisions on the allocation of material, e.g. if you are given 30 sterile test tubes, there is little value in designing an experiment that uses 25 of these to answer question 1, only to find that you need at least 15 tubes for subsequent questions!

Make sure you explain your choice of apparatus and experimental design. Calculations should be set out in a stepwise manner, so that credit can be given, even if the final answer is incorrect (see p. 223). If there are any questions that rely on recall of factual information and you are unable to remember specific details, e.g. you cannot identify a particular specimen, or slide, make sure that you describe the item fully, so that you gain credit for observational skills. Alternatively, leave a gap and return to the question at a later stage.

References

Anon. (1963) Tables of Spectrophotometric Absorption Data for Compounds used for the Colorimetric Detection of Elements, *International Union of Pure and Applied Chemistry*, Butterworth-Heinemann, London.

Billington, D., Jayson, G.G. and Maltby, P.J. (1992) *Radioisotopes*. Bios, Oxford.

Briscoe, M.H. (1990) *A Researcher's Guide to Scientific and Medical Illustrations*. Springer Verlag, Berlin.

Budavari, S., *et al.* (1989) *The Merck Index: An Encyclopedia of Chemicals, Drugs and Biologicals*, 11th edn. Merck & Co., Inc., Rahway, New Jersey.

Causton, D.R. and Venus, J.C. (1981) *The Biometry of Plant Growth*. Edward Arnold, London.

Clausen, J. (1988) *Immunochemical Techniques for the Identification and Estimation of Macromolecules*, 3rd edn. Elsevier, Amsterdam.

Collins, C.H., Lyne, P.M. and Grange, J.M. (1989) *Microbiological Methods*, 6th edn. Butterworth-Heinemann, London.

Diem, K. and Lentner, C. (1970) *Geigy Scientific Tables*, 6th edn. Geigy Pharmaceuticals, Macclesfield.

Dodds, J.H. and Robert, L.W. (1982) *Experiments in Plant Tissue Culture*. Cambridge University Press, Cambridge.

Eason, G., Coles, C.W. and Gettinby, G. (1980) *Mathematics and Statistics for the Bio-Sciences*. Ellis Horwood, Chichester.

Finney, D.J., Latscha, R., Bennett, B.M. and Hsu, P. (1963) *Tables for Testing Significance in a 2 × 2 Table*. Cambridge University Press, Cambridge.

Ford, T.C. and Graham, J.M. (1991) *An Introduction to Centrifugation*. Bios, Oxford.

Freshney, R.I. (1983) *Culture of Animal Cells: A Manual of Basic Techniques*. A.R. Liss, New York.

Geider, R.J. and Osborne, B.A. (1992) *Algal photosynthesis*. Chapman and Hall, New York.

Gerhardt, P. (ed.) (1981) *Manual of Methods for General Bacteriology*. American Society for Microbiology, Washington DC.

Golterman, H.L., Clymo, R.S. and Ohnstad, M.A.M. (1978) *Methods for Physical and Chemical Analysis of Fresh Waters*. Blackwell Scientific Publications, Oxford.

Green, E.J. and Carritt, D.E. (1967) New Tables for Oxygen Saturation of Seawater, *Journal of Marine Research*, **25**, pp. 140–7.

Grimstone, A.V. and Skaer, R.J. (1972) *A Guidebook to Microscopical Methods*. Cambridge University Press, Cambridge.

Kiernan, J.A. (1990) *Histological and Histochemical Methods: Theory and Practice*, 2nd edn. Pergamon Press, Oxford.

Lide, D.R. (ed.) (1990) *CRC Handbook of Chemistry and Physics*, 71st edn. CRC Press, Boca Raton, Florida.

Lillie, R.D. (ed.) (1977) *H.J. Conn's Biological Stains*, 9th edn. Williams and Wilkins, Baltimore.

Lüning, K.J. (1981) 'Light', in C.S. Lobban and M.J. Wynne (eds), *The Biology of Seaweeds*, pp. 326–55, Blackwell Scientific, Oxford.

Nobel, P.S. (1991) *Physicochemical and Environmental Plant Physiology*. Academic Press, New York.

Perrin, D.D. and Dempsey, B. (1974) *Buffers for pH and Metal Ion Control*. Chapman and Hall, London.

Prys-Jones, O.E. and Corbet, S.A. (1987) *Naturalist's Handbook 6: Bumblebees*. Cambridge University Press, Cambridge.

Reynolds, L. and Simmonds, D. (1981) *Presentation of Data in Science*. Nijhoff, Boston.

Robinson, R.A. and Stokes, R.H. (1970) *Electrolyte Solutions*. Butterworth-Heinemann, London.

Roitt, I. (1991) *Essential Immunology*, 7th edn. Blackwell Scientific Publications, Oxford.

Sokal, R.R. and Rohlf, F.J. (1981) *Biometry*, 2nd edn. W.H. Freeman and Co., San Francisco.

Stace, C. (1991) *New Flora of the British Isles*. Cambridge University Press, Cambridge.

Stahl, E. (1965) *Thin Layer Chromatography — a Laboratory Handbook*. Springer Verlag, Berlin.

Warburg, O. and Christian, W. (1942) Isolierung und Kristallisation des Garungferments Enolase, *Biochemische Zeitschrift*, **310**, pp. 384–421.

Wardlaw, A.C. (1985) *Practical Statistics for Experimental Biologists*. John Wiley and Sons Inc., New York.

White, B. (1991) *Studying for Science*. E. & F.N. Spon, Chapman and Hall, London.

Index

Index

Index